Devlin's

BOATBUILDING MANUAL

SECOND EDITION

Devlin's

BOATBUILDING MANUAL

SECOND EDITION

*How to Build Your Boat the
Stitch-and-Glue Way*

SAMUAL DEVLIN

Mc
Graw
Hill

New York   Chicago   San Francisco   Athens   London
Madrid   Mexico City   Milan   New Delhi
Singapore   Sydney   Toronto

1   2   3   4   5   6   7   8   9   LWI   28   27   26   25   24   23

ISBN   978-1-260-46767-3
MHID   1-260-46767-8

e-ISBN   978-1-260-46768-0
eMHID   1-260-46768-6

McGraw Hill is committed to making our products accessible to all learners. To learn more about the available support and accommodations we offer, please contact us at accessibility@mheducation.com. We also participate in the Access Text Network (www.accesstext.org), and ATN members may submit requests through ATN.

Contents

Introduction: The Magic of Building a Boat

Why build a boat? Your own boat, your dream, giving it life through your hands and heart and intellect.

For me, there is magic in boatbuilding. It begins with the dream. First, you imagine some beautiful place on the water. You feel the fresh wind in your face and the warmth of the sun, and then you imagine how the boat handles and moves in this environment. This dream has energy. Of course, the physical side of building this boat, the plans and tools and materials, are essential, but it's the dream that powers you through to the end, the actual hand-built creation that you're rowing or sailing or motoring through the water.

My own story illustrates such a dream.

It is the mid-1970s and a young Sam Devlin is working on a tugboat in the Gulf of Alaska. It's a wooden boat, probably built just before the start of the 20th century. I have been working six hours on, six hours off, for days now and have been trying to get used to the crazy schedule and somehow get enough sleep. One day I find myself off-watch in the galley of this old tug, drinking strong coffee from the pot that's always kept full on the huge, oil-burning stove. I pick up a magazine that the other deckhand has brought aboard and leaf through it. The magazine is called *WoodenBoat*; this is its start-up issue, and I'm suddenly intrigued. As I read it, a vision begins to form. There's a shop with a long workbench set below a bank of windows. The walls are rustic bare wood, dark with age, and the only light is the diffused daylight filtering through the windows. There are hand tools on the workbench, fragrant wood shavings on the floor, a crackling fire in a wood-burning stove, and next to the workbench, the inverted hull of a wooden boat. And I am there. Even now, more than 45 years later, I remember what a warm and inviting feeling this was and how it felt so totally *right*—that I belonged in the space of that dream and that it fit me.

I decided right then: This was the calling that would let me start every working day with energy and enthusiasm, creating useful things that have a life and spirit all their own.

Few things are more magical than transforming a rough pile of wood into a functional, graceful,

Figure 0-1. *My first tug job in Ketchikan, Alaska, was on the tug Amak with skipper Clyde Cowan.*

and beautiful boat. For me this is the essence of creative expression. It is like creating life. Anything that has a person sweat over it, bleed on it, curse at it, and experience joy with it is laced with the spirit of all that labor and emotion. That spirit forever inhabits the boat, and it is a living thing.

You may be wondering whether building a boat is within your capabilities. It's a good question to ask yourself, but there's only one way to learn the answer: Try it.

Try it.

I haven't a doubt that with the dream, the motivation, and the right design, anyone can build a wonderful boat. This book will be your guide. The Stitch-and-Glue method it explains is readily approachable and learnable. It will ask for patience and perseverance, but not high-octane math or professional woodworking skills.

You should be prepared for an emotional roller-coaster ride. Your energy and enthusiasm will soar to astounding heights and then possibly plunge to frightening depths—and then right back up again. There will be days when you gravely doubt the worth of the whole project. Expecting all this from the outset will prepare you to manage it. You'll take advantage of the highs and work like a demon

on the most challenging parts. But you'll have a list of some simple but interesting side projects on the boat, and when the emotional low rolls in, you'll tackle one of them and keep the momentum. You'll learn to pace yourself and stay refreshed and motivated, and you may find that this new skill adapts to other areas of your life. Just like an athlete in training, you are in control of tuning your body and your mind into being a boatbuilder.

Strive to keep the dream realistic and the project under control. Set time and budget constraints, and try to hold to them as you proceed through the building process. For every boat that I've built for myself, I've always kept in mind that it would someday have to be sold. I recommend you do this, too, even if it feels like you'd be auctioning your firstborn. This will help establish realistic limits on the actual cost and labor, and will keep you from getting stuck in the bog of perfectionism.

I recommend that you read through this whole book first, liberally underlining and scribbling marginal notes. Then as you begin each step on your boat project, reread the chapter about it. Also, collect and index relevant boatbuilding articles you find in *WoodenBoat* and other publications. There are many different ways to accomplish the tasks necessary to build a boat and many ideas that can enhance your project. Just keep that realistic budget in mind.

At this point, all you need in hand is your dream and motivation. The boat that will grow out of it will strengthen and reflect your inner spirit, and like me, you might find that building wooden boats is a very fine thing to do with your life.

In the nearly three decades since *Devlin's Boat Building* was first published, I have designed many dozens of new boats, and my excellent team and I have built many dozens more. Although our basic technique of Stitch-and-Glue construction has not changed, we have made numerous refinements and tested new ideas that streamline the process, such as frequently using staples instead of wire stitches in assembling the hull. In this thoroughly revised edition, I have incorporated a number of these refinements. I have also discussed some new materials and tools, added a chapter on building cabins and pilothouses, and described maintenance and repair in much more depth. Finally, writer Lawrence W. Cheek, a veteran of three Devlin boatbuilding projects that have been detailed in *The New York Times*, *Seattle Times*, and *WoodenBoat* magazine, has contributed a new chapter on Stitch-and-Glue building from the amateur's perspective. I believe you will find this new edition useful, and I hope it assures you that this dream is absolutely within your reach.

Figure 0-2. *The pinnacle of my design and boatbuilding career is the lovely motor cruiser Moon River, built for Ed and Susan Schulman of Seattle, Washington.*

CHAPTER 1

The Advantages of Stitch-and-Glue Boatbuilding

The differences between Stitch-and-Glue construction and other methods of building a wooden boat are significant. We can easily understand them by contrasting the structures of an early biplane and a modern jet airliner.

The biplane was made up of frames and spars over which was stretched a thin, membranous skin. The jet airliner's structure is actually much simpler, with a stressed aluminum skin rigidly attached to bulkheads and a smaller number of spars to create a homogeneous unit. A traditional plank-on-frame boat is structurally akin to the biplane. A Stitch-and-Glue boat more closely resembles the modern airliner; its monocoque skin and reinforcements bear the primary stresses, and its internal parts further contribute to the structural integrity while also defining the architectural spaces inside the boat.

Of course, I can't pretend to be a neutral observer, but I see so many advantages to Stitch-and-Glue construction:

- The initial construction is quicker and easier, uses fewer parts, and requires no or very simple building molds (the temporary forms around which the hull is shaped).
- The hull is extremely strong and rigid in relation to its light structural weight.

- Long-term maintenance is much easier because the epoxy sealing and epoxy/fiberglass composite sheathing form a complete waterproof barrier that protects the wood. On many other kinds of wooden boats, the paint attempts to assume this critical duty, but those coatings are not efficient, making frequent repainting necessary as the wood underneath swells and shrinks with changes in the weather and humidity.
- If the boat is to live on a trailer, as most small boats today must, the Stitch-and-Glue hull isn't subjected to repeated wetting/drying-out cycles. And the structure is so rigid that when the boat is bouncing along a country road, pieces aren't flexing and moving against each other, threatening to work themselves loose.
- From the designer's standpoint, Stitch-and-Glue invites a design to continually evolve, improve, and adapt, which is exactly what I do with my own design work. For example, my venerable Surf Scoter 22 design has recently grown a stretched pilothouse and wider beam for greater cruising comfort and rough-water stability. I can make changes like these at no cost except for a few hours' design work on the computer. Almost all other boatbuilding methods require expensive tooling: fiberglass production boats

need their full-form plugs and molds, and plank-on-frame and cold-molded wooden boats require complicated building setups and molds. Any boat design can benefit from refinement as people use it and give their feedback to the designer. Stitch-and-Glue construction makes it easy (and occasionally irresistible) to tinker and evolve.

When I began to get serious about boatbuilding, my first hurdle was finding shop space and buying the necessary tools. I was young and didn't have many resources, so it was a great temptation to toss my ideas about innovative boatbuilding and stick with plank-on-frame boats. Small boats can

be built with relatively simple tools, though they still require substantial skills. But my evolving ideas, along with a gradual accumulation of quality tools, steered me into the Stitch-and-Glue method. Today, after producing more than 400 boats in my shop, in sizes ranging from a 7½-foot dinghy to a 50-plus-foot motor cruiser, I'm still constantly amazed at how adaptable this construction method is and how it has enabled me to construct a life around designing and building boats.

When I began building boats in the Pacific Northwest, marine plywood was readily available in a variety of thicknesses and in sheets up to 16 feet long. To me, plywood has always seemed the most efficient expression of wood for building

Figure 1-1. *Polliwog is one of my very early designs and has been built in over 100 countries around the world. It was featured in my 1987 boatbuilding video, which has now had more than a million views on YouTube.*

things, so I naturally began visualizing how boats might be formed from these sheets. With my very first build, I began wiring (stitching) plywood panels together and fusing the joints and seams with epoxy resin and fiberglass cloth. I didn't know that other builders had started using a similar technology in Europe, Australia, and New Zealand; this was long before the Internet and its effortless global distribution of information. In retrospect, I'm glad I was unaware of those boats and designs. I was free to work out my own assumptions and refine my methods in isolation, without influence from others' prejudices. Maybe the most positive consequence was that it forced me to develop my own style, and I constantly worked toward creating fresh and interesting designs that didn't shout "plywood boat!" because they were slab-sided or simplistic. I came to believe that I could build any kind of boat—any style, any size, any intended use—with the Stitch-and-Glue method. After many years, I've confirmed this belief.

The more I worked with plywood, wire stitches, and epoxies, the more convinced I became that Stitch-and-Glue boats were indeed stronger, less demanding of maintenance, and easier to build than boats constructed with other methods. I grew certain that the method would come to be widely embraced in the boatbuilding universe. I even suspected that I might be able to make a living through it. But here's my most important discovery: I saw

Figure 1-2. *Lit'l Petrel is a nice little pram-bowed dinghy, suitable as a shoreboat for a much larger mothership.*

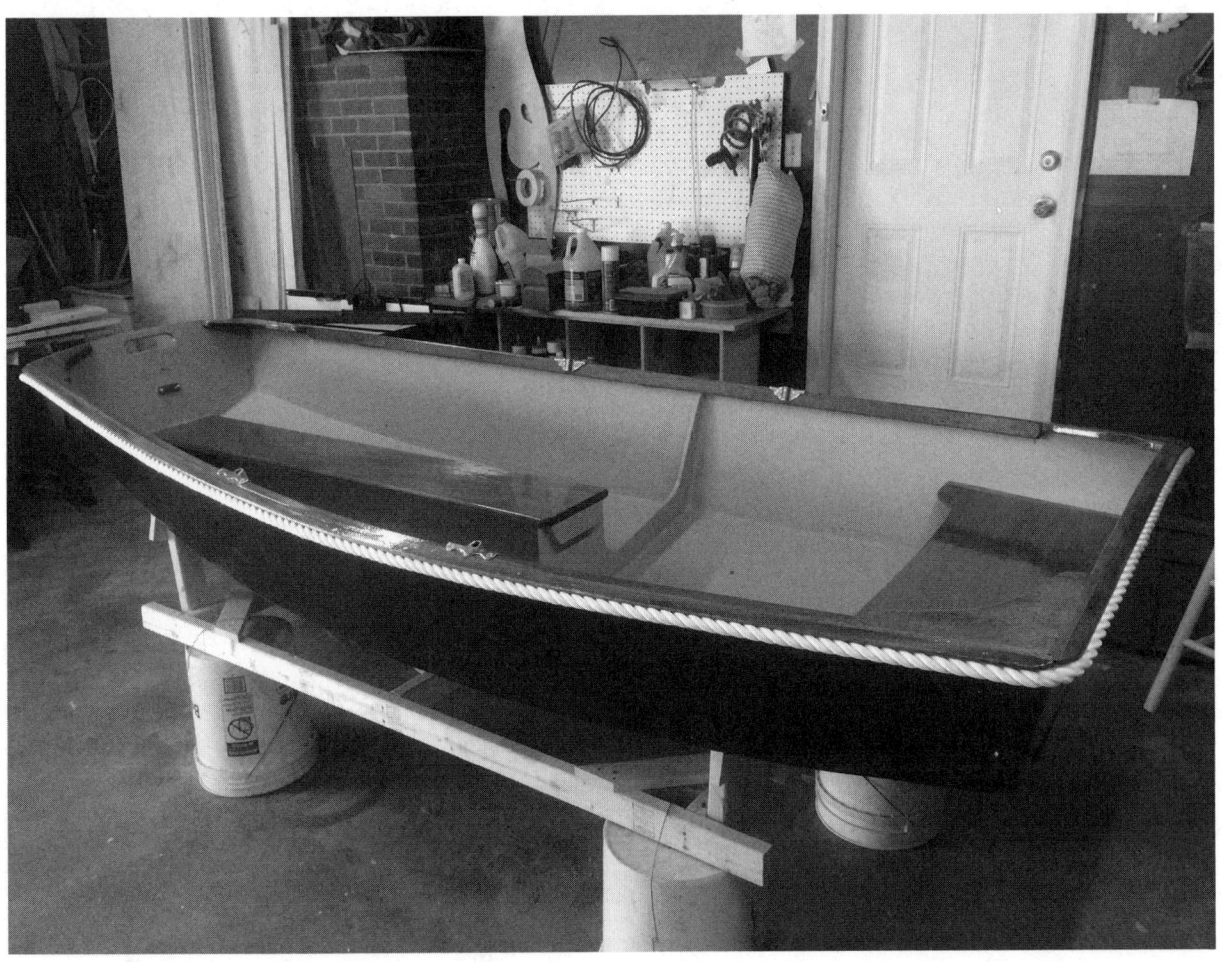

Figure 1-3. *If I make it to the gates of heaven, it is possible the guards might let me slip in because of the design of the Candlefish—simply a great and incredibly useful power skiff.*

that it had the potential to bring the pleasure of boatbuilding to anyone who wanted to try it.

I was probably fortunate to plunge into the boatbuilding world when I did, in the late 1970s. Two decades earlier, the first fiberglass boats had edged into the market, and through the late 1950s, 1960s, and 1970s, mass-produced glass boats of all kinds became an all-consuming tsunami. "Boatbuilding" essentially turned into boat assembly, and wooden boats—which celebrate the craftsman's hard-earned skills and provide infinite opportunities for individual expression—became almost extinct. But after three decades, time had ripened for a reaction. At least a few people had grown tired of the flood of similar-looking plastic boats and were ready for a more inspiring alternative. *WoodenBoat* magazine, launched in 1974, helped spread the word that a few scattered wooden boatbuilders around the North American continent were working in both traditional and innovative methods, and little by little, the boating public began to respond.

Certainly in the early days of my career, the fact that my boats were made of wood was an impediment to selling them, and the fact that they were *plywood* made it even worse. But as my design and building skills improved, a great number of the fiberglass production builders who had jumped into the market fell away, and consumers began to learn that fiberglass boats do not last forever. Gradually,

Figure 1-4. *I am surprised that there are not thousands of Pelicanos around the world, as they are so pretty on the water and very usable.*

wood/epoxy composite boats (a Stitch-and-Glue boat is in this category) picked up a marketing advantage as buyers became better informed. And amateur builders saw that the union of an ancient craft (wooden boatbuilding) with modern materials (plywood, fiberglass, and epoxy) made the home building of a fine, durable boat more accessible than ever. (A few DIYers might have been briefly tempted by the thought of building with fiberglass alone, but they would readily abandon it once they discovered that making the mold means essentially building *two* boats, only one of which will be usable. All-fiberglass boats make sense only when you're springing a fleet of identical hulls from that original mold.)

Let's undertake an honest assessment: *how* accessible is this boatbuilding method?

The most basic Stitch-and-Glue boat simply requires cutting out four or five plywood panels to form a hull, stitching them together with wire or staples, then coving and filling the interior seams with epoxy and fiberglass. When the epoxy cures, you pull out the wires and smooth the exterior in preparation for a sheathing with still more epoxy and fiberglass cloth. After the epoxy cures solid, you smooth the surfaces inside and out with a power sander. Reinforce the gunwales and add a seat or two, and you essentially have a complete boat, ready for paint and hardware. It's no more complicated than that.

(A)

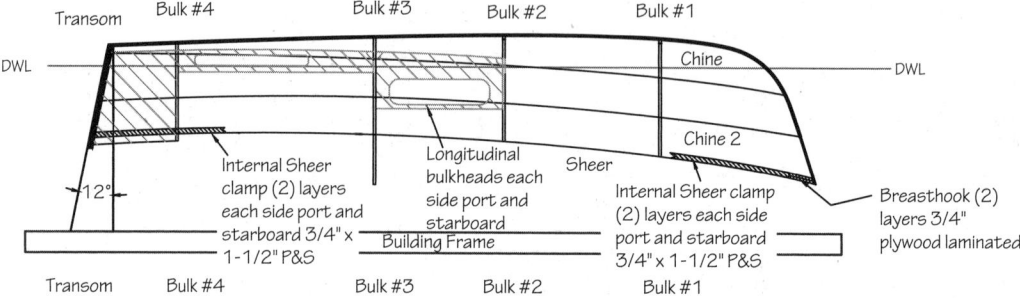

The Sequence for building first starts with a building jig or framework out of 2" x 6" boards as in Figure 1-5A. Mark out the interval of the bulkheads on the building framework's top surface. A Centerline is also necessary to help keep everything lined up. Now you can set your bulkheads up on the building framework and brace them to set vertically. Also set the longitudinals in place and set the transom up against them and the bulkheads. All the waterlines should be at exactly the same measurement from the top surface of the framework, and the centerlines of all the bulkheads must be in alignment. Now would be a good time to cut out a breasthook and place it on a pedestal at the proper angle and location. This will be the forward attaching points for the interior sheer clamp. The more accurate you are at this setup stage, the easier your building project will be.

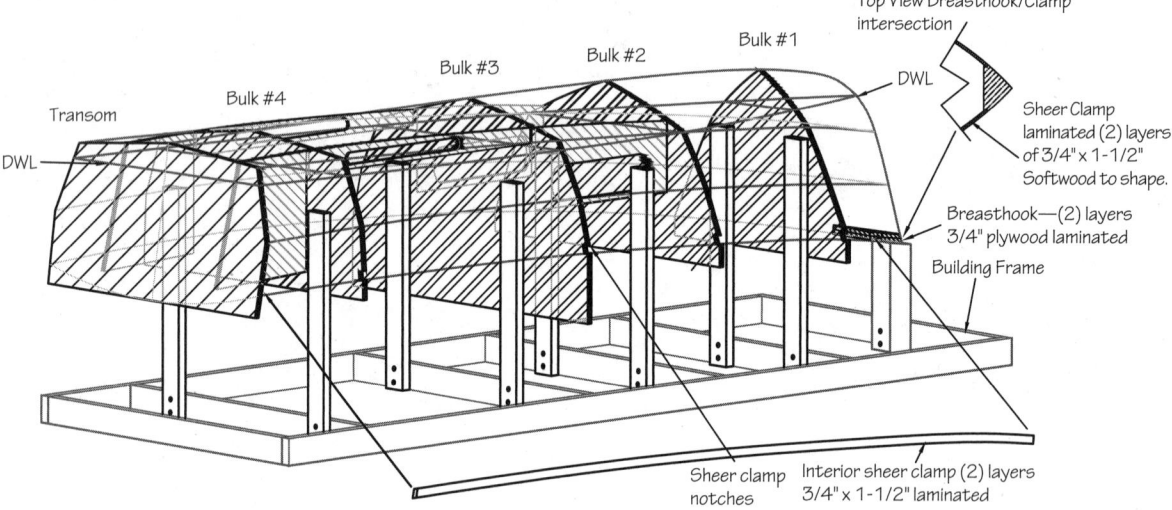

(B) Once everything is all lined up and set up properly, we are now ready to cut out the two bottom hull panels and Stitch them together. In Figure 1-5B you can see the two panels off the bulkheads but just about ready to come together. You will Stitch the two bottom panels together at the keel line or centerline of the hull face to face like the pages of a closed book. Then by opening up the panels (like opening up a book upside down) and laying them over the bulkheads, we are ready to apply those first panels to the setup. There should be a reference mark on the bottom panels that will show where you need to position the panels fore and aft on bulkhead #3.

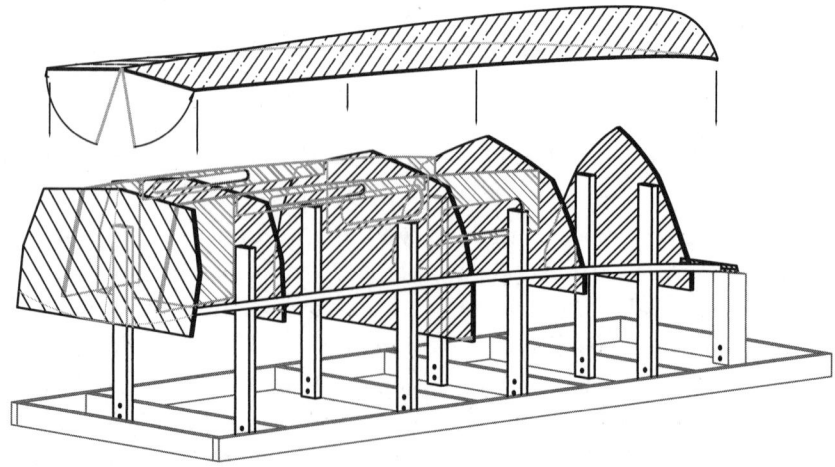

Figure 1-5A–1-5E. *Building a Stitch-and-Glue hull is a logical and approachable process, adaptable to almost any type and size of boat.*

(C)   Now that the bottom panels have been set into place, you are ready to Stitch the side panels into position (see Figure 1-5C). Start at the bow and Stitch or Staple (See "New Stitch-and-Glue Technology") the panels along the length of the chine from the Bow end to aft. You should have cut and beveled the double 45-degree angle on the inside edge of both of these panels at every edge of all of the panels except at the sheer edge. The double 45-degree bevel will allow the panels to lay alongside each other nicely, and the bulkheads will work wonders in keeping the panels all lined up and fair. At this point, don't fasten any of the panels to the bulkheads, but use them only as a framework for the panels. It's kind of like Stitching together a hat of the hull panels and then putting the hat over the bulkheads like putting a hat on your head.

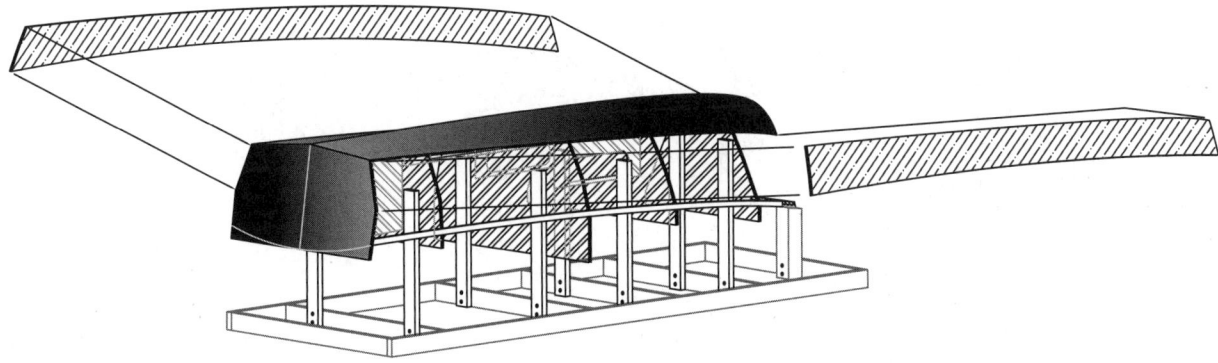

(D)   Now you are ready to Stitch the sheer panels into position (see Figure 1-5D). The bulkheads should have already been notched to accept the interior sheer clamps. Make sure sheer clamps have been entirely covered in epoxy resin before attaching them to the bulkheads. If you have not already done so, attach them before you stitch the sheer panels. Stitch the Sheer panels the same way you did for the Side panels.

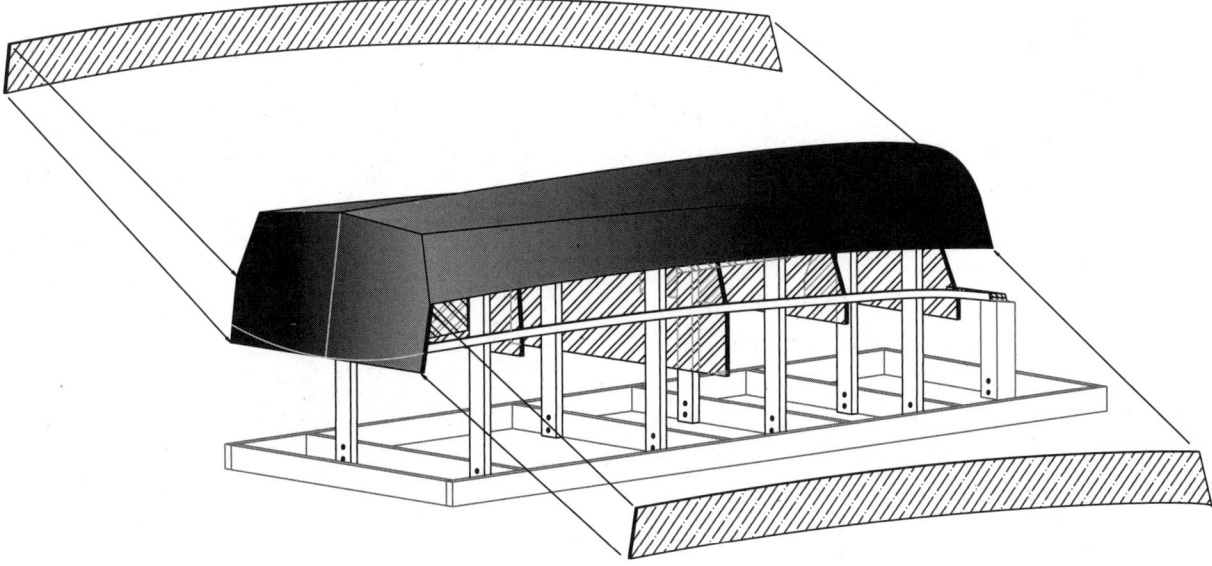

Figure 1-5A–1-5E. (Continued)

(E) All of the hull panels should be stitched and tabbed together now (glassing tabs done from the inside of the hull and welding together the panels into shape, also tabbing the panels to the bulkheads). Soak strips of biaxial tape in epoxy resin and tape the chines, keel-lines, and edges of the transom. These are the areas that concentrate wear, and the extra cloth will make your hull much more resistant to abrasion and wear. Next, sheath the entire exterior of the hull with 50"-wide dynel sheathing. The sheathing must get completely soaked in epoxy resin. It can either be put in place wet or you can place it over the hull, pouring the epoxy resin on it and spreading it around with rollers or squeegees. Make sure to thoroughly cover the sheathing. Underneath the hull, you need to glass tape all the seams that are easily reached and tab the other seams. The fiberglassing of the inside seams is pretty easy to do upside down on the bulkhead edges, but the chines and keel-lines are a bit harder to accomplish upside down, so what didn't get fully taped while the boat is upside down can get finished when the hull is turned rightside up.

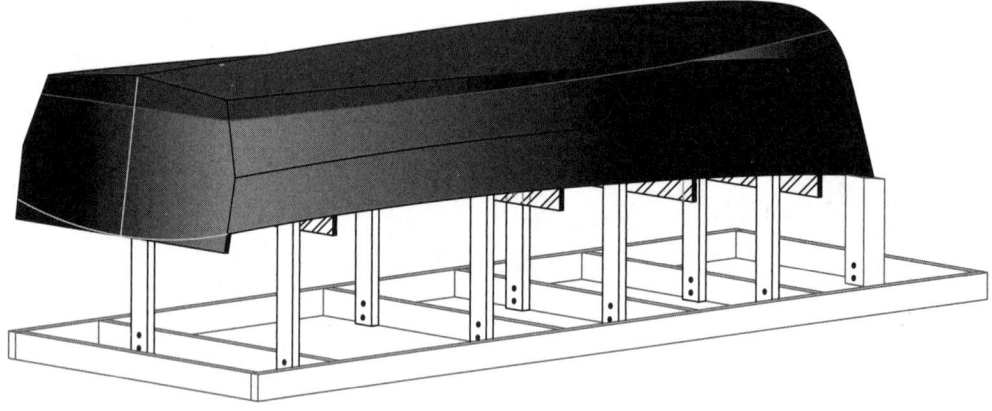

Figure 1-5A–1-5E. (Continued)

These basic boats (for example, the Polliwog, Lit'l Petrel, and Candlefish 13) do not require any traditional shipwright skill or advanced woodworking. Yet, the beauty of the Stitch-and-Glue medium is that it adapts to boats ascending the ladder of sophistication and complexity (the Pelicano 18 bass boat, the Arctic Tern 23 sailboat, and Godzilla 25 Harbor Tug). Amateurs have built these designs and many others very successfully.

I love plywood unashamedly. It utilizes all of wood's natural strengths while minimizing its weaknesses. It is a very efficient way to utilize wood resources, helping us to manage forests in a more careful and responsible manner. It's forgiving and easy to work with. And it will allow you to build a boat more quickly and easily than you might think.

Figure 1-6. *I designed the 23-foot Arctic Tern in 1985, and we have built several in our shop. Many more have been built by homebuilders around the world.*

Figure 1-7. *I am a tug nut and love designing and building tugboats that can be built either for work or pleasure cruising. The Godzilla 25 is one of many designs in that vein.*

Figure 1-8. *When pinned down and asked for my all-time favorite design, three-fourths of the time I would say it's the Blue Fin 54, this boat was built in Russia by a talented DIY builder.*

Figure 1-9. *My first build was the Egret design for rowing and sailing, and we still turn them out some 45 years later.*

Figure 1-10. *The Honker 20 was one of our favorite family boats when my boys were growing up on Puget Sound. It adapts to many uses and always keeps its occupants safe and sound.*

CHAPTER 2

Setting Up Shop

THE WORKPLACE

You need a dry, sheltered, and organized workplace to build your boat. Even if you're lucky enough to live in a climate of perpetually warm, sunny days, you will still need a covered area to build in, one that will keep your materials and boat-to-be dry, shaded, and protected through the duration of the project.

A backyard boatbuilding shed need not be permanent and expensive. A remarkable Canadian craftsman named Peter Gron, who built my 22 foot 8 inch Arctic Tern sloop design, began by fabricating a 16 × 24-foot shed in the form of a Gothic arch A-frame (Figures 2-1A and 2-1B). The building's ribs were paired 1 × 4 fir boards bent into an arch form with spacers (just like the open gunwales on a traditional canoe or rowboat) and were then walled with translucent polyethylene plastic sheeting. The pointed-arch form sheds rain and snow easily and is usually durable enough to see at least one boat through to completion (Peter's shed remained intact after six years). The major disadvantage of a plastic-walled shed is its lack of insulation, but this deficiency can be largely remedied with a double-wall sheathing. An outside and inside skin will also help to eliminate condensation and dripping.

How large a space do you need? Consider all three dimensions of the boat you plan to build. If you already have a garage or workshop, chalk the boat's outline on the floor, then allow at least 3 feet

Figure 2-1A. *If you don't have a shop already, why not consider building a simple and elegant shelter such as Peter Gron's?*

Figure 2-1B. *This type of shop provides superb daylighting for work with its translucent polyethylene skin.*

on each side and each end for moving around and working comfortably. Study the boat plan to determine whether you'll have adequate clearance under the ceiling and doorway through which the boat must eventually travel. If the boat is to be rolled out of the shop on a trailer, remember that its keel may be up to a foot off the floor when the boat begins its journey through the door. A large hull may also ask for considerable extra space alongside the edges of the hull when it's time to roll it over (see Chapter 21). Also, realistically consider the fourth dimension: time. Any boat project will tie up your garage or other workspace for a considerable amount of time and cause some disruption of normal household or shop activity. I once started building a very simple boat in my living room (at a time without the possibility of spousal debate) but quickly changed my mind as even the simplest procedures resulted in intolerable wood dust and shavings. So think through your workspace in advance, and take some time and pleasure in making it efficient and welcoming.

Good work asks for ample lighting, and a well-lit shop is a safer shop. If your workspace is an everyday garage, chances are that it has only minimal lighting, so it's worth considering hiring an electrician to add more ceiling fixtures (and several more electric outlets, while you're at it). Good boatbuilding needs a dry shop. If you don't have an enclosed work area, there's a greater chance that your epoxy work will not cure properly or that excessive moisture will stay trapped in the wood and eventually weaken the structure. Good health requires ventilation; you'll be using epoxy, solvents, paint, and varnish, so plan how you'll deal with chemical fumes when it's too cold to fling open all the doors and windows. Portable blowers available from your local hardware store can help keep these fumes from building to noxious levels.

Heat is more important than you might think. Epoxies and finishes demand curing time, and the warmer the ambient temperature, the more quickly they cure. Quick is good; delayed cure times invite contamination, dust, and insects to foul the surface. Your own comfort is important, too. Warm hands work better than cold hands, and if you are comfortable, your patience and attitude will be more attuned to the work at hand. A woodstove is a wonderful hearth-like presence in a boat shop, inviting the builder to fuel it with the day's wood scraps and take contemplative breaks in a comfortable chair beside the fire. However, in some areas such as my own Washington State, wood-burning stoves are now heavily regulated, and burn bans go into effect when air pollution rises above certain levels. The best alternative for shop heat is infrared heaters, which do not heat the entire air space (an inefficient prospect in a large shop) but rather the solid surfaces in the space. These heaters can be fired by propane, natural gas, or electricity. I first became aware of the miraculous efficiency of infrared heaters in my local car repair garage, a large building with lots of doors frequently opening as cars were shuttled in and out for service. Yet the shop space felt strangely warm and inviting. In a boat shop, likewise, this form of heating is safe and efficient, and it fits nicely with the Stitch-and-Glue method. For warm-weather work, a shop with large doors or many opening windows is the best bet for comfort. After years of living with a variety of shops, I have found that orientating to prevailing winds in your area and having doors at both ends makes boatbuilding and the builder's creature comfort much easier and more enjoyable. Opening both doors allows the wind to blow through the shop and helps to flush out some of the inevitably accumulating wood dust.

Your workspace should include ample room for your tools, and they should be organized so that they're easy to find and retrieve. I like to have my tools in well-organized storage cabinets or to hang them on open wall racks. A heavy, very sturdy workbench is an indispensable tool; preferably it should have two vises to double-clamp long stock. You can order a good-quality manufactured workbench from retailers such as Grizzly or Woodcraft for about $500 and up (and if you like, *way* up). Or you can build your own as a warmup project before beginning the boat. Workbench plans are easily available on the Internet or through magazines

Figure 2-2. *A boatbuilding project is easier and more fun if you have a sturdy and rugged work bench.*

such as *Popular Woodworking*. In my shop I have built tables at workbench height that are invaluable for placing stock at a level that avoids continual bending over and working on the floor. These are inexpensive and easy to construct with 2 × 6 lumber and simple plywood or particle board tops.

Your shop layout should be conducive to easy cleaning. Wood cutting, planing, drilling, sanding—in fact, practically every step in building a boat—generate a great deal of dust. It's worth considering an automatic dust collection system for your power tools, or at least for the sanders, which are the prime culprits in dust production. Whether or not you have such a system, I strongly recommend the discipline of cleaning and organizing the workspace at the end of every session. This will help you maintain both momentum and motivation. I've always found that I do some of my best work in well-motivated spurts, and if my shop is clean

and well organized, I can plunge into work with a clean slate and clean mind—rather than having to deal with the discouraging hangover of yesterday's debris. There is an interlocking matrix of cleanliness, organization, safety, and quality work; each supports and depends on the others.

A few more things are good to have in the shop: If you typically work alone when nobody else is at home, I recommend assembling a first-aid kit to keep handy in the shop. Although I am sure that you'll work carefully and mindfully, you're dealing with inherently hazardous tools and processes, and accidents happen. Some amateur builders have found it useful to station an old, not-quite-obsolete computer in the shop for ready Internet access. It's valuable for looking up advice on boatbuilders' forums and scouring catalogs for needed parts and supplies. An *old* computer is advisable because it's sure to get abused with dust and unfriendly chemicals. Finally, a small stereo is a great accoutrement to help make the work session pass pleasantly. I can always tell when I am happy and contented as I whistle to some pleasant music while working in the shop, and if I am happy and contented I tend to do very fine work.

The more comfortable and home-like your shop, the greater the chance you'll choose an evening of boatbuilding over an evening in front of the television. Still, you must be realistic about your time, money, and energy, and the shop is a means to an end, not an end in itself. I would rather see you embark on a boat project in a less-than-ideal shop than to get mired in the quest for the perfect workplace.

TOOLS

The avid boatbuilder will find no limit to the tools available and no bottom to the tool wishing well. Temper what you want with what you truly need, and then filter that list through what you can actually afford (or what you can clear through the tool-buying committee in your shop or home).

Here is my strongest advice: Spend good money on good tools. Rather than buying a cheap tool now, wait until you can afford the quality alternative. A poorly made tool is a terrible investment; it will frustrate you and keep you from improving your skills, and you'll want to replace it as soon as you realize how it's holding you back. On the other hand, a tool that's better than you are will reward your advancing craftsmanship. Professional-quality tools balance better, run smoother, last much longer, and are more enjoyable to use. If you're spending hard-earned money on the best material for your boat—which I recommend—it makes no sense to risk mauling it with inferior, less-than-sharp tools.

Your initial tool investment can be fairly modest if you're building a small boat, since a Stitch-and-Glue hull requires a minimum of tools.

The first consideration is whether you prefer hand or power tools. Hand tools will be less expensive, but your work will be much slower. I would consider three power tools to be practically indispensable for the amateur boatbuilder: power drill/screwdriver, sander, and some kind of power saw (discussion will follow). Many of the cordless tools now available—drills, drivers, and small circular saws—are wonderful to use, with good balance and a welcome absence of power cords draped over your work. I will often go out in the evening to work on a project, pacing myself by saying I'll work just until the battery runs out. For the home builder with limited time, this might be ideal. As with any other tools, the more expensive ones easily outperform the cheap ones.

Let's lay out a list of the basic essentials and then the dream list, followed by detailed considerations for some of the most important tools. In this edition I'm generally not supplying specific model numbers for tools since manufacturers frequently change them.

Basic Hand Tools
Assorted screwdrivers and driver bits for your cordless drill
Crescent wrench and/or or open-end wrench set
Pliers, lineman's type with side cutters
Hammer, 13- or 20-ounce
Rigging or pocketknife
Simple chisel set
Japanese-type pull saw or small hand saw
Scissors with 8- to 10-inch blades
Assorted C-clamps and bar clamps
Block plane
Sharpening stone
24-inch framing square or 48-inch drywall T-square
Compass (10–12-inch loose-leg dividers)
8-inch bevel gauge
Plumb bob
Level (electronic or spirit)
Reel-type chalk line
Tape measure
Architect's scale
Transparent water hose: 30 feet of ½-inch or ¾-inch (for water leveling the hull on its cradle)

(alternatively, a simple laser transit will do a much easier job of this task; these can either be bought or rented)

Dust respirator with organic vapor prefilters

Safety glasses

Hearing protection

Sawhorses: at least two, preferably with adjustable height 28" to 36"

Disposable gloves for working with epoxy (buy in boxes of 100)

Power Tools and Accessories

Sander-polisher, preferably variable-speed, with soft and hard backing pads

Palm sander or random-orbit finish sander

Sanding discs in 80-, 150-, and 220-grit paper

Circular saw

Jigsaw

Drill

Drill bit index: $\frac{1}{16}$- to $\frac{1}{2}$-inch high-speed steel bits

Driver and bits

Handheld propane torch with fuel cartridges

Shop vacuum

Miscellaneous Shop Materials

Waterless hand soap (e.g., Fast Orange) to remove epoxy or automotive hand cleaner

Tongue depressors or stirring sticks

Plastic squeegees for spreading thickened epoxy

Disposable mixing cups: graduated 13- to 16-ounce, for mixing epoxy and hardener

or minipumps for measuring and dispensing

Steel stitching wire and/or high-power stapler for wiring up hulls (see Chapters 12 and 13)

Pencils

Battens: $\frac{3}{4} \times \frac{3}{4}$ inch × 16 foot and $\frac{1}{2} \times \frac{1}{2}$ inch × 10 foot

Parrot-beak wirecutter pliers

Advanced Tools

Bandsaw, minimum 14-inch throat capacity

Table saw, 10-inch blade

Wood surface planer, minimum 12-inch width

Jointer, minimum 6-inch width

Handheld power planer

Stationary disc/belt sander combination

Cutoff saw, 10-inch blade (also called a chop saw)

Drill press

Router with $\frac{1}{4}$-inch and $\frac{1}{2}$-inch carbide round-over bits

Air compressor

HVLP (high velocity, low pressure) spray gun for painting

Drawknife

Wood rasps

Plug cutters

Mechanic's wrenches

Sledgehammer, short-handled

Compressed air stapler

Cold chisel set

Punches and nail sets

Ball peen hammer

Anvil

TOOL CONSIDERATIONS

Sander

The most used and abused power tool in my shop is an 8-inch sander-polisher. I consider this tool essential, even for the small-boat builder, in order to turn out a watercraft with a nice, uniform finish. Sander-polishers are similar to metal grinders but are lighter and run at slower speeds (1,500 to 3,200 rpm), which is an important consideration for boatbuilding. They turn slowly enough to hold the sandpaper and won't burn wood as quickly as the faster metal grinders. Used with a soft sanding pad and stick-on or hook-and-loop sanding discs, one of these will handle all your basic sanding/grinding needs, ranging from the broad expanses of the fiberglass hull sheathing to the interior epoxy fillets that bond the seams. With a deft touch, you can even do fine detail work with it.

When you begin shopping for sanders, you'll find a dizzying array of competing systems and sandpapers. In my experience, the stick-on discs seem to cut slightly more accurately as they adhere to the backing pad more closely and don't tend to

Figure 2-3. *Experienced Stitch-and-Glue boatbuilders consider the most important tool to be a rugged sander/polisher. This Makita is one of the best.*

ride up and over small, hard protrusions of epoxy and fillers as the hook-and-loop discs sometimes do. But hook-and-loop sanders are quite usable and will work very well for your shop.

What alternatives are there to the heavy-duty sander-polisher? In our shop we've purchased some of the German-made Festool sanders, which offer switchable rotary or random-orbit action and a vacuum dust-removal system that extracts dust through holes in the sanding discs. They are unbelievably expensive, but I am very impressed and have to say they seem to be worth every dollar. Festool's dust-removal system is so efficient it seems almost miraculous, and it allows long sanding sessions to be conducted safely and comfortably. Small palm and orbital sanders are relatively inexpensive and are designed to buff surfaces with small oscillations rather than grind them with swift circular action. By using coarser grits of sandpaper, one of these sanders might be able to replace the sander-polisher, although it will take more time to achieve the same results.

Block Plane

Every boatbuilder needs a good block plane. This elegant hand tool allows a woodworker to shave, smooth, or reshape wood with passes propelled only by arm strength. My father always had a simple-looking block plane in his toolbox,

and when I began boatbuilding, I was convinced I could do better. I selected the most complicated-looking and expensive model that I could find at the time, complete with knobs and levers for every imaginable adjustment. I was sure this plane would cut wood better than anything I had ever held in my hands before. As soon as I returned home, I re-sharpened the factory edge and placed a beautiful, straight-grained piece of yellow cedar in the bench vise. In my mind's eye, I could already see the long, smooth curls of shavings peeling off the block of wood. But the plane iron tore into the wood, digging in with every stroke. I adjusted the blade depth and tried again. Now it simply choked the slot between the blade and the body of the plane with wood chips—still no smooth shavings. No matter what I tried, this marvelous plane worked as though it had been designed by a fiend to gouge and mangle wood. After many hours of trying, I returned to the hardware store with tail between my legs and picked out a block plane just like Dad's—the original Stanley #118. I still have it.

Like wooden boats, high-quality hand planes edged to the brink of extinction, then came storming back as good craftspeople demonstrated that there was still a market for excellence. Most of the planes in my toolbox now come from Lie-Nielsen, a Maine company launched in 1981 with a philosophy of making top-quality hand tools with a "stubbornly local" attitude—they use Maine castings and Maine woods, and they don't outsource work overseas. I favor the low-angle block planes over the high-angle models (referring to the blade's angle of attack on the work surface) perhaps because of my home base in the American Northwest. The Northwest offers boatbuilders several useful softwoods such as cedar and fir, while the Northeast's supplies of hardwoods are more easily available (for more on the softwood/hardwood issue, see Chapter 4). Plane manufacturers tell us that low-angle models are more suited to slicing softwoods, while high-angle planes will remove hardwood material better with a scraping action. In your own toolbox, it might be good to acquaint yourself with the woods

Figure 2-4. *Festools are lovely, well designed, and very enjoyable to use. They are expensive but worth the cost.*

you will be using and make your blade-angle decision accordingly. With care, however, either kind of plane will work with most woods.

Saws

Building a Stitch-and-Glue boat will require saws to cut out hull parts with gradual curves, components such as rudders that have fancier curves, and occasionally a piece such as a thwart (a transversely mounted rowboat seat) that's all straight lines. If I had to get along with just two saws, I would choose a Japanese-style pull-stroke handsaw and a cordless circular saw. As you progress from simple to more complicated boats, you will undoubtedly find yourself acquiring a gamut of saws, including several types of handsaws, jigsaws, circular saws, bandsaws, and table saws.

You will be surprised at the versatility of a small circular saw; with proper technique you can cut almost any shape. If you're building a small boat, a 5½- or 6½-inch cordless saw will work very well. Use a good carbide-tipped blade with

Figure 2-5. *Important cutting tools for your toolbox from left to right: block plane, chisel, knife, multi-tool, scissors. Keep them all sharp!*

a high tooth count; the more teeth, the smoother the cut. I am right-handed, so I look for a saw that has the blade to the left of the motor. I can then view the workpiece and the cut line without having to peer over the top of the tool. I set the tool with the blade extending to just below the depth needed to cut cleanly through the piece. Then I hold my arm in, elbow against torso, and smoothly drive my arm forward, like a piston. The cordless saw's only drawback is a tendency to run out of juice during a long cutting session, but the simple solution is to keep an extra battery ready in the charger.

On a larger boat, you'll be cutting heavier stock and more of it, and my tool of choice is the corded heavy-duty worm-drive circular saw—again, with the blade on the left side.

When you are cutting sharper curves that can't be negotiated with a circular saw, nothing

will work as well as a good, well-balanced jigsaw. It can double as a scroll saw and a poor man's bandsaw. There isn't as much selection among cordless jigsaw models as with circular saws. But cordless tools are evolving, and I admit to a quiet dream of using cordless tools for all boat work short of those tasks needing large stationary tools.

While it is not an absolute necessity, after you've worked on your boat for a while you'll find yourself waking in the middle of the night, longing for a bandsaw. Once you've used one, you won't be able to imagine how you lived without it.

As with most other tools, a bargain-priced bandsaw is an investment you'll eventually regret. Light-duty benchtop bandsaws are available for under $300, but a good-quality, medium-power stationary bandsaw will cost between $1,000 and

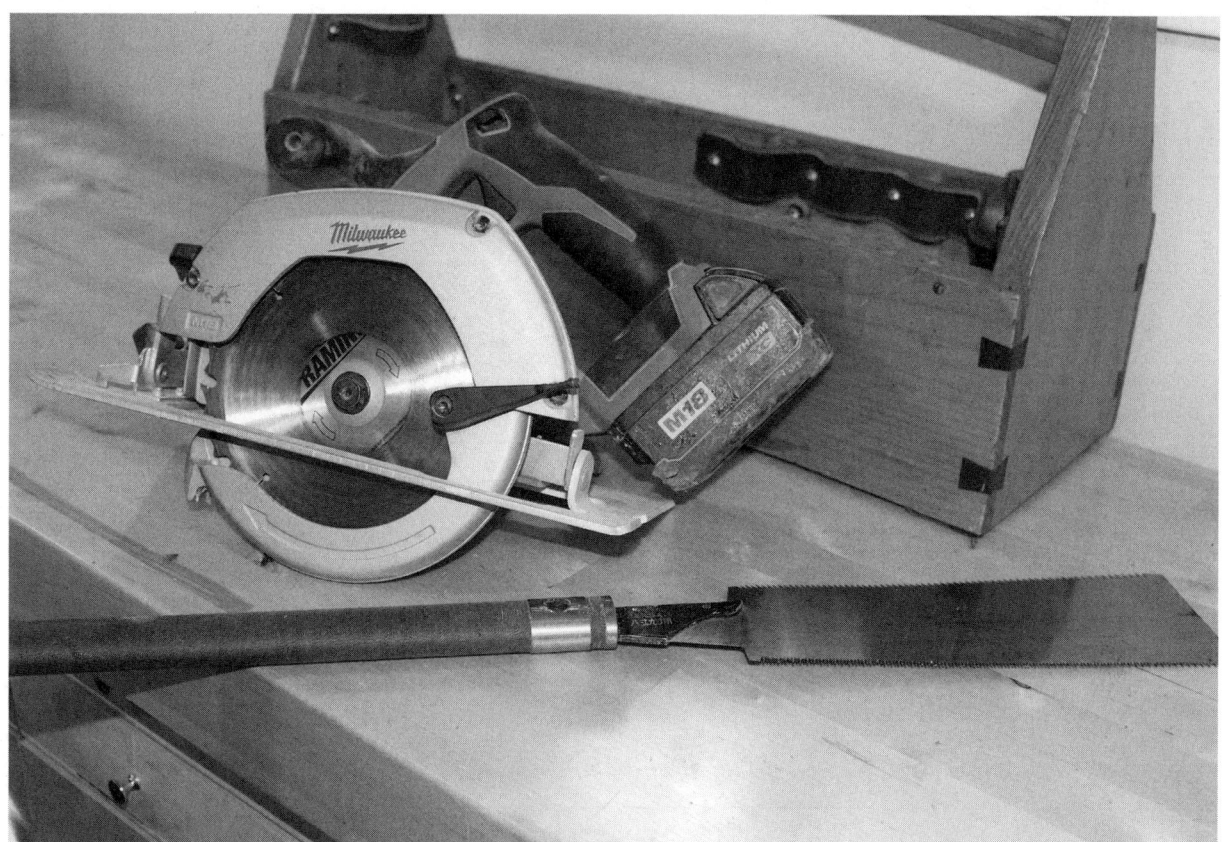

Figure 2-6. *Two of the boatbuilder's most important cutting tools are a cordless circular saw and a Japanese hand saw, which cuts on the pull stroke.*

$1,500. I have an older 14-inch bandsaw that has been my shop companion now for more than four decades, and I would feel lost without it. I judge a bandsaw based on the blade guides more than any other feature. Good ball-bearing-type blade guides ensure an easier, more precise cut than fixed blade-adjusting blocks. Any bandsaw is only as good as its blade, so once again I recommend against the everyday $20 blades found in typical hardware stores. Skip-tooth blades cost much more, but they will cut more cleanly, not bind in the wood when cutting a curving line, and last many times longer. Order them from a reputable mail-order outlet and eliminate one of the potential bogdown stations in your shop.

The larger the boat, the more likely you are to appreciate a table saw. It's useful for large pieces and precise, repetitive jobs such as cutting thin planks for the cabin sole. Buy a heavy-duty saw with as much horsepower as your shop wiring will permit. Carbide-tipped blades are worth their price, and a thin kerf (the blade's width) will minimize the amount of wood you turn into sawdust. Note that there are different types of blades for different cuts. Buy a good crosscut blade for plywood and a good rip blade for resawing dimensional wood.

Drill

A power drill is a must for Stitch-and-Glue construction. The good ones are well-balanced and long-lasting tools. Having two at hand is even better: You can keep a screwdriver bit in one and reserve the other strictly for drilling, saving time in swapping bits. Along with the drill, buy a drill index with bit sizes from $\frac{1}{16}$ to $\frac{1}{2}$ inch in increments

Figure 2-7. *You can build a boat without a bandsaw, but the work will be faster and more enjoyable with one.*

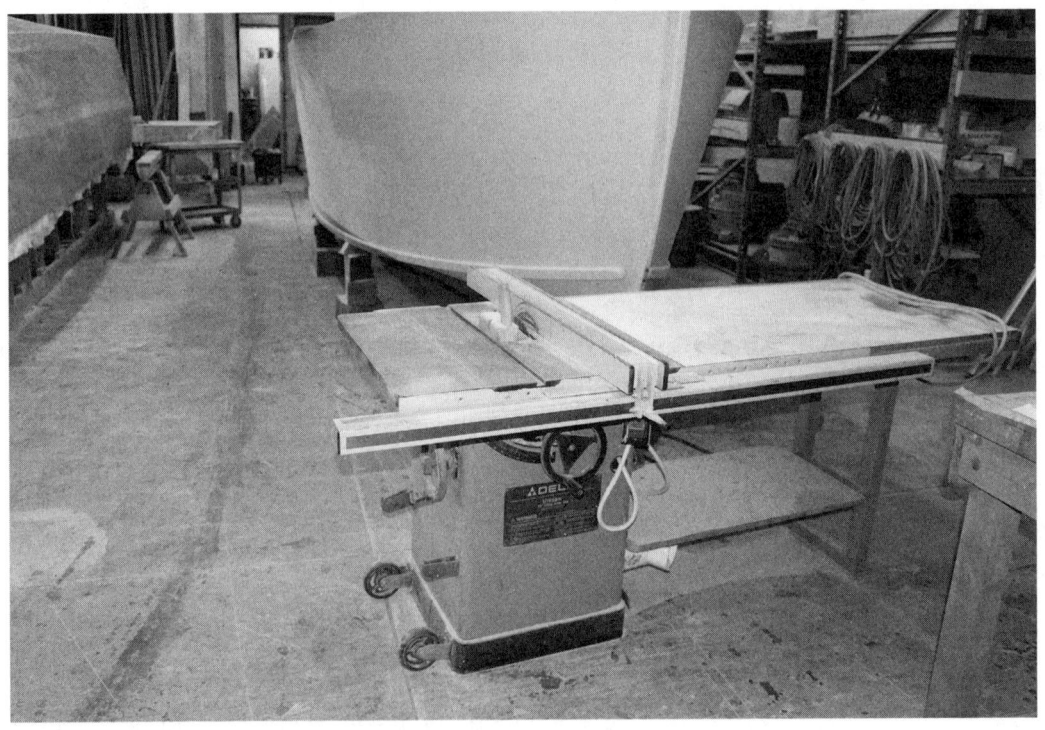

Figure 2-8. *Table saws excel at long, straight rip cuts, precise corner angles, beveled edges, and any operation that requires accurate, repetitive cuts.*

Figure 2-9. *Two more Stitch-and-Glue essentials are cordless drill and electric fencing wire.*

of $\frac{1}{32}$ inch. I prefer high-speed, steel twist bits for general work, but I have also used the more expensive bradpoint bits with good results. The latter are designed especially for wood and for boring flat-bottomed holes. You'll be countersinking most of the many screws you drive into a wooden boat, so a set of countersink bits—which bore a pilot hole and a countersink for the head at the same time—is a welcome time-saver. As with all these other tools, I advise buying the best-quality bits. They will bore cleaner and smoother holes, and your work will be of higher quality.

Hammers

Every boatbuilder needs a hammer even if you don't have many occasions to use nails in the boat itself. You'll frequently be building support frames and assorted jigs, and nailing is often the simplest way to put them together. I keep a light 13-ounce hammer for the more delicate work and a 16- to 20-ounce hammer for heavier pounding. In boatbuilding there is not much demand for the waffle-faced framing hammers, so stay with the smooth-faced variety. Look for good balance, preferably a wooden handle, and fine detailing and machining on the head. And try to avoid pulling out large, deeply driven nails with a wooden-handled hammer; you'll likely as not break its handle. Use a pry bar or crowbar for those tasks.

Clamps

Clamps are as vital as oxygen for the boatbuilder, and when you think you have far too many of them, it still won't be enough. Most all-around useful is the 12-inch bar clamp; it would also be good to have a few with longer reaches. C-clamps are also handy. An interesting trick with both types is to drill a small hole through the upper jaw and then thread a machine screw through it. When the clamp has to grip an awkward angle, the point of the screw will "bite" into the wood and hold it. When clamping parallel surfaces, the screw can be withdrawn. In a Stitch-and-Glue environment, clamps tend to get basted with epoxy as they're being used, so always wipe the moving parts with a rag wetted with solvent before the epoxy cures.

Sharpening Stone

Keep a good sharpening stone handy. The planes, chisels, knives, and other blades used in boatbuilding all need constant attention to function properly. Rather than sending them out for sharpening, learn to care for your blades yourself.

Many stones are available, but I much prefer the diamond type, which has a metallic-looking, hole-filled face glued to a chunk of plastic. Small diamond grains are embedded in the metal face, and the holes allow the minute sharpening residues to free themselves from the surface. This is the most versatile of sharpening stones, cutting virtually all metals with speed and accuracy. Water, usually the recommended lubricant, will rust the stones quickly unless they are wiped clean after each use, so I apply a light oil such as WD-40 instead. Kerosene also works. I mounted a stone on my workbench with screws to keep it from moving around, and I have made a box lid to keep shop dust from fouling its face.

Speaking of sharpening tools, it's nice to have a stropping leather to remove the burr after a workout on the stone. I keep a 24-inch finished-leather strop handy; a couple of swipes on it does wonders for any cutting edge. Any old-fashioned leather belt will double as a fine strop, and its buckle makes a handy fastener for belaying one end while stropping.

Other Tools

Find a good pair of pliers. Large lineman-type pliers are the best for cutting and twisting the wire you're going to use to assemble your Stitch-and-Glue hull. Buy the 9-inch size with a wire cutter built into the side of the jaws. It's important to buy pliers that feel comfortable in your hand, since you'll be using them a great deal in the wiring process. This is a top-priority tool and not a place to try to save money.

You'll need a small knife for myriad tasks. For years I used a Swiss Army knife, but after dropping yet another into the briny deep, I searched for a better alternative. I finally settled on a good rigging knife. This companion-on-my-hip had to be durable and hold a good edge, yet compact enough to fit my hand perfectly. Rigging knives are available at moderate cost from chandleries or in classier but more expensive versions from custom knife makers. I have a fine custom knife that's small enough to avoid that Daniel Boone look on my belt, and it's great not having to fish in my pocket for a folding knife.

You'll need a framing square for drawing the station lines during the lofting process. A simple L-square with measurement gradations or a 50-inch drywall T-square will work best for your layout lines, and of course either type can be used, along with battens to extend the station lines if you need the extra reach. I find that a straight, stiff, 50-inch wooden batten with a simple framing square will work for most purposes. These are simple and affordable tools.

Pick up a retracting tape measure at least 25 feet long. You'll find it's helpful to have one that's wide enough and rigid enough to extend

Figure 2-10A. *No boat shop ever had too many clamps.*

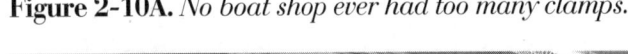

Figure 2-10B. *Nor can a boat shop ever have too many types of clamps.*

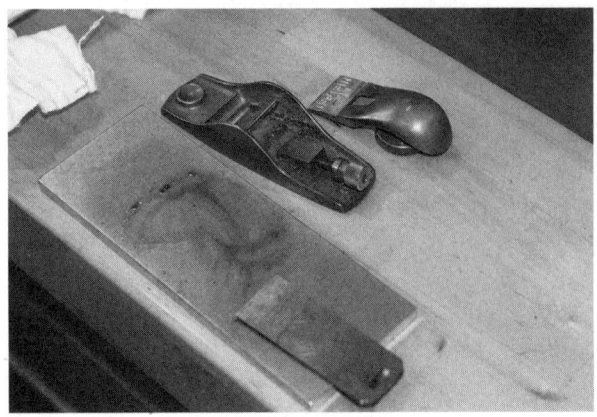

Figure 2-11. *You can't build a sharp boat without sharp tools, so sharpening stones are a fundamental shop essential.*

Figure 2-12. *Lineman's pliers will be needed for twisting and cutting wire stitches.*

10 feet and remain straight while unsupported. You will do a lot of chine-to-chine and sheer-to-sheer measuring to true up the hull. *Hint*: Take care to slow the blade when you are retracting it into the case. I have seen ends break off as a result of high-speed retraction. A 6-inch steel pocket rule, a 6-inch adjustable square, and adjustable bevel gauge are useful accessories, and they all fit nicely in a shop apron's pocket. Finally, once you buy a digital caliper—which is surprisingly inexpensive at $30 to $40—you'll wonder why you waited so long to make this small plunge.

You'll need a good compass. Buy one that has 10- or 12-inch legs and allows you to insert a pencil. You'll rely on your compass to offset lines, scribe the bulkheads, and fit various boat parts to curving surfaces. Try a used-tool dealer, since some of the best compasses are older models. An alternative might be to buy a set of dividers and tape a pencil to one leg. Although you'll need to re-tape each time you sharpen the pencil, you can get by with such improvisations.

To keep things level and true, add a plumb bob, a chalk line, and a spirit level or electronic smart level to your toolbox. I use my plumb bob for truing up the bulkheads and maintaining the vertical alignment of the hull during construction of the larger boats. Although a digital smart level or

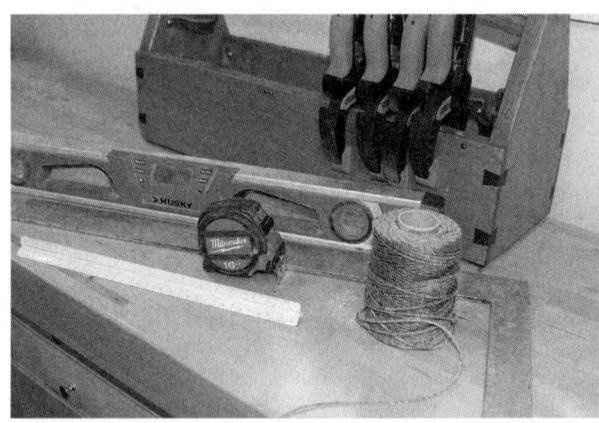

Figure 2-13. *Here are several tools to help you measure and keep things square and in proper position: architect's scale, tape measure, a spirit or digital level, framing square, and twine for snapping a chalk line.*

phone app asks for a bit of a learning curve, once mastered, it's a joy to use.

Get a pair of heavy-duty shears, not lightweight sewing scissors, to cut out cardboard or paper patterns, fiberglass cloth and tape, and peel ply. Some builders find it easier to use a razor blade or utility knife when cutting fiberglass tape, but I favor a good set of shears.

One of the most visible differences between professionally built and amateur-built boats is the

Figure 2-14. *A router and library of bits will help your project look more finished and more professional. Pictured here is a small, handheld router.*

detail in the finish work. That means shaped moldings, chamfered edges, and rounded corners on the boat's wood trim. A router will give you the ability to mill a variety of shaped wood parts. I rely on a heavy-duty, 3-hp plunge router mounted on a base bolted to the extension of my table saw. I also have a smaller and lighter router for handheld work.

There are three pieces of safety gear that you should consider vital to your health: respirator, safety glasses, and hearing protection. A well-fitted dust respirator should be part of your basic kit. Your boatbuilding career will be a short one if you don't take care of your lungs. While some builders may not be able to afford a complete dust collection system, you can certainly afford to spend $35 for a good organic vapor canister respirator. We've tried many types in our shop over the years, and our favorite is a 3M 6000 series—an affordable,

lightweight, and comfortable dust mask. It comes either with twist-on cartridges or with lighter weight twist-on prefilters. The newer GVS Elipse has many fans because it is designed to eliminate the breath-fog problem. I clean my face-piece in the dishwasher every week and replace prefilters at the same time. Safety glasses or goggles may seem like an inconvenience because they tend to scratch easily and breath-fog at inopportune times, but there are easy solutions. When they scratch, throw them away and dig out a new pair—they're not expensive. Now that I'm in my later years, I prefer the type that has a built-in magnifying lens so I don't need to fish for my reading glasses. To avoid fogging safety glasses, periodically spray and wipe the inside with anti-fog formula, the same stuff we Northwesterners use on our windshields. It works. For ear protection, get a pair of comfortable

Figure 2-15. *Safety gear includes disposable gloves to keep epoxy and other chemicals off your skin, earplugs to preserve your hearing, safety glasses to save your eyes, and a respirator with changeable filters for either dust or toxic organic vapor.*

ear defenders, and use them whenever you're working with your power tools. Comfort is important because it makes it more likely that you'll actually use them. One final thought: Buy the respirator, goggles, and ear protection at the same time so you can tell whether they'll all fit and function together.

Many other tools can greatly ease building, but certainly they are not required. Those tools worthy of special mention include a surface planer and an air compressor. If your budget will not stretch that far, you can always rent them. Start small with the basic tools, and if boatbuilding grows on you, so will your tool inventory.

I have found some of my best tools at garage sales and flea markets. Also, look for specialty-tool buyers in your area. They have experience and provide the best advice about the variety and quality of each tool. I have been thoroughly delighted by finds (in some cases, almost steals) at used-tool dealers. There are also tool reconditioners who specialize in older stationary power tools, which are sturdier—often heavier cast iron—and promise longer service than the newer, lightweight models. Shopping the alternative-tool market is good fun and well worth the time. Just don't let it interfere with your boatbuilding time.

CHAPTER 3

Selecting a Suitable Design

If this is your first boatbuilding project, it is wise to begin with a small, simple boat. Building confidence and skill is more important than building the boat itself. Many beginning boatbuilders have undertaken too large a project, only to stall or abandon the effort entirely. It is better to end up with a useful, pleasing boat that serves as a steppingstone to what you ultimately desire than a scrap pile of failed ambitions.

Think of a boat in terms of layers of expenditure. Parts and materials can exceed $15 per pound of a boat's dry weight, and the weight of a boat increases exponentially with its length. Labor hours rise exponentially right along with it. Larger boats also incorporate more systems, and each system raises the cash outlay along with new demands for the builder's skills. For example, an 8- or 10-foot rowing dinghy is a wonderfully uncomplicated watercraft that you can build for just a few hundred dollars in materials. It has no systems apart from the simple hull structure and oar propulsion. It can be transported on a cartop or in a pickup bed. But heat up the ambition, as we humans are known to do, and the dinghy grows to 12 or 14 feet. Now it needs a trailer. It gains a sailing rig and a centerboard, and a 2-horse outboard. Now it has three more systems, and each is a new thousand-plus-dollar layer of expenditure. If ambitions rise to 16 or 18 feet, the design may include a deck, a cabin, and an electrical system. And you're into serious boatbuilding.

Once you've exercised your chops on a small boat, you can hustle up the ladder of ambition with more confidence, a growing tool inventory, improved skills in using those tools, and a better idea of what you want and need in a larger boat. You may find a good use for the first boat as a tender for the bigger one. Or you may learn that you really appreciate the simplicity and intimacy that a small boat provides. Sailboat owners especially have a well-proven axiom: The amount of time actually spent sailing is inversely proportional to the size of the boat.

Here are the most important things to consider as you decide which design to build:

- What storage and transportation requirements need you consider? Do you have a place to store it out of the weather, and what size boat will this shelter accommodate? Will you need a trailer? Check cost and availability; used boat trailers are surprisingly hard to find and seldom inexpensive. While you're at it, check your state's boat-licensing requirements. Many states exempt muscle-powered boats, and some also give a free pass to sail- or motorboats under a specified length in the midteens.

- How will you use your boat? Fishing, picnicking, gunkholing, camp cruising, hunting, exercise? What capacity for crew and gear do you expect to need? What kind of water and weather conditions will you encounter? Do you want a beachable boat? Most designers address these issues in their design catalogs, and if they don't, a quick e-mail should get you an answer. If you'll be trailer-launching, consider the complications of preparing the boat for launch. A daysailer with a single sail and unstayed mast can be rigged in 10 minutes, but a small sloop with a stayed mast might take an hour. The time, effort, and complications of setting up the boat can have an adverse impact on how often you use it.

- What are your performance and comfort expectations? The dynamics of hull forms are complicated, but a few basic principles are easy to understand. Think of a hull form as a pendulum; the deeper its "V," the longer and slower the hull will swing. Conversely, the shallower the hull, the shorter and faster the swing. A wide or "beamy" hull will generally provide more stability; a narrow hull will be faster. For power boats you will have the choice of displacement, semi-displacement, or planing hull forms. A displacement hull's maximum speed is limited by its waterline length (see Chapter 27 for the mathematical formula). A differently shaped semi-displacement hull with adequate power will give you two to three times the speed of the displacement boat. With an obscene amount of power or a very light and properly shaped boat, a planing hull might reach speeds of four to six times the speed of its full-displacement sister.

- Do you love how the design looks? This is not a trivial issue. You're going to be investing a lot of time and effort as well as emotion in building it, and you will find it easier to sustain your motivation if you're in love. Most likely you will be living with this boat for a very long time, and it's important that it fit into your vision of what a proper boat should look and feel like. In this sense, boats are very much like clothing; what matters most is not how something looks on a hanger, but how it looks on you and how you feel in it. Beware of selecting a design *only* because it looks simple and easy to build. Very often those projects get stalled out in the building stage or are virtually worthless for resale.

You can cultivate your boat sense and tastes by reading magazines and books, and if you're lucky enough to live near a waterway, by watching how boats move through the water. With some study you'll start to become adept in the mental process of discerning and critiquing design. Whether a boat has a gently curved or straight sheer, a spoon-shaped stem or a plumb stem, you can develop an educated sense about its balance and proportion. As you study and learn, you'll begin to trust your design instincts. You might also sharpen your social instincts. As great as it is to plan on your spouse or partner sharing your boating dreams, this often

Figure 3-1. *The shape and angles of a hull determine a boat's performance on the water. Displacement hulls are slow, semi-displacement hulls are semi-slow, and planing hulls are fast.*

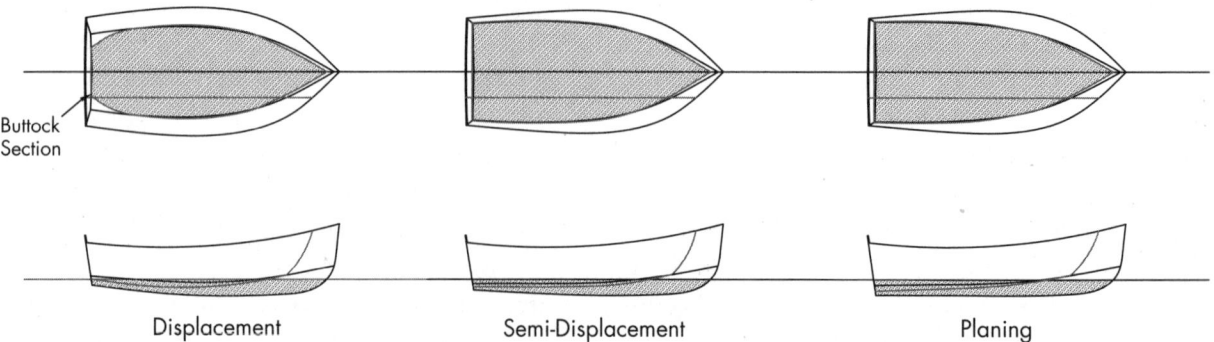

Buttock Section

Displacement   Semi-Displacement   Planing

does not prove to be the case and you will find yourself building and boating alone. The design you choose may well affect this social aspect, so think ahead.

When you've narrowed your choice to three or four different designs, it might be worthwhile to buy study plans for each one. Most designers offer study plans, often with the instant gratification of digital downloading, at nominal cost (Figures 3-2 through 3-4). These provide a clear picture of how the boat's components are laid out, what its structure looks like, and some of the basic dimensions. You won't have all the details you need for construction, but you'll be able to see yourself building and enjoying the boat, and you can chalk its outline on your shop floor to envision how you'll be working around it. Study plans also may give you a sneak preview of how clear and thorough various designers' full construction plans will be.

You can forage the Internet for wooden boat forums and amateur builders' blogs. You'll find many designs discussed and many projects well documented and illustrated. Thanks to this new world of information, there's never been a better time to become a boatbuilder.

Don't try to scrimp on the cost of the plans. A good set of plans is like a road map, guiding you through your boatbuilding journey. Poor plans will almost certainly ensure that you'll lose your way. Look for a classical, clean design, one with what I refer to as a "boat-like" look. Avoid designs that attempt to make a silk purse from a sow's ear or cram the accommodations of a 35-footer into a 25-foot hull. Such designs will just frustrate you in the long run and very well might cost the same as the larger boat anyway.

It may be that you won't find your ideal boat anywhere in the design catalogs, and you'll be tempted to buy plans for something that comes close and then modify them on your own. Some designers take an extremely dim view of this approach, and you should understand that deviating from the designer's intent is a ticklish business. But I have seen some incredibly innovative and thoughtful improvements that DIY builders made to optimize some of my designs, investing many additional hours in mockups to test the modifications. So I guardedly don't object, as long as the structure of the boat—its bulkheads, stringers, clamps, and other reinforcing members—is built according to the plans. This will ensure the craft's structural integrity and seaworthiness. Apart from this consideration, small cosmetic and functional variations will probably do no harm. Just understand that in boat design, every decision is a tradeoff. When you improve one characteristic, you sacrifice somewhere else. In a rather extreme example, if you raise the sheer a couple of inches, you may gain useful interior volume, but you also increase the boat's windage, raise its center of gravity, and add to its weight. But boatbuilding should be a dynamic and creative endeavor, and engaging your own mind and spirit into it is exactly what is needed. If you're successful, you will find this boatbuilding journey to be one of the most profound experiences of your life.

Boatbuilding can and should be a lifetime process, so it's never too late to start. I started as a young boy, first just by observing boats with my dad at the boat shows we would attend. We'd amble up and down the aisles, stopping only when a special boat would demand closer appraisal. My young and untutored eye would occasionally force me to stop and dwell on a boat that looked pleasing and just right. Soon I started to develop a sense of that certain "boat-like" look, and I could begin to envision how a particular boat would perform in the water, and why it might give its owner great satisfaction and pride.

I spent my childhood in Oregon's Willamette Valley where I saw the wonderful McKenzie River drift boats, and on the Oregon coast where I saw the graceful but rugged salmon fishing trollers. (I happen to own an 89-year-old version of the latter today.) To my young eyes these were beautiful shapes, and they were purposeful and functional. The drift boats had a strong sheer and an extremely rockered bottom that allowed excellent maneuverability in rivers treacherous with whitewater. The salmon trollers also had a strong sheer and

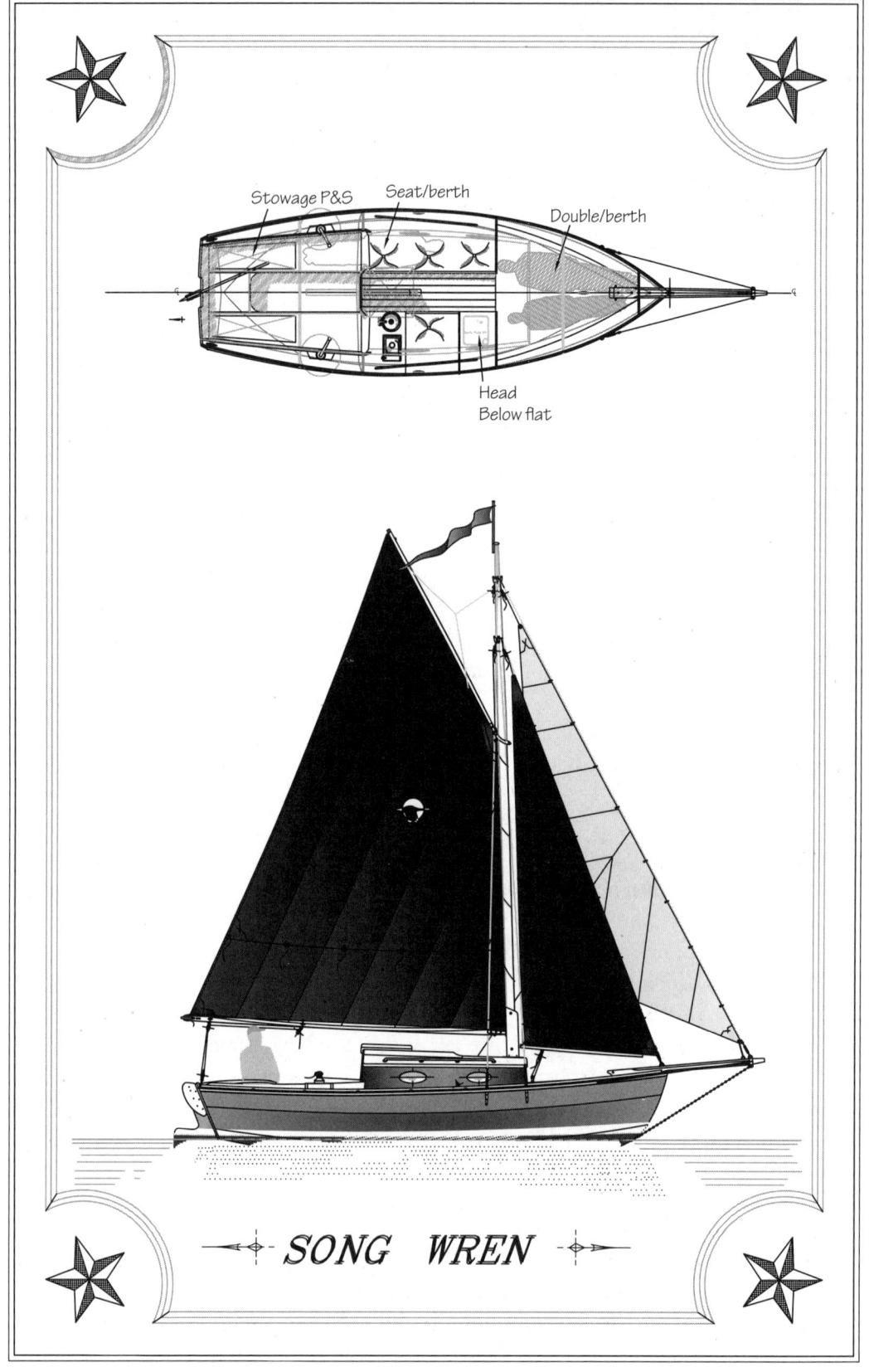

Stowage P&S

Seat/berth

Double/berth

Head
Below flat

SONG WREN

Figure 3-2. *Study plans are informative teasers showing what a boat will look like and a sampling of its construction, typically available for a fraction of the cost for the whole set of plans. This is the sail plan for the Song Wren 21.*

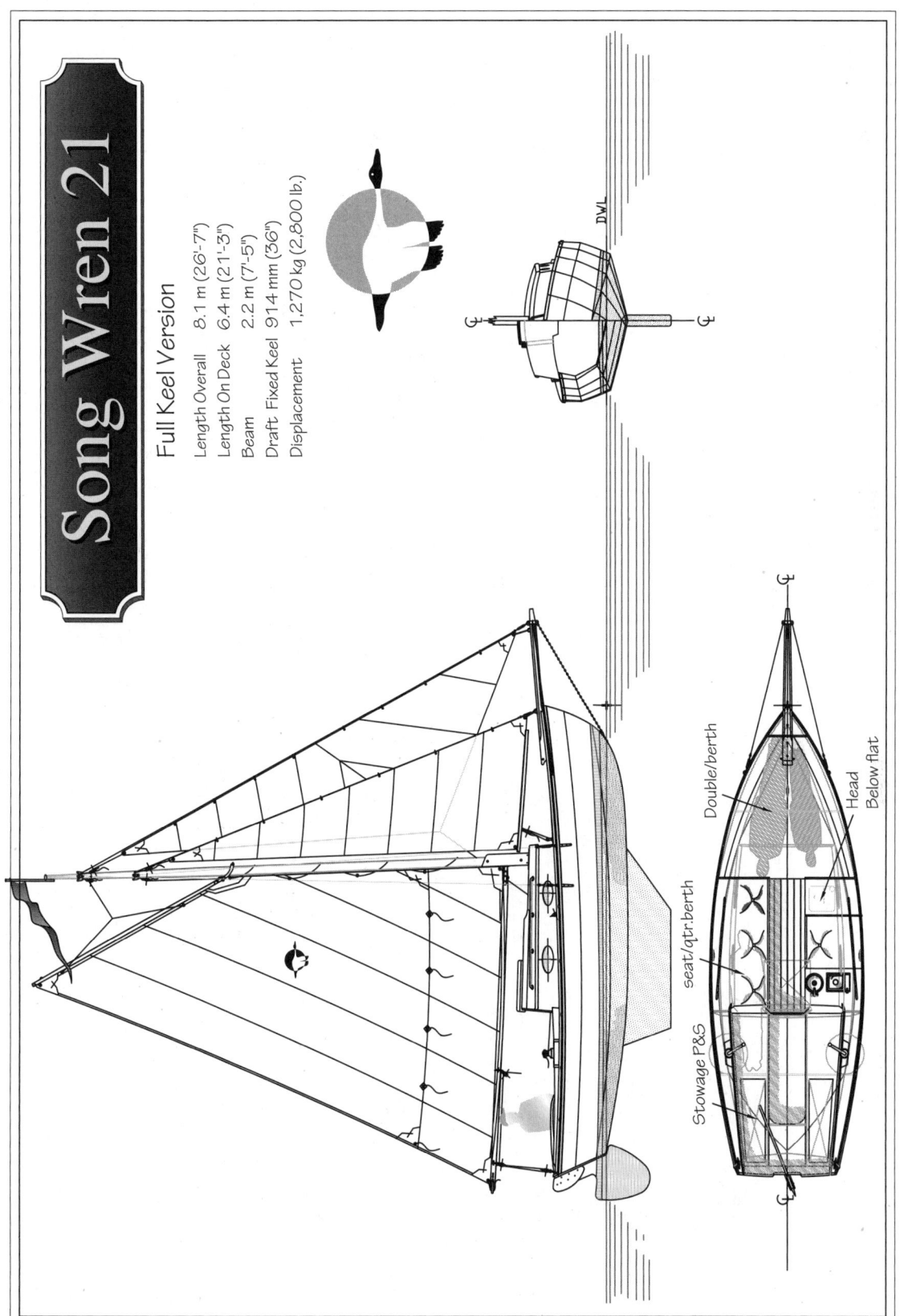

Song Wren 21

Full Keel Version

Length Overall   8.1 m (26'-7")
Length On Deck   6.4 m (21'-3")
Beam   2.2 m (7'-5")
Draft   Fixed Keel   914 mm (36")
Displacement   1,270 kg (2,800 lb.)

DWL

Double/berth

Head

Below flat

seat/qtr.berth

Stowage P&S

Figure 3-3. *The study plan package also includes these views of the interior structure and layout of the cabin and cockpit.*

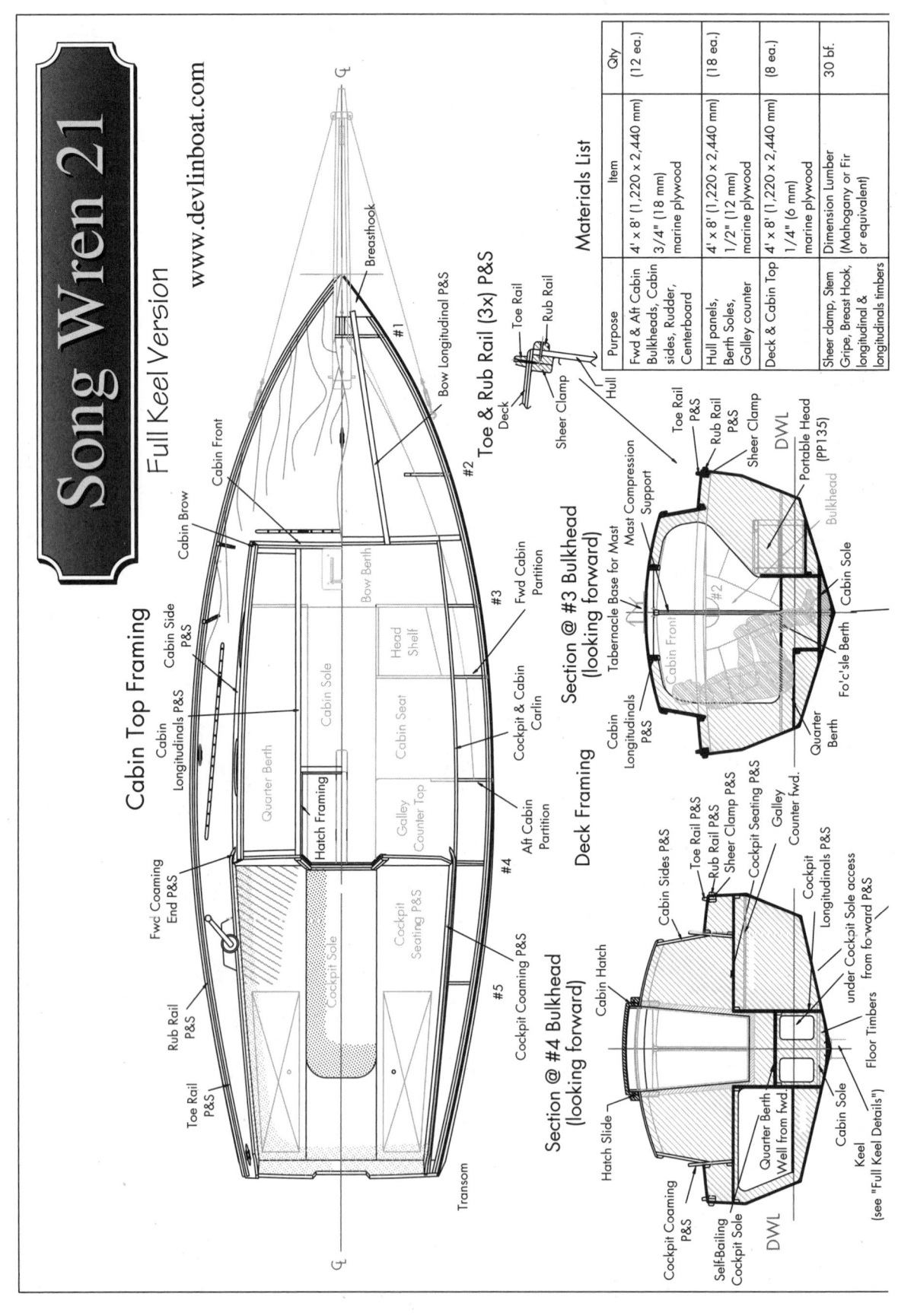

Song Wren 21

Full Keel Version

www.devlinboat.com

Materials List

Purpose	Item	Qty
Fwd & Aft Cabin Bulkheads, Cabin sides, Rudder, Centerboard	4' × 8' (1,220 × 2,440 mm) 3/4" (18 mm) marine plywood	(12 ea.)
Hull panels, Berth Soles, Galley counter	4' × 8' (1,220 × 2,440 mm) 1/2" (12 mm) marine plywood	(18 ea.)
Deck & Cabin Top	4' × 8' (1,220 × 2,440 mm) 1/4" (6 mm) marine plywood	(8 ea.)
Sheer clamp, Stem Gripe, Breast Hook, longitudinal & longitudinals timbers	Dimension Lumber (Mahogany or Fir or equivalent)	30 bf.

Cabin Top Framing

Toe & Rub Rail (3x) P&S

Section @ #3 Bulkhead (looking forward)

Deck Framing

Section @ #4 Bulkhead (looking forward)

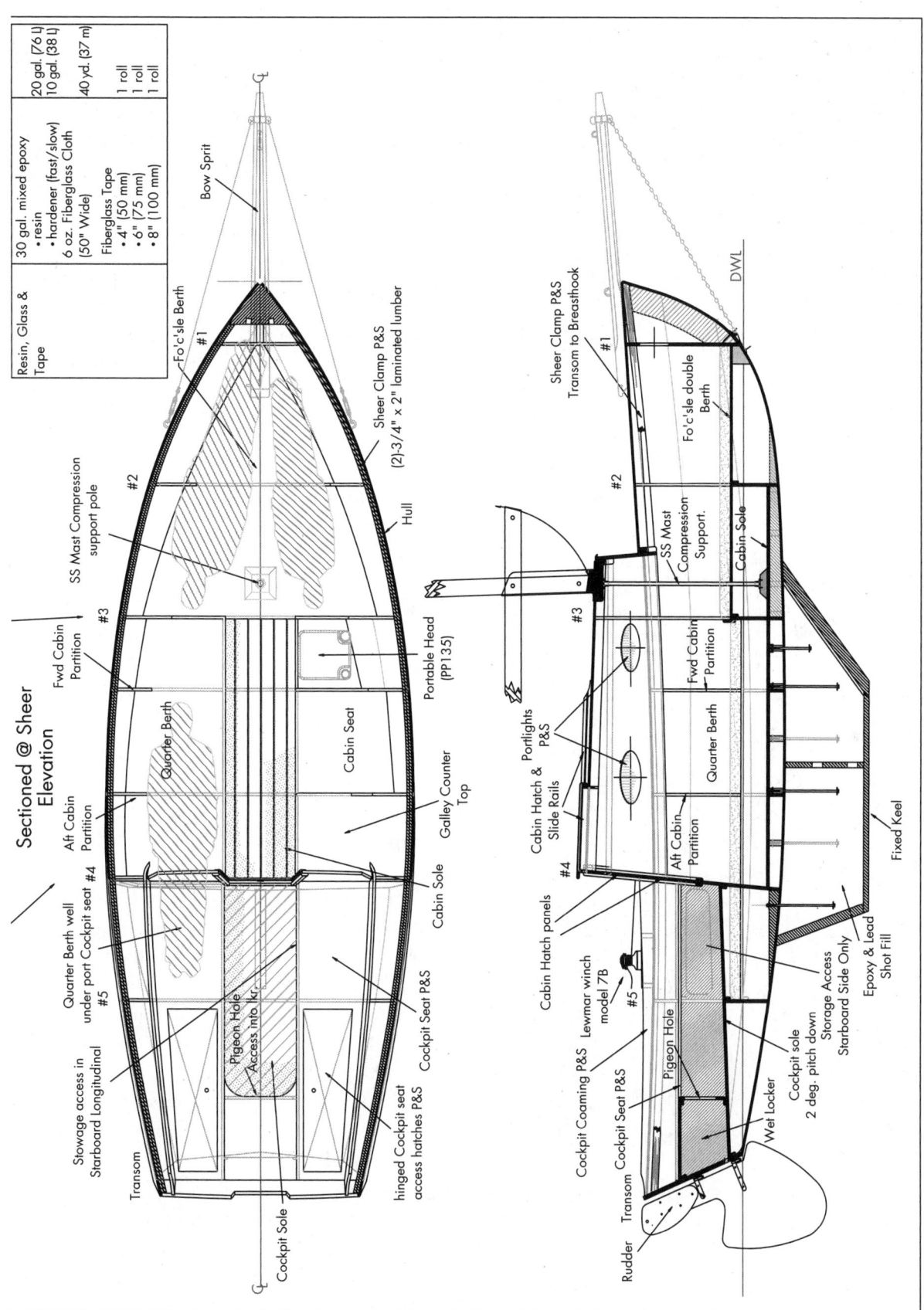

Figure 3-4. *Presentation drawings show the profile and topview of the boat's accommodations.*

Figure 3-5. *Growing up in Oregon, I saw many drift boats on the McKenzie and Willamette rivers. Used for drift fishing and built of marine plywood, they are highly evolved for their niche.*

Figure 3-7. *The North American West Coast has more than 3,000 miles of ocean exposure and has given birth to tough, seaworthy boats for commercial fishing and other work. In turn, these workboats have given much influence to my design and building eye.*

Figure 3-6. *These remarkably seaworthy salmon trollers have fished the Northwest Coast from mid-California to Alaska for more than a century. I have owned this one for more than 20 years.*

a high rise at the bow to tackle the rough Pacific surf where the rivers smacked into the ocean. In my youth, both types were always built of wood, the drift boats always with marine plywood.

It dawned on me at an early point that rough water is unforgiving of eccentric design and poorly constructed vessels. So the characteristics ingrained from my earliest observations were great strength in the shape and overall designs that would handle the roughest conditions.

Since those early years, my designs have been a mix of self-generated and commissioned inspiration, ranging from 7-foot dinghies to 65-foot sailboats and powerboats. All are wood and Stitch-and-Glue in construction, and all share the common heritage of coming from my heart and from my Pacific West Coast roots.

I'm often told that my boats are much different in real life from the way they look in their drawings, and I'm not sure whether that's an insult to my drafting or a compliment to the spirit of the boat. I'll hang on to the thought that it's a reflection of the boat's spirit, earned through the inspiration of its designer and the honest labors of its builders.

CHAPTER 4

Selecting Marine Plywood and Dimensional Lumber

MARINE PLYWOOD

Since Stitch-and-Glue boats are built of plywood, you'll need to know what quality of plywood to buy, how to judge that quality to ensure a first-rate boat, and where to locate top grades. Do your shopping at a **marine lumberyard** that offers accurate information, fair pricing, and a reputation for quality in both marine plywood and dimensional lumber. In many locales, you'll have to order online or by phone; look for sources in boatbuilding magazines and on the Internet.

There can be no compromise: The plywood *must* be marine grade, <u>**British Standard 1088**</u>. Plywood meant for house construction is much more susceptible to water damage and structural degradation. The veneers (the thin layers of wood that the plywood is laminated from) might be of a nondurable wood species, and the grade or quality of those veneers may not ensure a strong and safe boat. Your life may literally depend on the integrity of these materials. Sooner or later, nonmarine plywood will fail in some manner. You'll quickly spend

the few dollars you've saved—and much more— when it's time to repair or replace cheap plywood, and those many dollars and days of labor will be especially frustrating as the whole time you will know that it is your own fault.

The American system of grading plywood uses letters to designate the panel type. An "AA" designation indicates that both outside faces are the best grade, or "A" quality. In American grades of marine plywood panels, the interior plies must also meet certain criteria, such as the solidity of the veneers and the uniformity of species. This is important, since domestic marine plywood veneers can have fir, larch, or hemlock in their interior layers, and most observers can't tell the differences among species buried inside the plywood's sandwich of veneers. Fir and larch are desirable, but stay away from hemlock—it is neither stable nor durable enough for marine use.

After more than 40 years of experience building boats, I'm very biased toward using foreign plywoods that meet at least the BS 1088 standard;

the American grading system allows for too many defects on the veneers and in the interior plies. To put it simply, if you want your boat to last 30 years or more, use imported BS 1088 marine grade. In my early boatbuilding years, I used a lot of domestic fir marine plywood, most of it manufactured very near my original shop in Oregon. Often on a cool morning while walking out to the shop, I would notice dew on the sides of the boats stored in the yard, and on the dewy hulls would be several small, random, elliptical dry areas. These were shaped suspiciously like the repair patches I'd often noticed on the faces of domestic marine plywood. One day I drilled a hole in one such dry area, and lo and behold, it was actually a void in the interior plies.

You can probably guess that it wasn't too easy to drill holes into each of the dry spots on that boat

---

Figure 4-1A. *Top-quality marine plywood will carry a label or stamp certifying it meets the BS-1088 standard.*

to inject epoxy into the voids. Yet, if such holes or voids aren't properly filled and sealed, I would be just asking for moisture to work its way into the interior of the plywood and begin its dastardly work of swelling the fibers, delaminating the plies, and potentially and eventually rotting the wood.

You will pay for cheap plywood many times over. We once constructed two of my 19-foot Winter Wren sailboats side by side using the same epoxy, glass cloth, and work crew. One boat was built of less expensive fir marine plywood, the other of very expensive marine plywood from Holland (my first experience with BS 1088 plywood). Both boats were launched the same day, and both returned to our shop for refinishing during the same summer season about four years later. Both were repainted, revarnished, and checked very carefully for any potential problems. The inexpensive fir plywood boat had only one defect, a gouge where the outboard motor had hit the transom edge as it was being raised out of the water. The other boat showed no problems.

When both jobs were completed, I noticed something strange about the two work bills. The bill for the fir plywood Winter Wren indicated $46 more in materials and almost double the labor. The added material expense was for sandpaper and a small amount of epoxy for the transom-edge fix. Why the extra labor and sandpaper? The fir plywood was nowhere near as smooth and fair as the Dutch plywood, and it simply required more extensive sanding and preparation before repainting.

The owner of the boat built with the premium-grade plywood had spent an additional $900 for the Dutch plywood, but he saved more than $600 in labor for the first major refit of his boat. That savings, if you consider a refit at least every four years, would more than pay for the expensive plywood by the boat's sixth year. In addition, the beautiful grain of the European plywood allowed me to varnish the whole interior of the boat, while the interior of the fir plywood boat had to be painted because of surface imperfections.

Checking is another issue that is most common in fir plywood. These fine cracks, usually running lengthwise along the grain in the face of the plywood, allow moisture to enter the laminate. This is the greatest enemy of marine plywood and the biggest liability of plywood boat construction. Checking, and its attendant problems, may be the result of the plywood manufacturing process. Fir logs are thoroughly soaked and steamed before the veneering process, where rotary cutters peel the log's layers, forcing its curved surfaces to suddenly lie flat. That stress may manifest itself later as these checking cracks.

This checking problem can persist even through epoxy sealing and fiberglass/epoxy sheathing. Individual veneers and glue lines in the plywood may restrict the moisture problem to local areas, but even isolated areas are subject to swelling and contraction, and ultimately to delamination and failure. It is persistent enough to almost convince me against ever again using fir marine plywood in my boats.

Of the imported marine plywoods, I commonly use three species of African mahogany—khaya, sapele, and okoume. All three are available in a variety of thicknesses, and all are suitable for marine use. Hull panels, bulkheads, and decks can all be built with these. Okoume, which is a light salmon-pink color, is the lightest in weight and the least strong of the mahoganies, but I often use it as the base of my Stitch-and-Glue hulls. Khaya and sapele are both a bit heavier and finish up a much darker color when epoxy-sealed. Sapele's grain pattern is particularly beautiful and can make for spectacular brightwork when varnished. There is also a meranti marine plywood (this species of wood originates most often from Indonesia) available currently with very good strength but much greater weight than okoume, and the cost is affordable. You should keep in mind that species available today might only be marginally available 10 years from now, since wood is a dynamic resource and tends to come and go in availability. But the species mentioned above tend to be the major players in the world of fine marine plywood.

Because quality Stitch-and-Glue boat construction depends on top-grade marine plywood,

I often test plywood stock when it arrives at my shop (especially if it comes from a new or different manufacturer). I cut out 4-inch sample squares and without any epoxy sealing, boil them for 20 to 30 minutes in a small pan. Then I take them directly from boiling water to the freezer. When frozen solid, I remove them from the freezer and boil them again, repeating this cycle of heating and cooling at least three times. If your plywood can withstand this torture, your boat should last a long time.

You might also quick dry at low temperatures the test pieces in the oven to more closely simulate what will happen to plywood when it's in service on a boat. Boats experience wild swings of temperature and cycles of soaking and drying. They are trailered over hot highways, launched into cold water, then hauled and dried out again. Keep this constant punishment in mind when you select your materials, especially the marine plywood.

Although epoxies are wonderful products that have made modern wooden boats an economical possibility, they aren't quite the miracle potions some people believe. An epoxy-sealed and encapsulated boat is only as good as its core material. An epoxy and fiberglass skin cannot atone for an inherent weakness in the wood itself.

One final comment about plywood. Rot requires three things to flourish: a food source (which the wood under specific conditions eagerly supplies), oxygen (already present within the wood cells), and moisture. Moisture is the element most within our control. The internal voids in lesser grades of plywood can act like straws, drawing moisture to the interior. Any kind of break in the veneer, sealant, or sheathing—no matter how small—will also act as a straw. While you are assembling your Stitch-and-Glue boat, vigilantly inspect the plywood edges for voids or empty spaces of any type. Always be sure to seal all plywood edges and surfaces with many coats of epoxy to ensure maximum longevity and help prevent moisture invasion and veneer degradation. Plywood edges wick up a lot of epoxy, so they typically require two or three additional coats of epoxy to seal completely.

DIMENSIONAL LUMBER

Many components of the Stitch-and-Glue boat require dimensional (non-plywood) lumber—gunwales, skegs, keels, breasthooks, and thwarts, for example. A larger boat may call for dimensional lumber for sheer clamps, rubrails, sheerstrakes, bowsprits, masts, booms, deck beams, tillers, and cheekblocks (you will find definitions for all these structures later in this book), to name a few. There are two classifications for dry dimensional wood: *air-dried* and *kiln-dried*. It is imperative in any structure that will be encapsulated and sealed with epoxy (as in a Stitch-and-Glue boat) that all wood used be as dry as possible before sealing. You will want the moisture content to be between 8 and 12 percent in your project. Air-dried wood is usually better and more rewarding to work with because the high temperatures of the kiln-drying process can rob the wood of much of its suppleness and strength. You will find that kiln-dried wood is harder and can be more brittle, a distinct disadvantage in boatbuilding. On the other hand, economics and timing may dictate the use of kiln-dried wood since it is usually less expensive and more universally available than quality air-dried lumber. Planning ahead can and will help in your boatbuilding project: that batch of locally sourced wood, properly stacked and carefully air-dried to the 8 to 12 percent moisture content, will be your best friend when the time comes to build your dream boat.

Air-drying wood requires patience and forethought; the wood needs approximately one year per inch of thickness to dry fully. This drying needs to be done under cover and with spacers that allow each board maximum contact with air. During this lengthy process, much care must be taken to avoid warping or cupping, and the wood must be constantly monitored to protect against invading insects or rain moisture. Most lumberyards simply don't trouble themselves with stocking air-dried woods.

Furthermore, small lumberyards tend not to have a good selection of imported hardwoods, usually confining the bulk of their dimensional stock to

Figure 4-1B. *Air-dried lumber is best for wooden boats.*

domestic softwoods and a very few domestic hardwoods. While softwoods are useful for certain parts of your boat, you will still need to find a source of hardwoods for others. Depending on your locale, you might have some excellent indigenous hardwoods and softwoods. Research wood technology texts to find which local woods are durable and stable enough to make good boatbuilding materials. Here along the Northwest coast of North America, we have a lot of Douglas fir (a softwood), which is stable and durable and can be used for almost any part of the boat from keel to mast. To our dismay, most of the best clear, virgin-growth fir was logged years ago, and the highest-quality second-growth wood is being exported to foreign markets as fast as it can be logged. We also have some sources for

Yellow cedar and Port Orford cedar, which also are excellent softwoods and can be used for many parts of the boat. These are in short supply because they too are shipped overseas to foreign markets, but occasionally they can be available to the boatbuilder.

Investigate the boatbuilding magazines for lumberyards that specialize in boatbuilding woods. Indeed, these yards may be your best bet, since turning to an expert is always the way to avoid confusion and eliminate the costly errors of using the wrong type or grade of wood. Any lumber dealer that advertises in a national magazine will ship to you, and the shipping expense may not be as prohibitive as you might think.

Always look for clear-grained wood because knots and defects create untold problems. Often, dimensional wood stock will be bright finished (varnished) and visible on the completed boat. Clear stock will make your building job much easier and will result in less waste.

Especially if it is kiln-dried, look for lengthwise or cross-grain cracks and for any discoloring, which might indicate incipient decay. When you're buying from a distance, be specific as to your expectations. When you want clear stock, make sure the order-taker understands this requirement.

In most cases you will want to use the heartwood of the lumber and avoid the sap wood or the wood that comes from the edges of the tree. Sap and heartwood have distinct differences in strength and durability and usually are very distinctly colored and not hard to recognize. Almost always you will want to cut away any sapwood and keep only the heartwood for your boat.

Be sure to note the grain of the wood—flat, vertical, or mixed. "Flat grain" means that the grain lines (growth rings) are parallel to the wide face of the board. In a vertically grained board, the grain lines are perpendicular to the wide face. In dimensional lumber of square cross section, you can convert flat grain to vertical simply by rotating the piece 90 degrees (Figures 4-2 and 4-3). If you are bending gunwales with a cross section of ¾ × 1½ inch on a small boat, a gunwale with a vertical

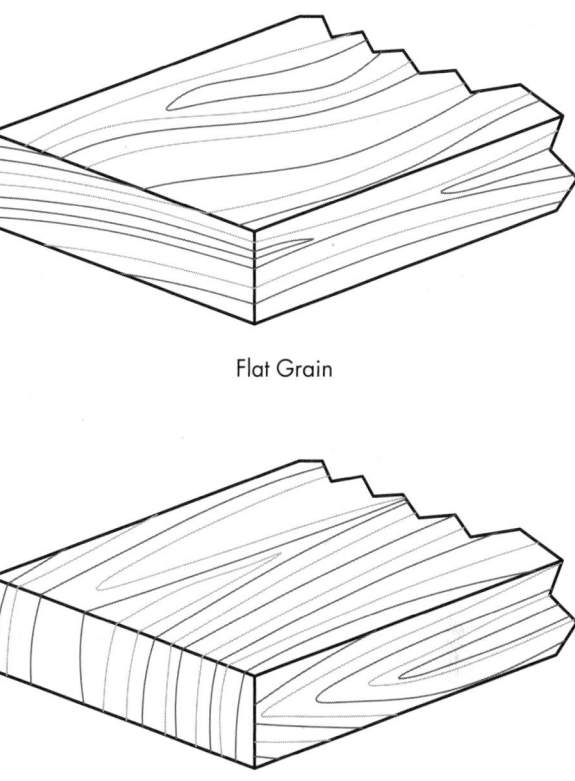

Flat Grain

Vertical Grain

Figure 4-2. *Proper orientation of the grain in your wood can add much strength to your boat.*

grain face will be more difficult to bend, but stiffer and stronger. A flat- or mixed-grain (anything less than grain at 45 degrees to the face) gunwale will bend into place more easily but won't be nearly as strong and stiff. When making a sheer clamp on a larger boat, where multiple layers of dimensional wood must be laminated to form the beam, flat-grained wood will bend into place more readily, fasten more easily, and do so without as much potential for splitting. Conversely, the vertically grained wood will be stiffer and contribute more to the structural strength of this critical area, but the fastenings that help to hold it in place while glue sets might by nature of the grain split the wood, and this would be a very awkward area to repair the splitting.

In the grand old-growth forests of our great-grandfathers' times, the trees were cut 8 to 10 feet from the ground to avoid the burls, butt growth,

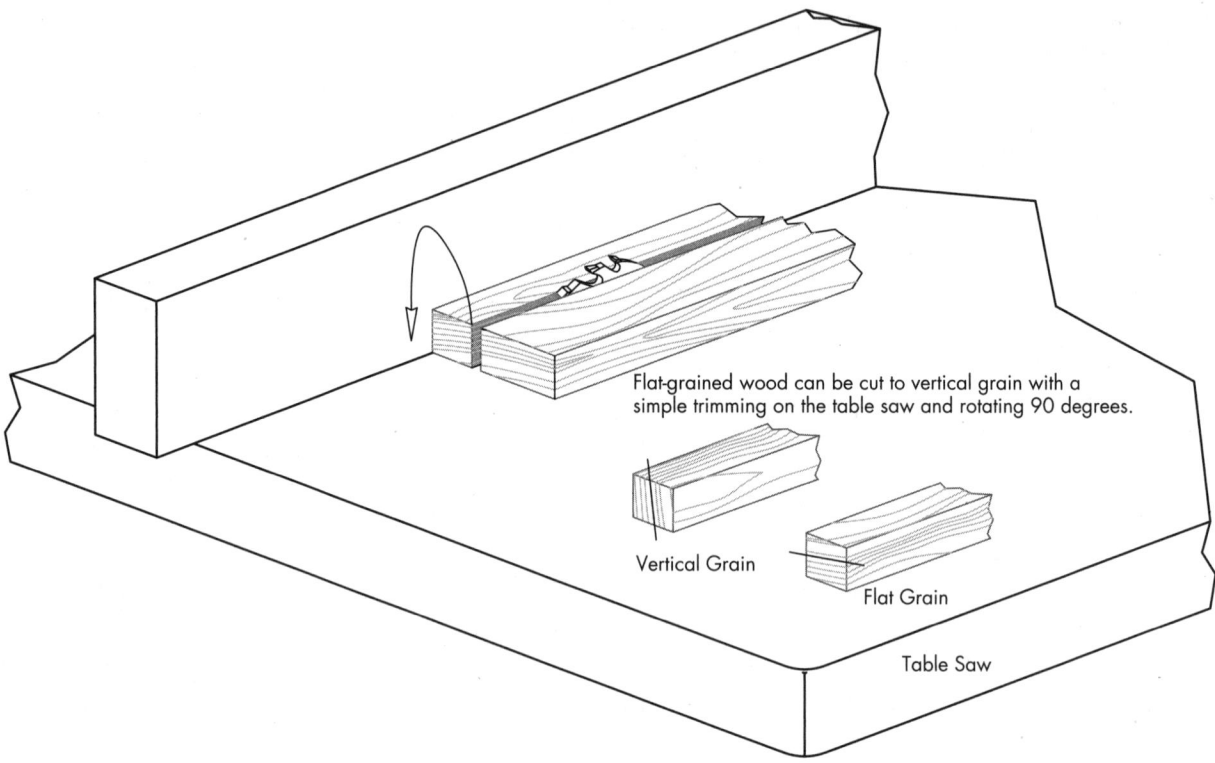

Flat-grained wood can be cut to vertical grain with a simple trimming on the table saw and rotating 90 degrees.

Vertical Grain

Flat Grain

Table Saw

Figure 4-3. *You can orient a plank for ripping on the table saw so that the wood will provide either more strength or more flexibility.*

and other irregularities at their bases. Often there was twisted grain at the base due to weight compression of the tree and other natural causes, and builders realized early on that when this wood at the base of the tree was dried, it tended to crack, twist, and warp. Big trees were plentiful, so they simply ignored the bottoms of the logs. Today, loggers cut as nearly flush to the ground as possible to maximize lumber footage, and butt- or twisted-grain woods end up in lumberyard stock.

To detect grain problems, sight the length of the board, and if the grain veers to the edge of the piece, you have grain runout (Figure 4-4). If it is pronounced, you can assume it is from a butt-cut log, and it may be a visually attractive but very unstable piece of wood. Once glued into the boat, these pieces will be more prone to cracks and joint failures that can be nearly impossible to repair. Use the same criteria for selecting dimensional lumber as you use for selecting marine plywood: Look for

quality, accurate lumberyard information, and fair pricing, but expect to pay a goodly amount for fine new stock.

If you're building on a budget, alternative sources can provide quality lumber, and if you have the time and inclination, you can score some great

Figure 4-4. *A plank with grain running off the edge is potentially weak and should be avoided unless using the wood in a nonstructural function.*

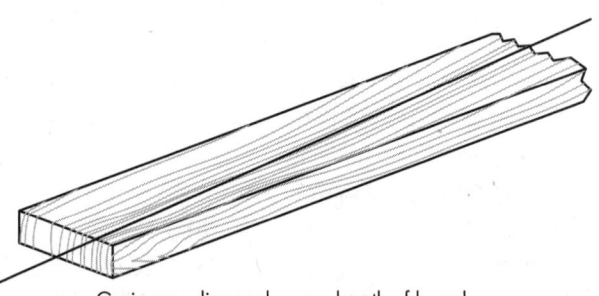

Grain runs diagonal across length of board

wood for your boat. Fine old wood salvaged from buildings, furniture, or other uses can be remilled and repurposed for use on your boat. And small, semiprofessional backyard sawmills are proliferating. They specialize in small lots or wood types that the huge commercial mills don't want to bother with, and they take the time to do it right. In the last couple of years, more than half the wood supplied to my own shop has derived from these sources. The best quality comes from small bandsaw mills. They are very accurate, and if you develop a good relationship with the sawyers, they will soon learn what type of wood you are looking for and how it might best be cut to optimize for boatbuilding. Unless you build yourself a small dry kiln (which can be done fairly easily with a plastic sheathed structure built around the stacks of wood and with a fan in the tent to keep fresh air coming in and exhausting the moisture-laden drying air), your best option is to air-dry this wood for many months. Salvaged wood from old buildings most likely is already very dry.

I often say, "Wood is God's way to keep me working," and it is not that far from the truth. It seems that every time I get a few bucks put away in the bank, my wood milling friends will call with a report of some trees that they are milling. A quick look at the logs and I am once again much richer for my wood/lot, but much lighter in the pockets for green cash.

Wood Types*

Douglas fir   Readily available, fir is light in color with a slightly reddish tone and a long, straight grain. It is light in weight relative to its strength.

_____

* Since I am from the Northwest coast of North America, I will focus on woods generally available in this area.

Figure 4-5. *A shop-built bending wrench helps to bend a sheer clamp into position.*

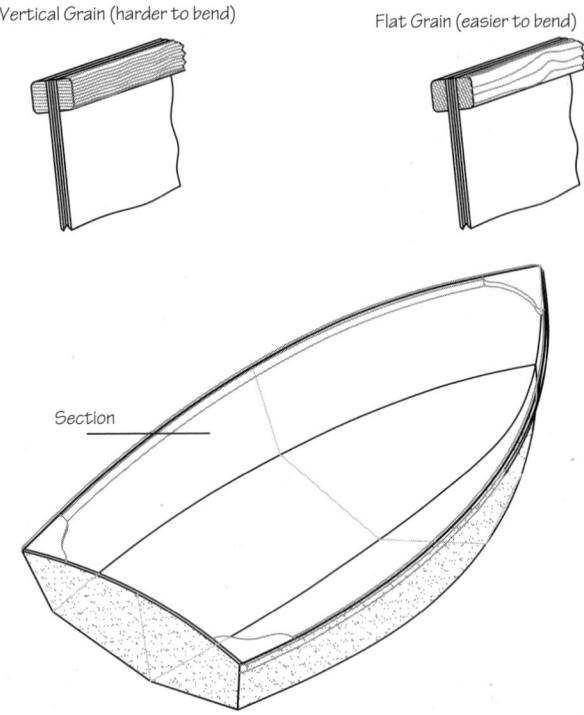

Vertical Grain (harder to bend)   Flat Grain (easier to bend)

Section

Figure 4-6. *Depending on the stiffness of the wood you are using, you might need to use flat grain to bend the sheer clamp around the bulkheads.*

For our purposes, it takes a fine finish and can be easily glued. You will find fir the best wood for keels, stems, stringers, deck beams, and clamps. It is also suitable for masts and spars (though heavier than Sitka spruce, our other best option for these).

Spruce   Forests of clear, old-growth spruce—most notably Sitka spruce—are still being harvested, especially in Canada and Alaska. This is absolutely the best wood for all spars, including masts, booms, gaffs, and bowsprits. Its major advantages are its light weight and extreme intergrain strength (meaning it is almost impossible to split lengthwise as the wood grains lock together to prevent a clean splitting, exactly the physical characteristics you want in spars). Spruce's blond color is similar to that of Douglas fir, though lighter in tone and not as reddish. When epoxy-sealed, spruce tends to yellow. For consistency of color, and if you want to finish that spar bright, you must take great care to seal the

wood evenly and then apply many coats of marine spar varnish to help protect the epoxy from ultraviolet damage (one of the few things that can cause epoxy to fail in its tenacious hold to the wood).

For spars, you will need long, straight-grained pieces of spruce, preferably air-dried. If you use kiln-dried spruce, check that the drying was consistent and that the wood is uniform in dryness and appearance. If you foresee that your building project will stretch over a long period, buy partially air-dried or green wood early on in the project, and carefully regulate its final drying, allowing the wood to stabilize while the building project is underway. When it's time to build the spars, your stock will be ready.

Mahogany   Used extensively in boatbuilding, the best variety available to us these days is Honduras mahogany from Central America and South America (it is plantation-grown in many areas these days). A durable, beautiful hardwood with dark red color and a distinctive grain that takes a bright finish, mahogany will never frustrate you. It glues superbly, seals with epoxy very well, and is always worth its premium price.

Also worth considering are the African mahoganies: sapele, khaya, and okoume. All three are similar in characteristics to Honduras mahogany, with sapele being the most striking in grain and color. Khaya and okoume are lighter in color and don't exhibit extravagant grain patterns, but the physical strength and characteristics of all these woods are very similar to Honduras mahogany.

Meranti mahogany (sometimes called lauan) is recognizable by its speckled grain and a range of colors from reddish-brown to pink. It is not a true mahogany but is more closely related to the cedar family. Its grain is quite fine, and it works wonderfully with hand or power tools. In the U.S. market, lumberyards usually stock meranti from the Philippines or Indonesia. With all the mahoganies, select the darker, denser varieties, and be cautious about using the light pink woods as they usually tend to be less dense and less durable.

Watch for wind shake (separation between growth rings) or jagged lines and cracks across the grain, especially in meranti (although it can be found in any available tropical hardwood). These defects can be traced to logging practices that allow trees to fall over each other and to clear-cutting, which eliminates protection from extreme tropical storms. It's easy to miss these defects until the piece is in place on the boat, so watch carefully. There is nothing so miserable as having spent hours shaping and installing a piece and then discovering its tiny, and fatal, defect. Very often when working with dimensional woods, I will plane smooth the planks that I intend to use and wipe them with a rag wet in isopropyl alcohol. This will show me if there is wind shake or other defects before I have spent hours fitting them into place in my boat. This tip has very often saved much aggravation.

Teak Teak was once a great wood for exterior surfaces, particularly rubrails, sheer guards, handrails, toerails, decks, and seat thwarts. Teak is heavy and durable and will withstand more abuse and neglect than almost any other wood. It is a dark chocolate-brown wood, sometimes with a beautiful and pronounced grain. But old-growth teak is just about extinct. Most teak available today is overpriced and plantation-grown, with large, thick growth rings and nothing like the physical properties of our grandfathers' teak. Another negative is that it does not glue well. Use it if you must, but I have had very good results using sapele, purpleheart, and other species for the exterior pieces that I made with teak in the past. Don't be hesitant to try something new as the world of commercial wood evolves. There are excellent online reference sources that provide precise information on various wood species' weight, workability, rot resistance, strength, and other characteristics.

Cedar Two species of cedar are popular boatbuilding woods. Both are lightweight and have good strength. These woods hold mechanical fasteners well and have excellent durability. My preference is Port Orford cedar. A native of the Northwest, it has a honey-blond color and a sweet, pungent smell that I never tire of while working it. Durable and rot-resistant, it glues well, takes a finish readily, and like fir, has many uses. Virtually any part of the boat could be built from Port Orford cedar. In boatbuilding heaven, I'll make boats of Port Orford cedar plywood.

Alaska yellow cedar has similar qualities, but its aroma is more pungent, often reminding me of juniper berries. Both Port Orford cedar and Alaska yellow cedar are members of the cypress family (not true cedars) and have all the desirable characteristics of cypress: durability, good gluing ability, and high strength-to-weight ratios.

Western red cedar, often used for fencing and house siding, is widely available in the West. Although it is light in weight and wonderfully rot-resistant, it tends to be brittle and splits along the grain very nicely (not normally a quality you want in a boatbuilding wood). You might use it for seat thwarts in a small skiff or dinghy, but I would avoid it for most structural uses in our Stitch-and-Glue boats.

Oak White Oak is a hardwood that runs very light beige to nut-brown in color. It is medium to heavy in weight, has high strength, and is potentially durable—though sometimes I have seen this wood rot with frightening speed. I think this problem may be related to the season of harvest. Oak harvested during the spring, when the sap is flowing at its maximum, seems to have a much greater tendency to rot. If the tree is harvested in the fall or winter and is seasoned properly, I would wager that the wood might last forever.

White oak is used in areas of the boat where extreme strength is needed. Gunwale clamps and rubrails are the two most usual places. I also favor it for tillers, which need to be as stiff and as strong as possible. Its glueability has been questioned, but I have never had any trouble. As with teak or any oily wood, it's best to wipe down white oak with acetone and a clean rag before gluing, and mechanical fasteners should be used to assist the bond to other surfaces.

Red oak possesses many of the same physical characteristics as white oak, although it isn't anywhere near as durable. Its high cell porosity allows moisture to penetrate, thus making it susceptible to decay. I would **not** use it for anything on a boat.

There is one principle I keep in mind when I am selecting oak: When examining a pile, if you find even one piece of wood with rot or decay in it, avoid the lot! An old boatbuilder told me this early in my career, and it has been good advice.

Purpleheart Purpleheart is one of the woods that I have had very good luck in using in my own shop, especially in the last 20 years as its availability has improved. Its physical characteristics and uses are appropriate for virtually all the exterior wood of the boat: rails, rail-caps, handrails, guards, and trim, to name a few. This wood is extremely hard and works well with both power and hand tools, and once you get it shaped into its final form, its stability is its best quality. It has three disadvantages: It is very heavy—close to the same weight as water; it is difficult to glue properly; and I am very put off by its garish purple color. But if you apply an oil or alkyd-based finish, it almost instantly oxidizes to a lovely color much like a very dark red mahogany. Thus finished, purpleheart looks very nice against nearly any color that you might paint your hull. For gluing it, first wipe the faying face (the side to be glued to the boat) with acetone, and then seal it with thinned epoxy, which will help prevent moisture from penetrating. You can then use thickened epoxy, or bed in polysulfide caulks for the gluing, but always back it up with mechanical fasteners. I have found that sealing with 10 percent thinned epoxy will help give a good base for final finishes, either several coats of marine spar varnish or Sikkens Cetol for protection.

Specialty Woods These woods may be used on any boat, particularly to decorate and to highlight customized interior work. My advice is to limit the variety of woods. Using too many different varieties in the interior can be confusing and detract from rather than accent the vessel's structure. Always consider how well any particular wood will glue, and if you are in doubt add mechanical fasteners along with the glue to help keep the structure together.

In summary, I use the finest grades of imported marine plywood, and I prefer fir or Port Orford cedar gunwales and clamps; mahogany floor timbers, deck beams, and interior trim; spruce spars; oak for tillers; and purpleheart for exterior guards and trim. Once again, my concern for design simplicity leads me to caution against using too many different woods. Simple is better.

And for heaven's sake, don't fool with the natural color of any wood. Staining diminishes its ability to seal with epoxy, and the color Nature gave the wood is always the best face to leave on it.

Figure 4-7. *The saloon of a Sockeye 45 illustrates how simple, nonfussy wood treatments can be truly elegant.*

CHAPTER 5

Handling Epoxy

EPOXY RESIN AND HARDENER

While Stitch-and-Glue boat construction was possible before epoxy, this chemical prodigy has let us build larger boats of much higher quality and greater strength. Before epoxy, hull joints had to be reinforced with layers of fiberglass cloth and polyester resin (the normal fiberglass boat-type stuff). The bonding quality of the polyester resin was limited greatly by the builder's experience with the polyester chemicals and the arbitrary mixing ratio of the resin and catalyst. The durability of the boat also depended heavily on proper maintenance and greater care of the boat in storage. If those early boats were left at a moorage or out in the rain without cover, the plywood quickly absorbed water and swelled. When the wood dried, it shrank, and after thousands of these cycles, the polyester resins would lose their bonds to the wood. Many early boats suffered from resin/cloth failures, resulting in exasperating fixes to remedy the resulting problems.

Epoxy's greater bonding strength, its capability of sealing the entire vessel, and the increased durability opened a wide range of new possibilities. Boats could be made much larger. They could be left at a moorage without fear of absorbing water, their structural integrity was enhanced, their maintenance issues were greatly reduced, and their life expectancy increased exponentially.

Epoxy is a multipurpose material for boatbuilding. It is used as a coating to seal all of the plywood and dimensional wood surfaces and as a glue for the structural joints. It can be mixed with wood flour or other fillers to make strong structural fillets.

Epoxy is also used in combination with fiberglass or other synthetic cloths to reinforce seams (literally welding the structure) and sheathe exterior surfaces. This hull sheathing of fabric and epoxy is critically important for three reasons: It protects the plywood against moisture intrusion, provides a hard skin that resists abrasion, and adds to the hull's strength and stiffness.

The downsides: Epoxy requires precision when mixing the resin and hardener, it can be toxic if improperly handled, and it is expensive.

Epoxy is a two-part adhesive. The resin component is a clear, syrupy liquid, while the hardener is typically thicker (more viscous) and can be anything from water-clear to the color of honey. These two fluids must be mixed in the exact ratio specified by the manufacturer. This ratio can differ greatly from one epoxy system to another, so follow the manufacturer's directions precisely. A deviation of as little as 5 percent in either direction can undermine the final physical properties of the cured epoxy. An error of 10 to 20 percent can prevent it from curing at all. And once you have mixed the two fluids in the correct ratio, they must be fully and thoroughly stirred, scraping the surfaces of the mixing container and moving all the liquid around until you are sure of a complete blend.

When mixed and cured at room temperature, epoxies undergo an exothermic reaction, generating heat as they cure. They need this heat along with some heat from the environment to become a solid. Ambient air temperature affects the speed with which the epoxy sets up or, as I say in my shop,

Figure 5-1A. *Epoxy, wonderful material that it is, still needs careful handling for best results.*

"kicks off." The optimal temperature for almost all the room-temperature-cured epoxies is around 75°F. Lower temperatures mean a slower cure rate; higher temperatures mean faster.

To counter this effect, epoxy manufacturers formulate various "speeds" of hardeners. Along the U.S. Northwest coast, temperatures stay fairly moderate. Consequently, I use a fast hardener through much of the year, switching to a slower formulation in the two or three warmest months. The resin formulation is always the same; only the hardener varies. When you're early in the epoxy learning curve, use a slow- or medium-speed hardener. As you gain experience, you may want to move into the fast range to speed the curing process. In a production shop where speed is essential, we typically want to use the fastest or hottest hardener possible.

Under average conditions, epoxy with medium hardener takes roughly 24 hours to reach an easily sandable state, but this varies with temperature. In summer, when temperatures range between 80° and 90°F, I have been able to epoxy a small project in the morning and work with it the same afternoon.

With epoxy's self-heating characteristics, it is not uncommon to have a container of it smoke or even foam up and "flash off." The smaller, more confined, and narrow-mouthed the mixing container, the hotter this exothermic reaction. When a batch of epoxy overheats in your container, remove it from your workspace immediately. Use caution: The epoxy can become hot enough to melt a plastic container! And don't breathe the fumes. Move it outside and allow it to cool off. It will solidify in a matter of a couple of hours and then can be disposed of in the trash (Figure 5-2).

If I find the epoxy flashing off or boiling, I mix smaller amounts and work faster, or I switch to a slower hardener. It also helps to pour the batch into

Figure 5-1B. *WEST System is one of several brands of quality marine-grade epoxy resin. Decide on one brand and use it consistently.*

Figure 5-2. *This container had too much epoxy in too confined a space; it was smoking before we (quickly) took it outside the shop.*

a large, flat container with a lot of surface area to dissipate heat. The more confined the container, the faster the self-heating process; and the more unconfined the container, the slower the self-heating process. A painter's roller tray is a great way to slow the heating down and is also easy to use for rolling epoxy onto the parts of the boat you are working on. This can add considerably to the epoxy's working time.

You must be patient with epoxy when working in an unheated space. In winter, when the temperature in an unheated shop hovers in the 40s, an epoxied surface can remain slightly tacky even after 24 hours. It is even possible in extreme cold that epoxy itself cannot generate enough internal heat to cure and will stay in a tacky limbo forever. I have used a kerosene space heater to warm the shop and accelerate the rate of cure, but I have found, to my dismay, that the slightly incomplete combustion of kerosene leaves a minute residue in the air that can foul the material surfaces, interfering with the epoxy's ability to bond. Both propane and natural gas space heaters make water vapor as a combustion by-product and thus create similar problems. If you need heat in your shop, one of the best options is an infrared electric or gas quartz heater as I described in Chapter 2. Other cold-weather solutions are to keep the epoxy and hardener themselves warm, thereby slightly assisting the natural

warming that happens when they are mixed; and keeping the boat itself warm with a small space heater underneath the upside-down hull. The rule with epoxy is that as long as it either self-heats or enough external heat is available for a sufficiently long time, room-temperature epoxies will cure to a solid.

None of the suppliers of boatbuilding epoxies actually manufactures its own raw resins and hardeners. Each supplier formulates its resins and hardeners by mixing base stock material with various additives that vary the epoxy's curing properties and performance. I strongly recommend using one of the epoxy systems that has been designed for marine use and the marine environment.

Among the marine epoxy systems available today a very rough rule can be applied: the greater the proportion of resin to hardener, the harder the cured epoxy product. Epoxies that use two parts resin to one part hardener are generally a bit more flexible than those using five parts resin to one part hardener. Mixing the two components is certainly simpler with a 2:1 system, which is what I use. (Imagine the potential for mistakes at the end of a long day when you must calculate and measure the amount of 5:1 hardener for 12 ounces of resin—2.4 ounces, to be exact.)

If you are confused by suppliers' claims as to the physical properties of their systems, I recommend you experiment on your own. Ask other boatbuilders you've met for their recommendations. Try to settle on one epoxy system that will meet all your needs. Consider product availability and access to technical help. Select a manufacturer that supplies adequate technical manuals; the more specific the manual, the better. Some formulators have taken the additional step of demonstrating how to use their products in specific boatbuilding applications, and they maintain a staff of experienced technicians for telephone assistance. A high level of service is a distinct advantage.

Since epoxy is a critical component, don't shop for low price—buy the best available. Most of the epoxy failures in boatbuilding have been with cheap systems or those not really designed for marine applications. The better epoxy systems come from reputable companies that have their own research staffs constantly testing and striving to ensure a high-quality product that will always work in a marine environment.

You can measure the epoxy and hardener with either disposable graduated measuring cups (my preferred option) or with calibrated dispensing pumps that mount individually on the resin and hardener containers. There are also separate mixing pumps into which you decant epoxy resin and hardener into individual containers on the pumps themselves, an expensive but usually reliable dispensing method. In my shop we exclusively use the graduated cups because the individual pumps are too slow when mixing large quantities, and we use the 2:1 resin to hardener ratio systems to ease measuring and mixing issues. For building a relatively small boat in a home shop, the container-mounted pumps may be fine. To keep from making a mistake in the proportions, set up your resin and hardener containers so that you always pump in the same order, and never change it.

I routinely check the epoxy work of the previous day when I open the shop in the morning. If a surface seems tacky or not properly cured, the culprit is almost always improper mixing. A shortage of stirring is a more likely cause of failure than mismeasuring. *You cannot stir too much!* Be sure

to scrape around the sides of the container, then scrape the mixing stick on the container lip and follow that with another round of stirring. There is no way to determine the adequacy of mixing by looking at it, as both components are for the most part clear. The only proof is finding a fully cured job the next morning. Colder shop temperatures will make the epoxy more viscous, which can affect the accuracy of a pump metering system. If the pump dispenser is clogged, use chemical-proof gloves with long, sleeve-covering gauntlets to clean all the parts with solvent, preferably denatured alcohol. Before reassembling, lubricate the moving parts that won't have physical contact with the epoxy, being careful not to contaminate the resin or hardener sections of the dispenser with the lubricant. After the plungers are fully reloaded with the liquid, use a measuring cup or a weight scale to be sure that the correct individual amounts of resin and hardener are being dispensed. As you will see, cleaning pumps is a fussy and messy job.

Figure 5-4. *After more than 40 years of boatbuilding I have concluded that mixing pumps were invented by the Devil. Hand measuring with graduated cups or a weight scale gives more consistent results.*

Figure 5-3. *Graduated cups are good for accurate mixing and faster than pumps when you need larger quantities.*

Figure 5-5. *Stationary metered pumps are used by some professional shops. They work very well when they are working well but are really difficult to troubleshoot and fix when they quit working well.*

All measuring devices must be kept clean to function properly. They will tend to gunk up, collect dust, and build up residue until they won't serve their intended purpose accurately. During his annual visit to my shop, the local fire marshal always comments that these buildups on dispensers are a fire hazard, and no amount of arguing seems to dissuade him. Given all the hassles with dispensers, you might see why I think graduated cups work best because we simply throw them out when they're past their prime.

SAFETY

The most controversial aspect of epoxy use is the matter of safety. Because epoxy plays such a key role in Stitch-and-Glue boatbuilding, it's important to have a clear understanding of it. There is no

way around it: *The improper use of epoxy can be hazardous to your health*. But constant vigilance and continuous care for safe and proper use will minimize the hazard. Boatbuilders using normal precautions and staying safety-minded at all times can use epoxy with the best of results while fully protecting their health.

My strongest advice is to keep epoxy off your skin. Prolonged contact with the resin and hardener can cause an allergic reaction—sensitization—in some people. Once sensitized, the slightest contact with the resin and hardener, their fumes, or even sanding dust from epoxy that hasn't fully cured can trigger a reaction.

Epoxy vapors can be hazardous at high concentrations or if you are already sensitized. High ambient temperature, an unventilated room, or spraying the epoxy—which I don't recommend in any circumstance—all can increase the fume concentration to the point where you need a respirator. Keep in mind that there are no active solvents in a marine epoxy system, so there is really not much aroma you can detect in the vapor. Never breathe the sanding dust of partially cured epoxy; it will cause severe respiratory irritation. It's always wise to use a sander with a vacuum attachment.

Keep epoxy off your tools, and always wear gloves that protect wrists as well as hands. For the big epoxy-intensive projects such as sheathing the hull, a disposable Tyvek suit is advisable (Figure 5-6). I know of three examples where boatbuilders threw caution to the wind and suffered the consequences. Two were first-time builders, but one was a professional who should have known better. The common denominator was failure to use proper gloves. The professional worked at another boatbuilding shop and was a reckless fool in all aspects of his life. He refused to use gloves and would plunge his hands into acetone at the end of each job to clean off half-cured resin. While using urethane paints, he would refuse to wear even the simplest dust-filter mask, let alone an organic-vapor respirator or a fresh-air system. Predictably, he experienced lung damage from the urethane paint and spent several days spitting up blood. In addition, the exposure to the epoxy caused a rash on both wrists and his forehead that resembled a reaction to poison oak. The rash would disappear after five or six days if he stayed clear of epoxy, but as soon as he walked back into the shop, it reappeared. In the end, he had to give up boatbuilding with epoxy altogether, and the last I heard of him, he was at work in a tin can factory.

When I consulted with the two amateur builders, we traced their reactions to the cleanup process. Most gloves available to boatbuilders are adequate for epoxy but will never stand up to cleanup solvents, especially the more aggressive acetone or lacquer thinner. The fingertips of the gloves are weak, and after normal use, the solvents can leak through the glove to the skin. In both cases, I found that uncured epoxy had repeatedly been allowed to stay in contact with the builders' hands. Over time, each experienced skin sensitization. When cleaning up, discard the thin latex gloves you used for epoxying and don slightly heavier and much sturdier solvent-proof gloves.

And then there's "Devlin's Law," a variant of Murphy's Law. After a goodly amount of experience, I have identified three natural temptations that you will experience when working with epoxy. *Once you have epoxy on your gloves, you WILL*

Figure 5-6. *Disposable gloves are essential for protecting your skin during any epoxy work, and I recommend Tyvek coveralls in addition for the big and messy jobs.*

have an itch on your nose, your eyes WILL need to be rubbed, and you WILL begin to sweat and need to wipe your brow. I guarantee you'll experience these urges, and just as surely, if you succumb to temptations, you will experience some nose or eye sensitization due to epoxy exposure.

There is simply no alternative to constant vigilance: using safety gear, working as cleanly as possible, and not getting epoxy on your skin. Keeping Devlin's Law in mind, note that one reason for wearing a canister respirator—apart from the fumes and dust—is to keep yourself from scratching your nose. After more than 45 years of using epoxies almost daily, the only reaction I notice is a slight constriction of the throat during extended use. When I use a respirator with organic vapor cartridges, I never experience even the throat irritation.

Of the two epoxy components, the hardener is the more toxic. Keep this fact in mind, particularly when cleaning the hardener side of your dispenser. Extreme caution should also be used when sanding partially cured (green) epoxy surfaces, as may happen in the winter in an unheated shop. Always wear a respirator and protective clothing, even if it's only street clothes that are laundered daily, and cover all parts of the body likely to come in contact with uncured epoxy. If you insist on keeping your beard, a full-hood, powered-respirator fresh-air system may be the only answer, since regular cartridge-type respirators will not seal properly over a beard.

I have seen instances of almost magical acts of reverse gravity in which epoxy or its resin and hardener components splashed *up* into a boatbuilder's eyes. In each instance we had to rush the victim outside to a water hose for a lengthy flushing of his eyes, and then rush him to the emergency room where the doctor repeated the process—not something anyone would do by choice. Wear safety glasses or eyeglasses at all times. If your eye protection is uncomfortable, at some point you'll find yourself working without it—and that's when accidents happen. These days I routinely need reading glasses for close-up work, so that's what I wear in the shop. Whatever kind of glasses you settle on, wear them constantly in the shop so that you get used to them.

The bottom line, my macho friends, is to respect these chemicals; just because the hazards are invisible does not mean they are absent. If you are apt to disregard such hazards and won't adopt a fervent attitude about safety, then I advise building your boat using traditional methods, and stay away from Stitch-and-Glue construction.

And even the protection can't be taken for granted. I've also seen a worker develop nasty-looking, painful hand rashes as a reaction to latex disposable gloves, which in his case was probably a reaction to the talcum powder in them. He was fine after he switched to non-talc gloves over soft, lightweight cotton liner gloves.

Always shower after a work session; it will help keep your body clean and healthy. Launder your boatbuilding clothing daily. Wearing epoxy-encrusted clothes day after day just continues the exposure to uncured resin or hardener.

BASIC EPOXY TIPS

I always try to use the hottest (fastest-curing) hardener system possible, taking into account the weather and temperature. I have found that simple trial-and-error experimentation with various hardeners helps me identify the appropriate combination for the day's temperature and the job at hand. Keep records for which works best under specific conditions.

It is also important to store the resin and hardener at a consistent temperature, especially during the winter. The better you control the epoxy's temperature, the surer and more consistent the cure rates will be.

A way to keep the resin and hardener at a uniform temperature is to store the dispensers in a heated box. In my shop we use old refrigerators with 100-watt incandescent light bulbs for heat. With the compressor off and the light bulb rewired to stay on, the sealed interior stays at about 80°F.

The freezer compartment also makes a handy place to store gloves, stir sticks, mixing cups, fillet squeegees, and other epoxying paraphernalia. If you don't happen to have a spare old refrigerator, or you didn't stock up on now-obsolete incandescent bulbs, a quick warmup project (so to speak) for your boatbuilding could be a simple plywood box with an inexpensive Golden Rod (a low-wattage storage dehumidifier-heater) installed inside it. With a little experimenting with a thermometer, you can tailor a heat source to give you consistent heat.

Thin coats of epoxy applied with a roller or squeegee take more time to cure than thicker applications such as a taped epoxy-and-fiberglass hull seam because the heat dissipates so easily from a thin epoxy coating. When seal-coating panels of plywood with thin coats of epoxy, you can apply an external heat source such as a heat gun or heat lamp to warm the entire coated surface. This will also help level the epoxy coating so that your surfaces are smoother and require less labor to prepare for paint or varnish. Use caution, however, when applying heat to the first coat of epoxy on a dry wood surface. It's quite easy to lower the viscosity of the epoxy sufficiently to drive a small amount of it into the wood grain. While this doesn't compromise the sealing process, it may create finishing problems because the thin epoxy can displace air bubbles (out-gassing) from the wood grain, causing the epoxy to cure with a pocked surface and necessitating another sealing coat. You can avoid this problem with a slight, very even application of heat.

Particularly when humidity is high, a coating of epoxy may develop a foggy or waxy appearance called amine blush. This is a chemical reaction among water vapor, carbon dioxide, and fluid epoxy, and it will interfere with the bonding of anything you try to apply on top of the epoxy—paint, varnish, or further epoxy coats. Happily, this film is water-soluble, so all you have to do is wash the affected surface with clean water and wipe first with a Scotch-Brite pad and then lint-free cloths.

But be careful! Moisture must not be allowed to drip or pool onto any unsealed plywood edges.

Epoxy also seems to magically appear on every tool, clamp, and boatbuilding device in the shop. The glop can rear its ugly head anywhere. Its appearance may not bother you initially, but soon this residue will diminish the usefulness or precision of your tools. That's why it's important to work cleanly and methodically. Epoxy solvents are eternally controversial, but here's my rundown:

Acetone is a very effective solvent. Its disadvantages are extremely rapid evaporation, risk of fire from high concentrations of its vapor, and the fact that stray smudges or spills will ruin painted surfaces and some plastics, such as clear Plexiglass and Lexan. But for gluing oily woods it excels and it does cut or thin mixed epoxies very effectively.

Lacquer thinner is very much like acetone but evaporates slightly more slowly, giving more working time with the solvent itself. Like acetone, it is also highly volatile. Using it around paints or plastics will be disastrous. But it does work very well for cleaning tools.

Acetone replacement solvents such as Bio-Solv are not nearly as volatile and are quite effective. They are expensive and will dull paint, but they are somewhat more environmentally benign than acetone. These products are very smelly and can permeate the whole shop.

White vinegar has its fans, and it has the advantages of being cheap and entirely harmless, as far as we know. But it's not as effective as these other solvents.

Denatured alcohol is the closest thing I know to a universal solvent. It's inexpensive, environmentally friendly, and it won't hurt painted surfaces or you (as long as you don't try to drink it). It is not nearly as fast in cleaning epoxy as acetone, but it is usable on a wide variety of epoxied surfaces and is easily available at your hardware store.

Paint thinners, or mineral spirits as they are often called, are useless as epoxy solvents. They will work very well for cleaning up polysulfide caulks and sealing compounds such as 3M-5200 or Sikaflex—more on these later.

Whatever solvent you use, keep it away from your skin and most definitely your eyes. Develop the discipline of wiping tools and shop surfaces clean before stray epoxy cures to a rock-hard crust. If one works nervously under pressure, it may be difficult to work with epoxy, but being careful, mindful, and methodical will go a very long way toward controlling the mess.

FILLERS

Fillers are used to modify epoxy for various applications, from fairing compounds to structural adhesives. Fillers are always added after the proper epoxy resin/hardener mixture has been thoroughly blended. You'll find a great variety of epoxy fillers on the market, including specialized formulations for everything from high-strength hardware bonding to easy-sanding fillers. Except for wood flour, they are all quite expensive. In my own shop I have found that a great variety is unnecessary. I have narrowed down our needs to only three basic fillers: wood flour, CAB-O-SIL, and microballoons. These adequately answer a variety of boatbuilding needs and help to keep the overall material cost of the hull down. I might additionally have a small amount of high-density filler on hand for some hardware bonding, but this amount is so small as to be insignificant.

No matter which of the following fillers you're using, you should seal it after curing and sanding with a coat of epoxy. All fillers are porous to some degree, and the epoxy sealing will help to seal up that porosity and to protect the joint or patch.

Wood flour fits the bill for Stitch-and-Glue boatbuilding because it creates exceptionally strong joints. It is both a bulking and a thixotropic filler, which means it will make a thickened epoxy that tools easily into joints but also stays in place once you've finished the application. Wood flour is little more than the finely ground sawdust commonly used in the baking industry as a cellulose

Figure 5-7. *Precutting and saturating fiberglass tape in a shop-built glassing box minimizes mess and epoxy waste.*

filler for breads (fiber additive, if you must know) and in the manufacture of wood putties for home construction and cabinetry. It is finer than the sawdust made by your table saw or belt sander. Wood flour is available from most epoxy suppliers and many chandleries, and in most cases is not expensive.

Wood flour mixes uniformly with epoxy, making a thick paste. In my shop, we use a mix of CAB-O-SIL and wood flour in $\frac{1}{3}$ ratios to make the fillet material smoother and easier to work. For the right consistency, add the filler until the shininess of the epoxy has disappeared and the mixture

Figure 5-8. *The workhorse of fillers is wood flour/epoxy mixed to the consistency of thick peanut butter.*

looks and feels like thick, creamy peanut butter (Figure 5-8). I use this paste to create the base for our Stitch-and-Glue composite seam wherever plywood sheets meet in the hull structure, and to cove (fillet) the joints for all permanent interior structures. Whenever I attach a cleat or shelf, I use the paste on the faying surfaces to glue the components in place.

By its nature, wood flour is the perfect partner to join plywood panels and the wood components of the boat because it shares the same cellulose composition and provides a close color match. I find it disconcerting to see a well-built boat with purple microballoons or white microspheres glaring out from every seam. If you plan to finish your interior bright, wood flour is the only way to go.

You might ask, as many do, "Why use a filler at all?" It's nearly impossible to cut all components for a perfect fit, and there will be slight gaps between the surfaces being bonded together. Clear epoxy will run out of the joints or fail to fill the gaps and voids. Thickened epoxy will eliminate both problems while creating a joint that is actually stronger than the pieces being joined. If your joinery is perfect, then by all means show it as such, but you can create a fine-looking boat with structural fillets that's probably stronger to boot.

CAB-O-SIL is a white powder that is used primarily as a thixotropic additive (Figure 5-9). It helps prevent epoxy from sagging when applied to vertical surfaces, and it can enable you to apply a thicker sealing coating of epoxy in a single pass, whereas

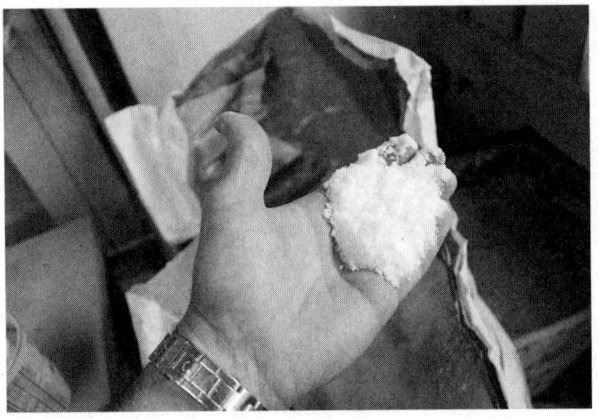

Figure 5-9. *We typically mix CAB-O-SIL about 50-50 with wood flour for a smooth filleting blend.*

Figure 5-10. *Microballoons mixed into epoxy form a thick purple paste that can be sanded much more easily than wood-floured paste.*

unthickened epoxy might require several coats to achieve the same depth and thickness. When blended with wood flour, it's not colored enough to take away from the wood flour's natural color. This is a versatile filler that can be combined with other fillers to create custom blends tailored to each specific task.

Microballoons   When filling low spots with a thickened epoxy fairing compound in preparation for the final painting of the boat, wood flour has a couple of drawbacks. It doesn't sand easily, and its density is greater than necessary for a simple filling/fairing operation. In such cases, I prefer an easier-to-sand compound made with epoxy and microballoons. Microballoons are lightweight phenolic spheres, light purple in color, and when mixed with epoxy, they make a dark purple filler that is easy to sand and sculpt (Figure 5-10). I think its color is unsightly and distracting in boat interiors unless the surface is painted, but I do like its ability to hold a feathered edge when sanded and to fill an unfair surface more

rapidly than a wood flour mixture. **Microspheres** are small, white glass spheres that are almost microscopic. They can be used in addition to or in place of microballoons. The physical properties and uses are about the same (Figure 5-11).

Finally, there are other epoxy fillers that can be real time-savers. West System's 410 Microlight fairing compound is 30 percent easier to sand than microballoons, mixes into the epoxy faster, and has a tan color similar to wood flour's natural color. System Three's QuikFair is a similar compound. These are less dense and easier to sand than microballoons, and while they are more expensive, they surely save sanding and fairing time (Figure 5-12). They are not suitable for structural filleting (see Chapter 14). After sanding Microlight or QuikFair, reseal the surface with epoxy to help eliminate any difference in porosity (which would foul a paint job quickly).

Figure 5-11. *Microspheres are similar to microballoons but are white in color.*

Figure 5-12. *Microlight is a very low-density epoxy filler, good for fairing large areas of a hull.*

CHAPTER 6

Epoxy-Fabric Reinforcement

By itself, a skin of epoxy on wood provides a barrier to water intrusion and some abrasion resistance. When epoxy and a synthetic fabric together sheathe a wooden form, they add to the wood a great deal of stiffness and resistance to bending and breaking, as well as an even more formidable moisture defense. The fabric also helps to regulate the thickness of the synthetic skin on the plywood, creating a matrix that holds the epoxy. Fiberglass cloth is the most common and versatile fabric for Stitch-and-Glue boats, and you will be using it to reinforce epoxied and filleted joints, sheathe the outside of the hull, and strengthen panels such as rudders and cabin tops.

Fiberglass comes in various forms for boatbuilding, including **woven** and **knitted** cloths and a random-fiber mat that resembles a coarse felt. The first two types are used extensively for Stitch-and-Glue construction. The woven cloth (and for our purposes in this book, any woven fiberglass fabric is a cloth) is particularly suited for use over composite joints (i.e., any joint reinforced with resin, fabric, and a fillet material) in small boats. But large boats mean greater stresses converging at the joints, so knitted biaxial fabrics are layered with the woven cloth to create even stronger seams.

Figure 6-1. *Woven fabric tape comes in 2- to 8-inch widths and is very useful for smaller Stitch-and-Glue boats.*

Figure 6-2. *Woven cloth sheathing is available in 38-inch, 50-inch, and 60-inch widths. Weights of 4 to 6 ounces (per square yard) are typically used for sheathing hulls and decks.*

In all Stitch-and-Glue boats, these strong epoxy/fiberglass composite joints replace the dimensional wood chine logs and frames of a traditional plywood boat. Mat cloths are not very useful for Stitch-and-Glue boats as there is so much sizing applied to make these mats that the epoxy will not saturate easily or stick very well to the resulting mat cloth.

To understand how fiberglass makes a boat's joints stronger, you should understand how the fabric is constructed. Fiberglass is made from continuous filaments of glass, which are drawn or pulled from molten glass through precise, multi-holed bushings. These tiny filaments are combined into strands. Depending on the type of fiberglass, there may be from 51 to 1,224 filaments per strand. Thinner strands are called threads, and thicker strands with more filaments are yarns. Fiberglass is available in a variety of forms and will be characterized by these properties: the number of yarns per inch in each direction, the weight of the fabric

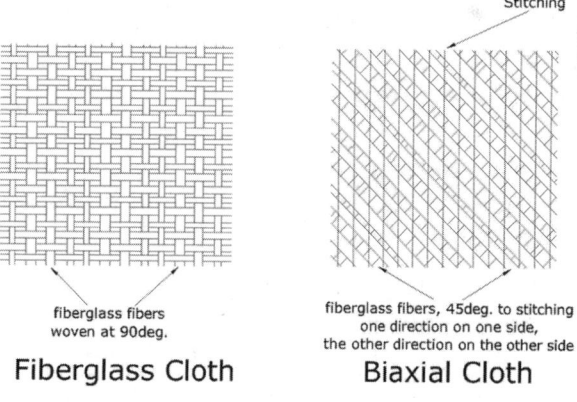

Figure 6-3. *Fiber orientation comparison of woven and biaxial cloth.*

in ounces per square yard, the thickness in thousandths of an inch (mils), yarn construction, weave style, and finish. Most fiberglass fabrics are coated with lubricants or sizings so that the filaments won't fly away during the high-speed weaving process and to ensure a uniform appearance. After the weaving is done, the fabric is heat-cleaned to remove most of the lubricant. Unfortunately, the heat also greatly reduces the original tensile strength of the glass strands. All woven fiberglass products share this problem; if the strands need to be woven, lubricant is required, but the heat used to remove the lubricant diminishes the fiber strength.

It was only a matter of time (decades, actually) until someone asked, "Why weave it?" In answer, knitted fabrics appeared, which for our purposes means biaxials and triaxials in cloth/roll form. Knitting machines easily handle the glass yarns without lubricants. The greatest boon to boatbuilders, besides maintaining the original tensile strength of the glass filaments, is that knitted fabrics can orient the fibers at 45°/45 axis if desired, rather than only 0/90° axis as with woven fabric. With this orientation in a tape cut from a roll of biaxial cloth, 100 percent of the fibers cross the joint, giving it far greater reinforcement and taking full advantage of the strength of the cloth across the axis of a Stitch-and-Glue joint. Triaxials are biaxial cloths with a backing of mat cloth in the matrix, and as explained above I don't think these work well with Stitch-and-Glue construction. I don't recommend using them.

Knitted fabrics are rougher in texture than woven fabrics. If used alone, they don't finish nearly as smoothly as woven cloth. The usual solution in Stitch-and-Glue construction is to use a woven cloth as the final or top surface layer over the knitted layers beneath. For the interior seams of a Stitch-and-Glue boat, multiple layers of knitted cloth tapes covered with a layer of woven fiberglass tape significantly speed up the glassing of the hull and bulkhead seams. But when it comes to fiberglassing the upper portions of the inside of the boat—the more visible areas and places where you might want a bright rather than painted finish—biaxial or knitted fabrics are unsuitable, and woven tapes make for a more sightly, though slightly less strong, joint.

Factory-produced woven fiberglass tapes, widely available in 2-, 4-, and 6-inch widths, are tempting because they seem so easy to use—until you try them. The edges are woven with a tight sizing thread, which eliminates the irritating problem of unraveling at those same edges. However, this sizing will cause the fabric to pucker when you try to wedge the tape into the inside of a compound-angle joint. And these compound nooks and crannies are most of what we have in boats, such as the bow, chines, and bulkhead-to-hull seams. One solution is to make a cut or dart into the fabric edges every 8 to 12 inches. A better solution is to make your own tape strips. Unroll a couple of yards of wide woven cloth on a clean workbench surface. Lay a straightedge along it at about a 10- to 20-degree angle to the long side and cut strips with a razor knife in whatever widths you need. Handle the strips with care to avoid unraveling, and you'll find that they conform very nicely to any kind of joint you have, and without the woven tapes' puckering at the edges.

When designing a Stitch-and-Glue boat, I examine each seam and joint for the level of stress it will experience to determine the amount of filleting and the number of layers of fiberglass it will require. The higher the stress, the more layers of fiberglass tape necessary for adequate strength. For small dinghies and skiffs up to perhaps 15 feet, I specify a minimum of 12 ounces of cloth in total, which typically means two 6-ounce layers. A 15- to 20-foot boat might require 18-ounce reinforcements and so on up to 42 ounces for boats approaching 50 feet (the largest we've built to date). In boats larger than the smallest dinghies or skiffs, I use layers of biaxial fiberglass tape topped with a woven fiberglass finish layer to smooth the joint.

I have enormous confidence in joints built like this. I have indulged in some destruction tests of mocked-up hull sections, and when extreme force is applied, they usually fail at the edge where the tape ends, rather than at the joint where the two pieces of wood come together. And when compared to

Figure 6-4. *A keel on a 22ft Surf Scoter design is being glassed in place with layers of biaxial cloth tape run the length of the joint between the keel and the hull over a fillet of woodflour/epoxy.*

glued and screwed joints of dimensional wood, the composite Stitch-and-Glue joints test out at eight to ten times as strong. These joints are fully structural and have the unique advantage of minimizing the interfaces in the joints. By this I mean that *all* the joint is structural with less of a perceived break in transitioning from one area of the joint into another. Thus, the hull itself really functions as a one-piece component, with no discernible breaks between one part and another.

Print-through, the tendency of the cloth weave pattern to telegraph or show through the final topside paint, is an eternal problem. The first and most effective way to minimize it is to use white or light-colored paint, as most production boat manufacturers do. Lighter paints contain more pigments to filter out the UV rays of the sun, and light colors absorb less infrared radiation in direct sunlight, so the underlying epoxy matrix does not heat up and deform as readily. Nobody listens to my advice on this score. The vast majority of boats built to my plans over the years have sported nice, bright, and troublesome red, blue, green, and black topsides. It's a seductive dream to imagine a brilliant blue wooden boat that pops out from the crowd of refrigerator-white production boats at the local marina, and this siren always sings loudly to the boatbuilder.

In my shop, we've been battling the print-through curse for years, and we've gradually

Figure 6-5A. *Cutting your own tapes from a roll of cloth will give you smoother joints and works well for the larger building jobs where you need a variety of widths and lengths.*

Figure 6-5B. *Biaxial tape cut and being saturated in a glassing box.*

learned ways to mitigate it. First, selecting one of the harder epoxy formulations (Chapter 5) seems to reduce the tendency to print-through. Next, post-curing the epoxy will help. If you build your hull in the winter, the infrared shop heaters I recommended in Chapter 2 will further harden and assist in the curing of the epoxy, with no effort at all on your part. If it's summer, you can put a coat of dark primer on the finished hull and move it outdoors in the sun for a week or two for a solar cure. In whatever way you cure it, the longer the epoxy has to set or cure before painting, the less noticeable the print-through. A lighter, tighter weave (such as 4-ounce instead of 6-ounce) fiberglass cloth used for sheathing reduces print-through, but it will also reduce the panel stiffness. Two other ideas are additional coats of primer, or an outer layer of Dynel cloth over the initial layer of fiberglass cloth. The latter is the sheathing schedule that we use most often on larger boats.

For exterior sheathing, you do have choices other than fiberglass cloth, such as Dynel, Xynole, Kevlar, and even carbon fiber cloth. In recent years, I have increasingly tended to use Dynel or Xynole (these two are almost exactly the same in application) on my decks. These two cloths will flex and recover from a sharp blow, whereas glass

fibers may fracture. Decks take a lot of sharp blows, such as anchors being dropped on them. For the same reason, I like Dynel/Xynole for the deck-to-hull and cabin-to-deck joints because it may be better at absorbing the structural stresses imposed on these critical structures during their life cycles. Dynel/Xynole holds more epoxy in its weave than a comparable weight of glass cloth. Because of its stretchability, it may be more trouble to smooth out over the more complicated shape of a boat's bottom, but I have found it easy to use and without many vices. My only complaint is its limited availability. At this writing, it is available from marine suppliers Defender Industries and Jamestown Distributors. Note that, unlike fiberglass, Dynel does not become transparent when wet out with epoxy. For whatever surfaces you insist on finishing bright, fiberglass remains your only choice.

Kevlar is bulletproof—literally, since it's known for its use in bulletproof vests—so it can be effective where very high impact resistance is required. If you're planning on driving a high-speed power boat in shallow, rocky water, you will want to sheathe the hull in Kevlar. When you do, be ready for a real job—it presents much more hassle than fiberglass, Dynel, or Xynole. Kevlar

Figure 6-6. *Various fabrics you can use in building a Stitch-and-Glue boat, left to right: mat, peel ply, Kevlar, 6-ounce glass, 12-ounce biaxial, and Dynel.*

Figure 6-7. *This boat is being sheathed with the dry method. First, 6-ounce glass cloth is draped over the upside-down hull.*

Figure 6-8. *The cloth is best smoothed down with the heels of your palms. Try not to pull and stretch it.*

is difficult to wet out, you can't cut it with normal scissors or razor knives, and you must add a layer of Dynel or fiberglass over it before you can sand overlaps and seams smooth. Vacuum bagging works well for Kevlar, but that introduces quite another set of technologies and hassles. Sometimes when building a larger boat with cold-molding of extra plywood panels on the hull to build up skin thickness, we strategically apply Kevlar sheathing as an interlaminate, fastening the outer plywood panels over the wet and not fully cured Kevlar layer. Staple or screw fastening of the cold-molding/plywood layer will apply the uniform pressure necessary to complete a structural laminate. We use slower hardeners to give us more time to work before the epoxy starts to set up and take special care with overlaps and extra thicknesses. These interior layers of Kevlar really use the cloth effectively and allow some of the nuisance of working with it to be minimized.

In my opinion, there are few reasons to use extremely expensive carbon fiber sheathing on a Stitch-and-Glue boat. However, in building a 28-foot Onyx sailboat recently, we found that we needed to add substantial stiffness to the rudder without making it any thicker than the largest standard gudgeons (1½ inches) we could obtain. Carbon fiber proved to be the ideal solution as a sheathing for the plywood-cored rudder; it added an amazing degree of stiffness with very little extra material.

The bottom line is that fiberglass is the most readily available and least expensive exterior sheathing material, and with proper care in application, it will perform very well to strengthen and protect your wooden boat. When preparing to paint or varnish, be sure to buy finishes that are epoxy-compatible. If you buy from a reputable marine supplier, they will know what works. You'll read more on this in Chapter 25.

Figure 6-9. *Pour mixed epoxy over the fiberglass cloth and squeegee it around in a figure-8 pattern, making sure the weave becomes fully saturated.*

Figure 6-10. *A foam roller is also a very good tool for evenly distributing the epoxy.*

Figure 6-11. *Here the glass cloth has all turned transparent, which shows that the epoxy has saturated it. Now trim the edges off and let the epoxy set up a bit. When it's tacky, roll on another coat to build up the epoxy to proper thickness.*

CHAPTER 7

Scarfing

Your Stitch-and-Glue boat must always use full-length plywood panels for the initial stitching of the hull. If you need to make longer sheets of plywood out of shorter sections, that process is called *scarfing*. This is a fundamental technique in boatbuilding, particularly in any of the methods of construction that employ plywood. You can also use it for joining dimensional lumber into longer lengths for masts, booms, sheer clamps, and rubrails. It's not something you need to dread or fear. If you digest the instructions in this chapter and take proper care when setting up your scarf, you will do it successfully the first time out.

A scarf joint simply joins two or more shorter pieces of lumber or plywood lengthwise to make a single, longer piece. Standard marine plywood sheets measure 4 × 8 feet, with some limited availability in 4' × 10' or 5' × 10' sheets. The 10-foot panels will yield a boat of roughly 9½ feet maximum in overall length, allowing for side curvature. If your ambition stirs you to build a boat larger than that, you will need to scarf. (Some suppliers offer pre-scarfed 16-foot panels for a premium price. I don't endorse them because the supplier often pays no attention to matching grain and color to achieve uniform panels, and there's no independent oversight to guarantee the quality of the scarfs. And finally because scarfing is a basic boatbuilder's skill, you might as well master it.)

The classic scarf joint is what I call the "slash" scarf (Figure 7-1). The two plywood ends are sawn or planed into long bevels and glued together with epoxy. There is a linear mathematical relationship between the length of the bevel slope and its holding strength. The longer the bevel (or you might say the flatter its angle), the more gluing surface it allows, and so the stronger the resulting joint. A longer scarf also facilitates a more uniform bend in the scarfed section of the panel, making for a smoother hull curvature and shape. The minimum bevel length to panel thickness ratio is 8:1, and a practical maximum is 12:1. If you're scarfing ¼-inch plywood, an 8:1 ratio translates into a bevel 2 inches wide, and the 12:1 ratio gives a total bevel width of 3 inches.

Figure 7-1. *An 8:1 ratio of length to thickness is the minimum for scarfing, and the flatter the slope, the stronger the scarf.*

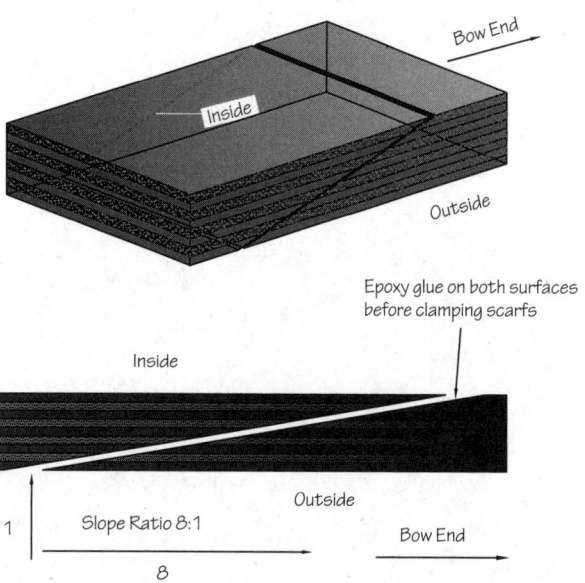

There are several methods of scarfing panels. The most all-around useful method (and the one that uses the fewest specialized tools) is the "staircase" technique (Figure 7-2). Basically, you stack your pieces of plywood in a staircase fashion, staggered on the interval that creates the slope ratio you've chosen. Use extra scrap pieces to form the bottom and top steps; these sacrificial pieces will absorb the tendency of your tool to roll off at the extremes of the bevel. Firmly fasten all the pieces together (either screw them together well clear of the cutting area or use large clamps). The best tool to use initially is a handheld power planer to "knock off" the steps of the stack, creating an even, smooth, uniform bevel, but for a small job, a good long-bed hand plane might be used. You are simply cutting off the tops of the steps until you reach the apex of the intersection with the step below. Work methodically and accurately, but the steps themselves will help you keep everything aligned and even. A belt sander or disc grinder will refine and finish the job. When the lines formed by all the exposed plies are straight and equivalent in thickness, you have a good scarf bevel.

Take the stack apart, being very careful not to damage the delicate thin edges of the bevels. Reversing directions on the sheets or boards will show if the cuts are accurate and neat.

A creative alternative for making long panels, particularly when building a boat larger than 30 feet, is to buy your plywood in thin sheets and laminate two or three layers together with staggered

Figure 7-2. *Stacking all the panels you will need to scarf is the quick and simple way to go.*

Figure 7-3. *Making the first pass with the power planer.*

Figure 7-5. *The scarf stack here is completed. If the lines of exposed inner plies are straight, then your scarf is clean and accurate.*

Figure 7-4. *Further passes continue cutting the edges off the stair-stepped panels.*

butt joints. For instance, three laminations of ¼-inch or four laminations of 4 mm would produce excellent planking stock for the hull of a large boat. Vacuum bagging is the best way to assure even clamping pressure (you wouldn't want to be using fastenings for this method) while the epoxied layers cure. Laminating panels works quite well, especially for creating long panels wider than 4 feet.

Still another method is the brainchild of John Henry, who has invented a scarfing attachment that bolts onto the base of a Makita handheld power planer. A snap to use, the attachment can be adjusted to set the planer's knives at various angles. This tool can be set to cut scarfs on plywood as thick as ¾ inch or 18 mm. It is accurate and cuts very smooth bevels that will need no extra dressing or sanding to be ready to glue up.

The West System epoxy people had in the past developed a scarfing tool that attached to the base of a circular saw. The Model #875 Tool is now out of production, but you might find a used one on eBay or some such site. These only cut thru-scarfs in ⅜" or 9 mm marine plywood, but I have cut 12 mm or ½" panels with just a bit of dressing with a sharp block plane.

The 21st century has now provided us with a mechanized and computerized way to make scarf joints, and most boat kit suppliers are happily using it: the puzzle scarf, so called because its joints look like giant jigsaw puzzle pieces fitting together (Figure 7-10). A CNC (Computer Numerical Controlled) router cuts the mating pieces with machine precision, and the builder simply paints the end grain with epoxy—first unthickened, to soak in, then a second thickened application—to fuse them together. It is always wise to reinforce the joint with a layer of glass tape wider than the puzzle legs on both sides. To keep the added thickness of the fiberglass from causing a hard-to-fair

Figure 7-6. *Power planer and grinder are the basic tools for cutting scarf stacks.*

bulge on the outside of the hull, you can carve a very shallow channel across the width of the panels to contain the glass and epoxy with a grinder, or a hand power plane with the blade set very shallow, or hand planes. Avoid the risk of weakening the panel by making sure the channel does not cut the same path on both sides. If your reinforcement tape spans 8 inches on one side, allow 12 or 14 inches on the other.

If you're building from scratch and don't happen to have a CNC machine in your garage, you can farm out the work: Many commercial CNC shops will cut the puzzle scarfs on full plywood sheets for you. One such shop in my area charges $80 per sheet for the work. If you're building a 15-foot boat, that would likely require the puzzle treatment on at least six 4 × 8 foot sheets for $480. If you truly fear or dread scarfing, it may be worth the cost to you. But I've seen many beginning boatbuilders who

blanched at the idea of undertaking a scarf and then discovered to their surprise that it wasn't really that difficult. In any event, you may later be undertaking scarfs on smaller pieces of dimensional lumber—for rub strips or sheer clamps extending the full length of the boat, for example—so it remains my recommendation to conquer your fear of it.

Whichever method you use, you will need a level work area with good ventilation for gluing up full-sized sheets of plywood. You can use a patio, shop floor, or even a loft, but make the surface as absolutely level as possible. Before gluing the scarfs, cover your work surface with plastic sheeting, preferably the thicker 6 mil variety, to prevent excess epoxy from fouling it or sticking it to other surfaces. Prepare all needed tools and materials before you start because once the bevels are coated with epoxy, there is no time to waste as you position and clamp the panels. Coat the cut bevel edges

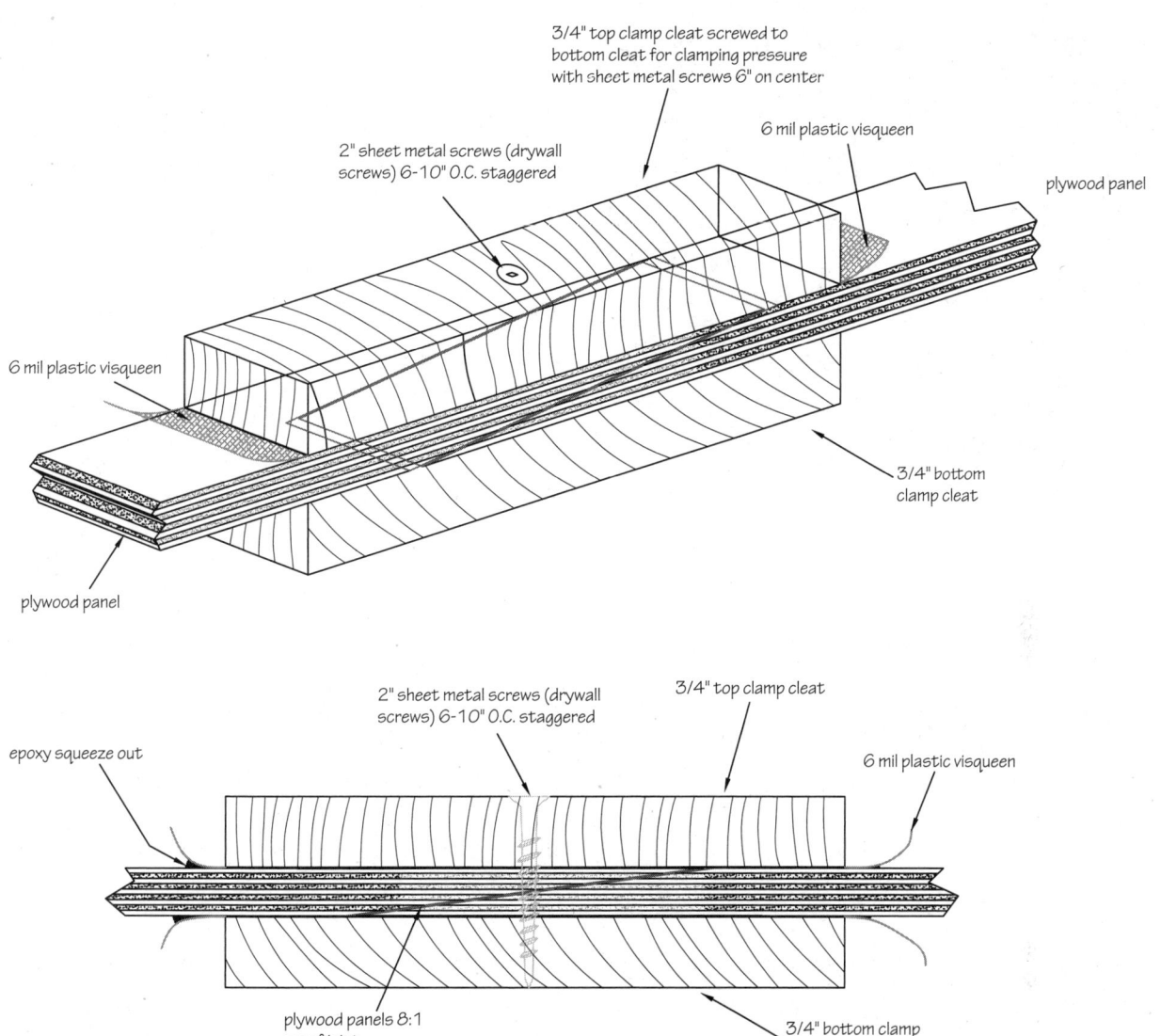

Figure 7-7. *Scarfs are cut and glued up between two stiff boards to help squeeze the plywood together. Don't forget the plastic film to keep epoxy from sticking the boards to the plywood panels.*

with unthickened epoxy 15 to 20 minutes before assembling the scarf joints. This allows the epoxy to soak into the end grain, avoiding the possibility of a weakened, epoxy-starved joint. Just before final assembly, recoat the mating edges with epoxy thickened slightly with CAB-O-SIL. The thickening will help fill any small gaps left by slight imprecisions in your scarf bevels.

Plastic sheeting or wax paper is a good thing to keep within reach. If you're going to stack two or more scarfed panels on top of each other, separate the glued areas with either kind of sheeting so that you don't create accidental laminations. Let the sheeting extend well beyond the bevels, as epoxy may ooze and creep out of the glue joints.

A note of warning: When using lighter-gauge plastic sheeting or wax paper like this, I have often found that after the epoxy sets solid and I am taking my stacks of plywood apart, the plastic or paper

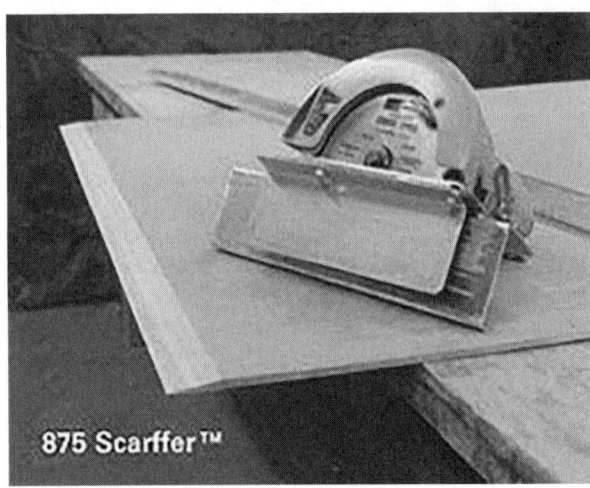

Figure 7-8. *Though now out of production, the Gougeon Scarfer model #875 was a good and cheap way to scarf up to ³⁄₈-inch plywood.*

Figure 7-9. *The John Henry scarfing jig will help the planer maintain a consistent angle as it cuts the bevel.*

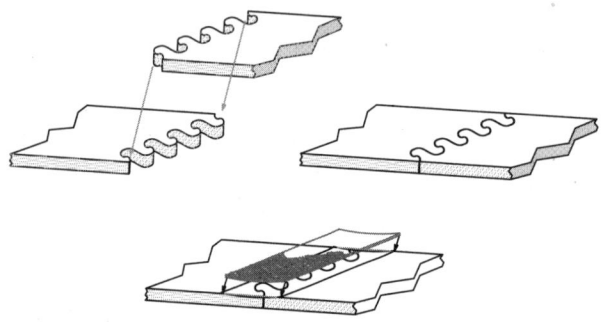

Figure 7-10. *Most kits now use CNC-cut puzzle joints for painless and perfectly accurate scarfing.*

has accidentally crept into the scarf joint itself. Watch out! It is very difficult to fix this problem.

Imprecision cannot be tolerated in the alignments of one panel to another. The long sides of the panels you're creating must be perfectly straight. Use a long straightedge or a string stretched the full length of the panel to check the alignment as you mate the pieces. At the same time, check the vertical alignment of the joint by laying a small, stiff wood batten across the joint. The transition from one panel to the next must be perfectly smooth and not change in level. There should be no slop or over- or underbite as you cross from one sheet to the other. When you're satisfied, tap a few small nails or staples into the scarf to hold the alignment temporarily while you weight or clamp the scarf. These fastenings will help hold both the vertical and lengthwise alignments.

You can apply the necessary clamping pressure for gluing up your scarfs in several ways. The simplest method I have found is to use weights or props from an overhead beam. Any kind of weights will work—hand dumbbells, if you don't mind ducking your workout for a day; or if you're building a sailboat and have already obtained lead shot for your ballast, scoop it into zip-lock bags for versatile glue-setting weights. I have found that placing weights on stiff, straight 2 × 6- or 2 × 8-inch boards spreads the pressure more uniformly across the joint or stack of panels. Be sure to have a layer of plastic to avoid sticking this board to your marine plywood.

You can also make a very effective clamp by placing a pair of stiff boards over and under the scarf joint—don't forget the plastic sheeting!— then screwing long drywall screws through the top board, scarfed panels, and into the bottom board. I would space the screws 6 to 10 inches apart in staggered rows. After the joint has cured, you will have the minor extra step of patching the fastening holes with epoxy/wood flour paste and sanding the surface smooth.

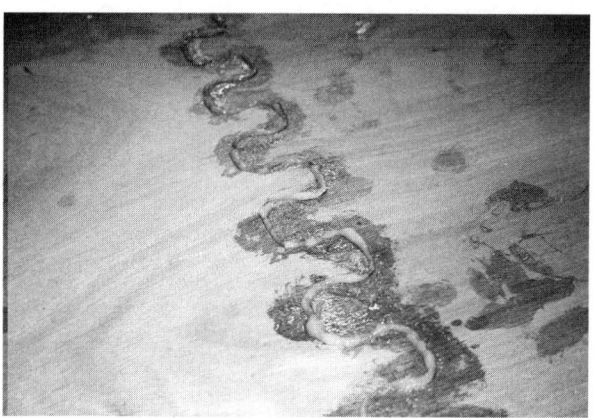

Figure 7-11. *Use plenty of thickened epoxy painted on all the faying surfaces of the puzzle joints to ensure a strong scarf.*

Figure 7-12. *A good squeeze-out of thickened epoxy is what we want to see when the joint is pressed together.*

Figure 7-13. *Place a stiff, straight board over the joint with 6-mil Visqueen plastic sheeting to keep it from being glued to the plywood panels.*

Figure 7-14. *Now place lead weights ("gravity clamps") along the board to hold the joint firmly in place while the epoxy cures. Alternatively, a few temporary screws could be used between boards on both sides of the panel for good squeezing of the joint.*

Figure 7-15. *All the puzzle joints are now glued up, but the job isn't finished—reinforcements must still be applied.*

Figure 7-16. *Sand the puzzle-joined section smooth on both sides.*

Allow a couple of days for the epoxy to cure fully; then sand the excess glue from the scarf joints. The joint may not be quite perfect, but that can still be remedied. It's important to understand the stress loads of panel joinery in principle. If I conduct a destruction test of a scarf joint, it will always fail either at the start or the stop of the glued-up bevel. These are the weakest parts of the joint because even though they are glued together, the interface of the grain of the wood layers is interrupted. If my vertical alignment was a little sloppy (i.e., an under- or overbite of the panels), these points will be weaker still. But I can compensate by laminating one or more pieces of fiberglass cloth tape across the scarf, extending the cloth edge well beyond the weakest points. This effectively distributes the stress over a wider area of the panel. As with puzzle scarfs, you can dish out a slight concavity to accommodate the fabric and epoxy.

Figure 7-17. *An 8-inch sander with 80 grit sandpaper held slightly elevated on one side will carve a shallow valley across the joint to contain the glass reinforcement.*

Figure 7-18. *On this puzzle joint we are applying two layers of glass tape over each side. The first layer is narrower than the puzzle outline itself.*

Figure 7-19. *The second layer of glass tape overlaps the puzzle outline and is laid wet-on-wet over the first layer.*

Figure 7-20. *Use more of the 6-mil plastic or peel ply with plastic over that and more boards and weights to smooth out the glassing. This needs to be done on both sides of the scarfed panel.*

CHAPTER 8

Lofting

For a Stitch-and-Glue boat, lofting simply means taking the boat's designed panel shapes and drawing them, enlarged to full size, on full-length sheets of marine plywood. As in traditional boatbuilding, this lofting process is the most important stage of building. The final shape of the boat is dictated by the panel shapes; at stake are a pleasing appearance, efficient hydrodynamics, and overall performance. Any unfairness in the underwater panels will quickly affect the coefficient of drag and change the way the boat moves through the water. If the lines of the hull are in any way crooked, uneven, or unfair, the result will be an ungainly appearance and a boat out of proportion, and all your hard work will be for naught. While some designers provide optional full-size patterns on dimensionally stable mylar (not a particularly easy way to transfer the shape to your wood pieces), the lofting process for a Stitch-and-Glue boat requires such a small amount of time that little is to be gained from full-sized patterns.

Think of a Stitch-and-Glue boat as if it were a banana. If you were to peel a banana, eat the fruit, and then reassemble those same peels in the same order that they had originally, you would have a banana shape again. Furthermore, if you traced the shape of each flattened segment of peel onto cardboard or kraft paper, you could then cut out and assemble a paper model of the banana. In essence, this is exactly what a boat designer does with a Stitch-and-Glue design. He or she imagines the boat's shape as a collection of large peels, or component parts, with their edges forming the sheer (hull-deck junction), chine, and keel lines of the

boat. In lofting the Stitch-and-Glue boat, all you are doing is taking the designer's outlines for the peels (panels) and drawing them full-size on flat plywood sheets.

Drawing the full-sized panels is a simple procedure. For the simplest shape, a V-bottomed hull, there are only two side panels, two bottom panels, and a transom. With larger designs, the lofting requires drawing full-sized bulkheads and additional interior members to strengthen the hull athwartships (side-to-side) and longitudinally (fore-and-aft). Any one of these many panels or interior parts can drastically affect the final symmetry and aesthetics of the hull, so each panel or part must be lofted and proofed as accurately as possible. We will go into proofing in Chapters 12–14.

When lofting, if the boat is small enough to make it practical, I prefer to lay the plywood on sawhorses to avoid working on my knees. With bigger boats, lay the plywood on the floor atop wooden spacers with enough height above the floor or base to allow you to cut the shapes out without your circular saw blade hitting the substrate. Whichever method you prefer, the panels must be level and flat. If the floor is uneven, begin by laying out and leveling battens (sticks) every 12–18 inches. Again, you need to avoid interference with the blade's path when you begin to cut out the parts. If you're leveling the floor, I have found that the spray foam insulation sold in home improvement stores can provide a good semi-adhesive to aid in holding the boards in position. Just spray around the edges of the boards and the foam will freeze them in position. Accuracy in

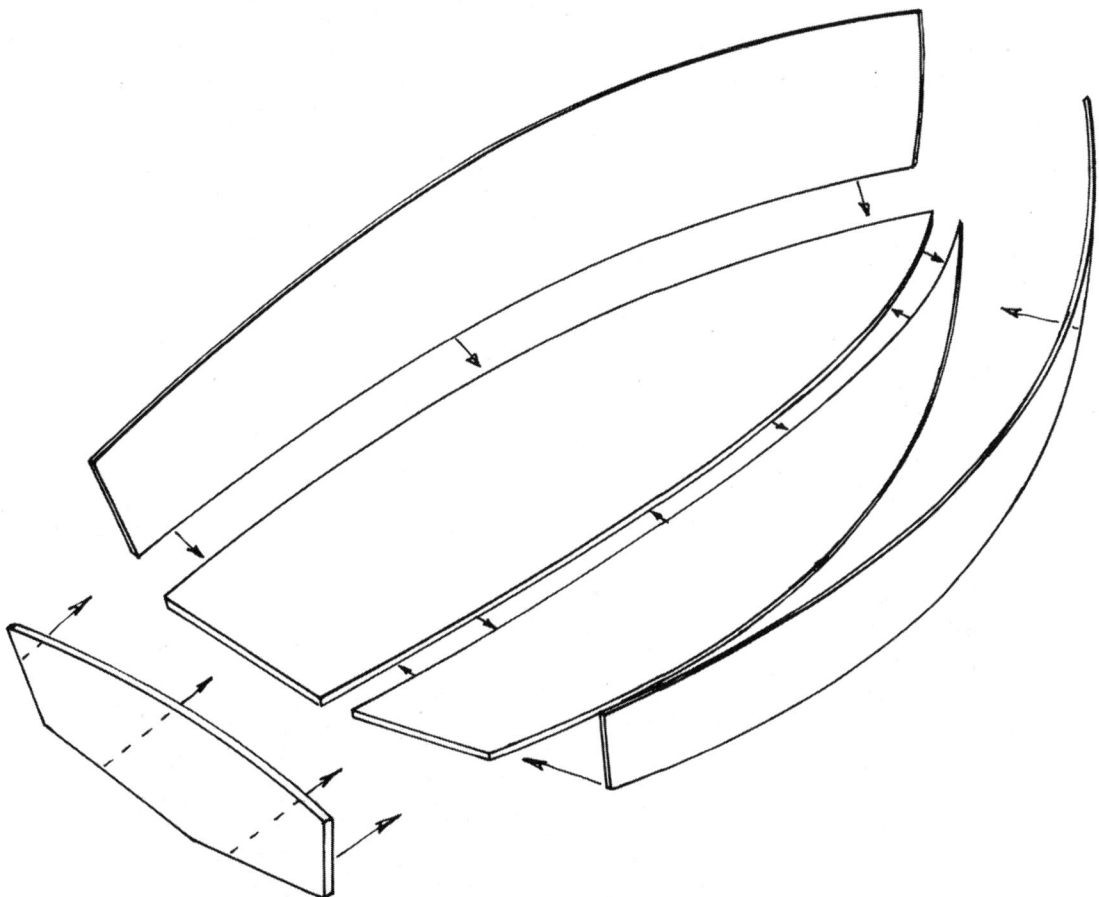

Figure 8-1. *The simplest Stitch-and-Glue hulls have only four or five panels. These illustrations show them in exploded form and fitted together for a complete hull.*

lofting depends partially on how level the panels are while you mark the station and dimension points, so be sure to set up properly, flat and uniform.

Your next step is to draw in the layout station marks on the panels. These are usually athwartships stations drawn perpendicular to the long side of the panels at some regular interval. Most often they are laid out at 12-inch or 300-mm stations along the plywood's long edges. The particular interval the designer has used makes no difference, as long as you are faithful when determining and setting the dimension marks.

When setting station marks, I most often use a drywall T-square with a 50-inch leg to mark the full width of the panel; this speeds up the process. If you have only a carpenter's framing or L-square, simply

extend one leg by attaching a straight wooden batten. Double-check all station marks for accuracy, and mark the numbers on them corresponding to the station numbers on your plans. Remember, all we are really trying to accomplish is to draw in full size what the designer has drawn in small scale on the building plans.

Once the station marks are satisfactory, begin to project the dimensioned intersection marks onto the panels. If your plans were hand-drawn or are older designs, the measurements will be given in the boatbuilder's traditional notation. The rule to remember is *feet-inches-eighths*—the universal three-numeral designation for boatbuilders locating exact dimension points by measuring distances along given station marks. Hence, 3–4–2 on a

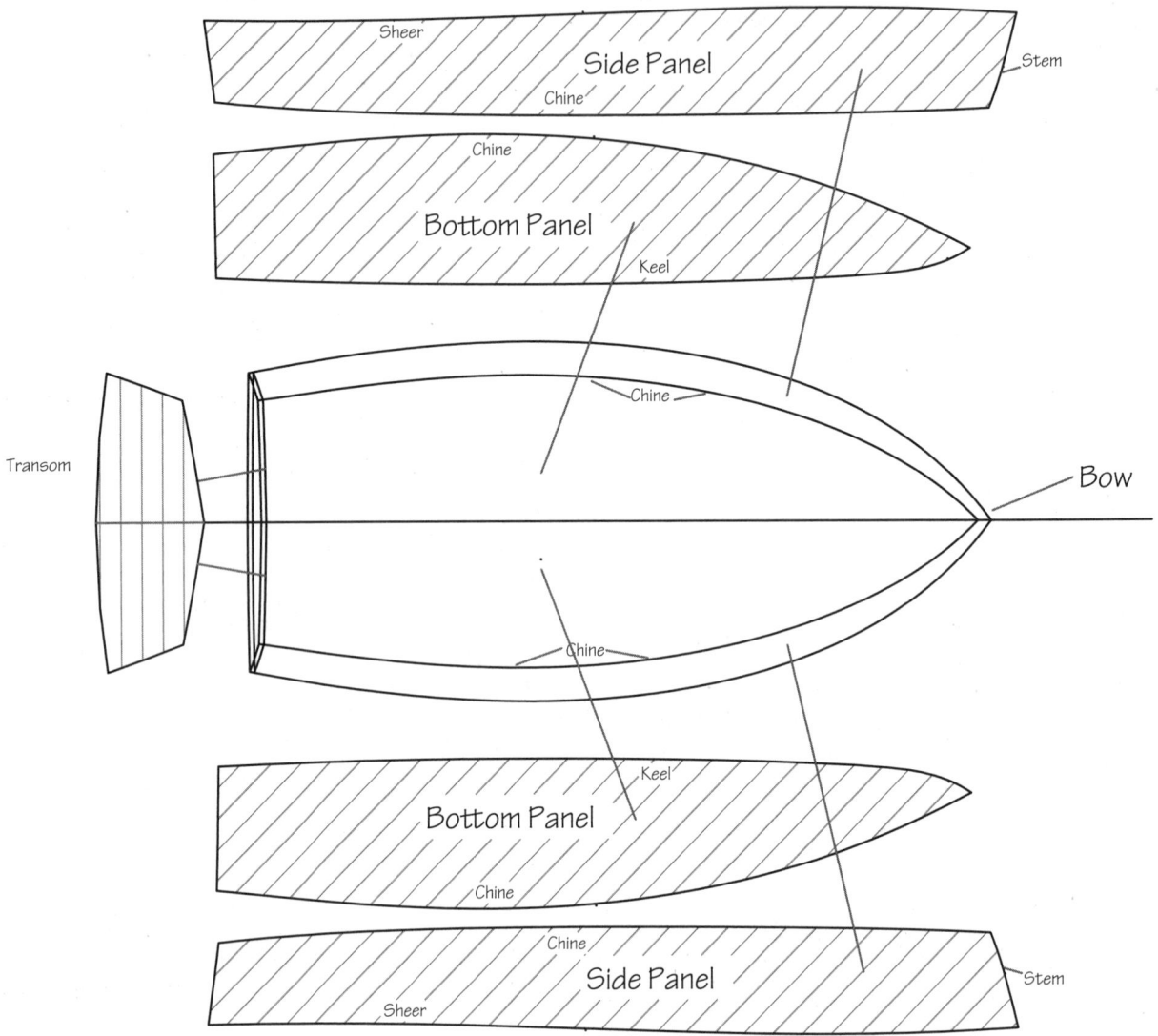

Figure 8-2. *The exploded and laid-flat parts of a simple single-chined Stitch-and-Glue boat.*

blueprint translates to 3 feet 4 inches and 2-eighths of an inch, from the long edge or baseline on the plywood panel. The third numeral is eighths, never fourths, or halves, or anything else. The only additional variable is a plus (+) or minus (−) sign; if these signs should appear, it translates to plus or minus a sixteenth of an inch. Thus, 2–5–5+ translates to 2 feet 5-11/16 inches. If you find the system confusing at first, it might help to go through the plans and pencil in "translations" in conventional notation. However, the more you use the triple numeral system, the more second nature it will

become. It was developed when boats had a table of scantlings and allowed more accuracy than if you had a bunch of other measurements.

If your plans have been drafted on a computer, the dimensions will be given in the usual imperial or metric measurements as there is no practical way to easily determine dimension using the feet-inches-eighths rule.

Boat by boat, I have been working through my design catalog and have added a metric option for the plans. Certainly, metric measurements simplify things and reduce the likelihood of errors. If enough

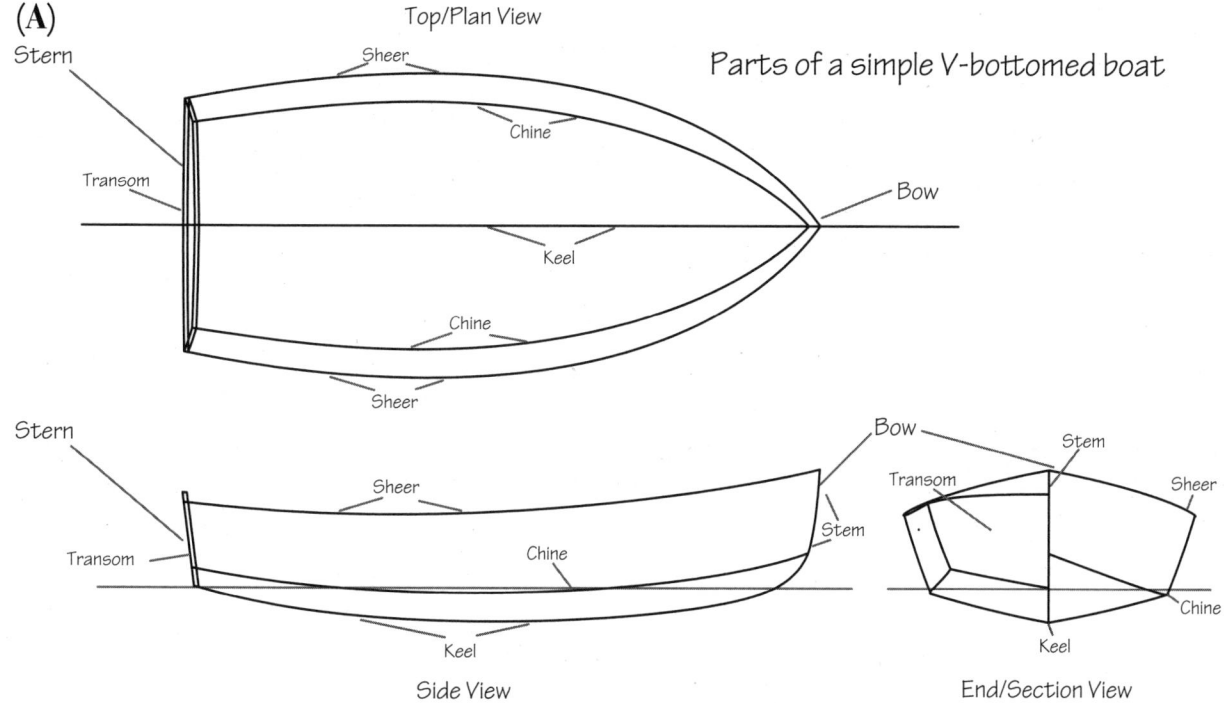

Figure 8-3A. *The parts shown on a single-chine Stitch-and-Glue boat.*

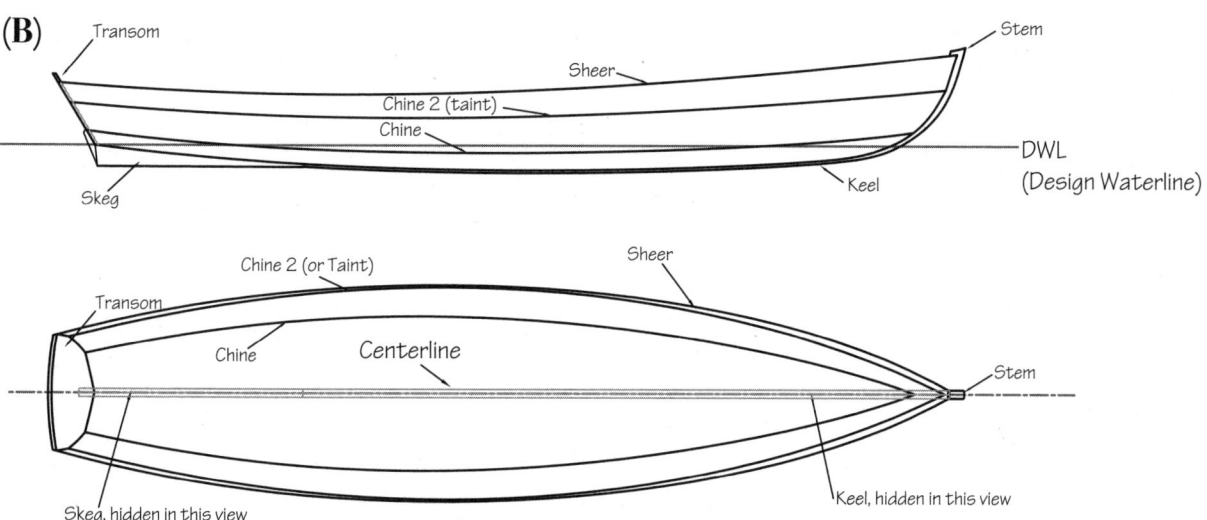

Figure 8-3B. *A multi-chined design has two or more panels on each side.*

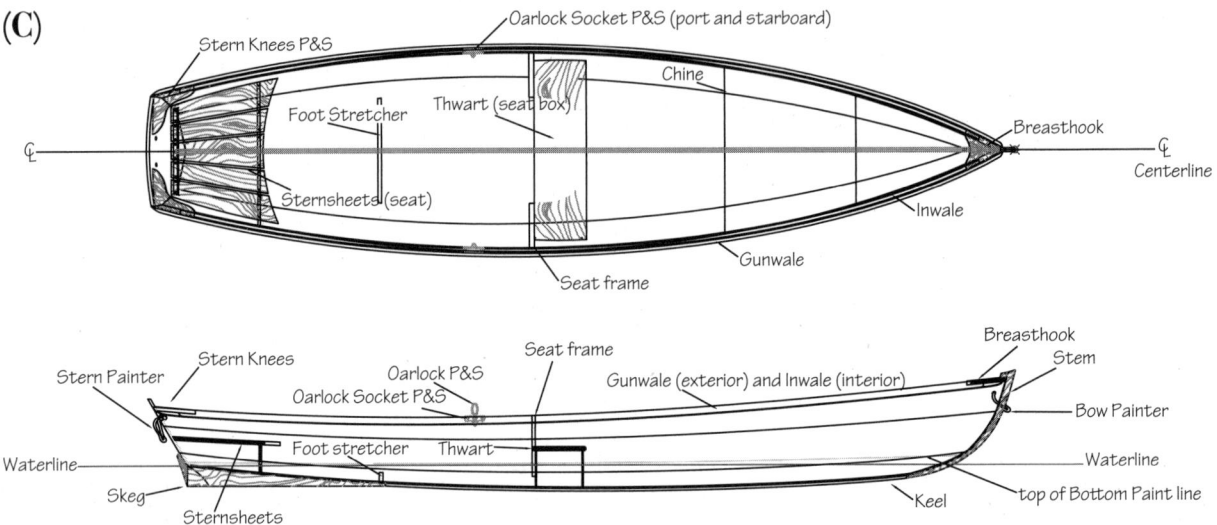

Figure 8-3C. *Here are top and side views illustrating the parts of a small rowboat.*

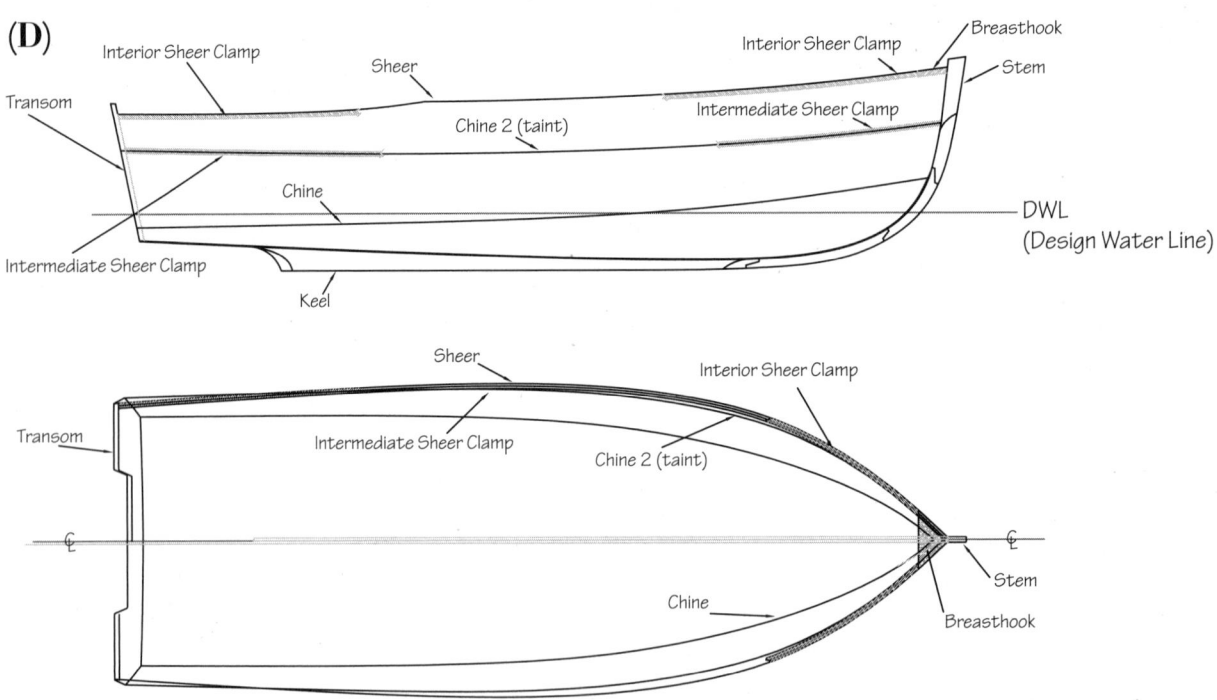

Figure 8-3D. *The parts of a multi-chined larger boat.*

wooden boatbuilders make the switch to the metric system, perhaps we will drag the United States along with us. Don't tell anyone, but I hope I am too old to have to completely make the transition to metric. Old habits die hard.

With all the dimension points drawn on every station in place and double-checked against the building plans, the next step is to place a batten along the points to draw the curved lines (the edges of our hull panels, or banana peels). First, drive small brad nails (4 penny 1½-inch bright finish nails are a few bucks a box and are quite useful around the boatbuilding shop) on each of the marked dimensions. Then lay a batten adjacent to all the nails and dimension marks. Avoid the temptation of nailing through the batten. Not only would this ruin a perfectly good batten, but it would also immediately affect the true curvature of the line by not allowing you to adjust for fairness along the length of the batten. Good battens of clear, vertical-grain wood with uniform flexibility are hard to find, so think of your batten as a tool. You would never dream of nailing through the wooden handles of other tools; *do not nail through your batten*. When lofting, you want the curves to be as smooth as possible, without bumps or hollows, and you'll need a batten that will flex uniformly throughout its length. The smoothness of a batten's curvature will be affected if one section is either stiffer or more flexible than the rest of the batten. I prefer to use softwood for battens, and I have several ranging from a ¾ × ¾-inch batten about 18 feet long to a couple at ½ inch × ½ inch × 10 feet long. One as small as ⅜ inch × ⅜ inch × 6 feet long will serve well for curvier sections of the panels. A trick is to set more of the small brad nails on the backside of the batten to help hold the spring of the batten in its position along the original marks. Again, remember to not nail through the batten itself.

Step back after aligning the batten and move from end to end of the panel, sighting the curvature along the batten's full length. Take note of any flat areas in the curvature, and look for bumps that might be caused by improper positioning of a nail or by the earlier mismarking of a dimension point. If something is grossly out of line, it will be apparent at this time, and rechecking the measurements will fix it. There are lots of marks, so mistakes can happen. Catch them now, before you pick up your saw (Figure 8-4).

If the curve is a severe one, you may need to set additional nail guides to achieve a fair line; then again, if the curve is soft, you'll need fewer nail guides to allow the batten to flex smoothly.

In the end, the batten has to find its proper level of tension. The bow section of the plywood panels, where the curves are more pronounced, are the most difficult and critical. Here you must be absolutely faithful to the designer's dimensions and not attempt to average or straighten out lines.

When you are satisfied that the dimension marks are correct and that the batten is flexing smoothly over its full length, it's time to connect all the individual dimension marks with a continuous drawn line. Keep the pencil at a constant angle to the batten over the full length of the curve. Use your free hand to hold the batten down to avoid shifting or pushing it with the pencil. Repeat the procedure and mark all the edges of the panels. When finished, pull out all the nails and get ready to cut your marked panels.

You now have the outlines of your panels, albeit in only two dimensions. Your next step will be to cut out the panels with a small circular saw or jigsaw. Obviously, this step requires care, considering the money and scarfing time you've invested in your plywood panels. Good light is essential; vacuum dust extraction is helpful. Practice following curved lines on a few scrap pieces of plywood with your saw of choice until you feel reasonably assured. Set the depth of the cut to just barely clear through the plywood. And when you finally cut into your expensive marine ply, aim to the *outside edge* of the line you've drawn. After you finish sawing, it will be easy to trim the edges precisely to their true marked lines with a block plane and sanding block.

Figure 8-4. *In this photo, the batten is not pressing tightly alongside the brads placed at the intersections with the lofting station marks.*

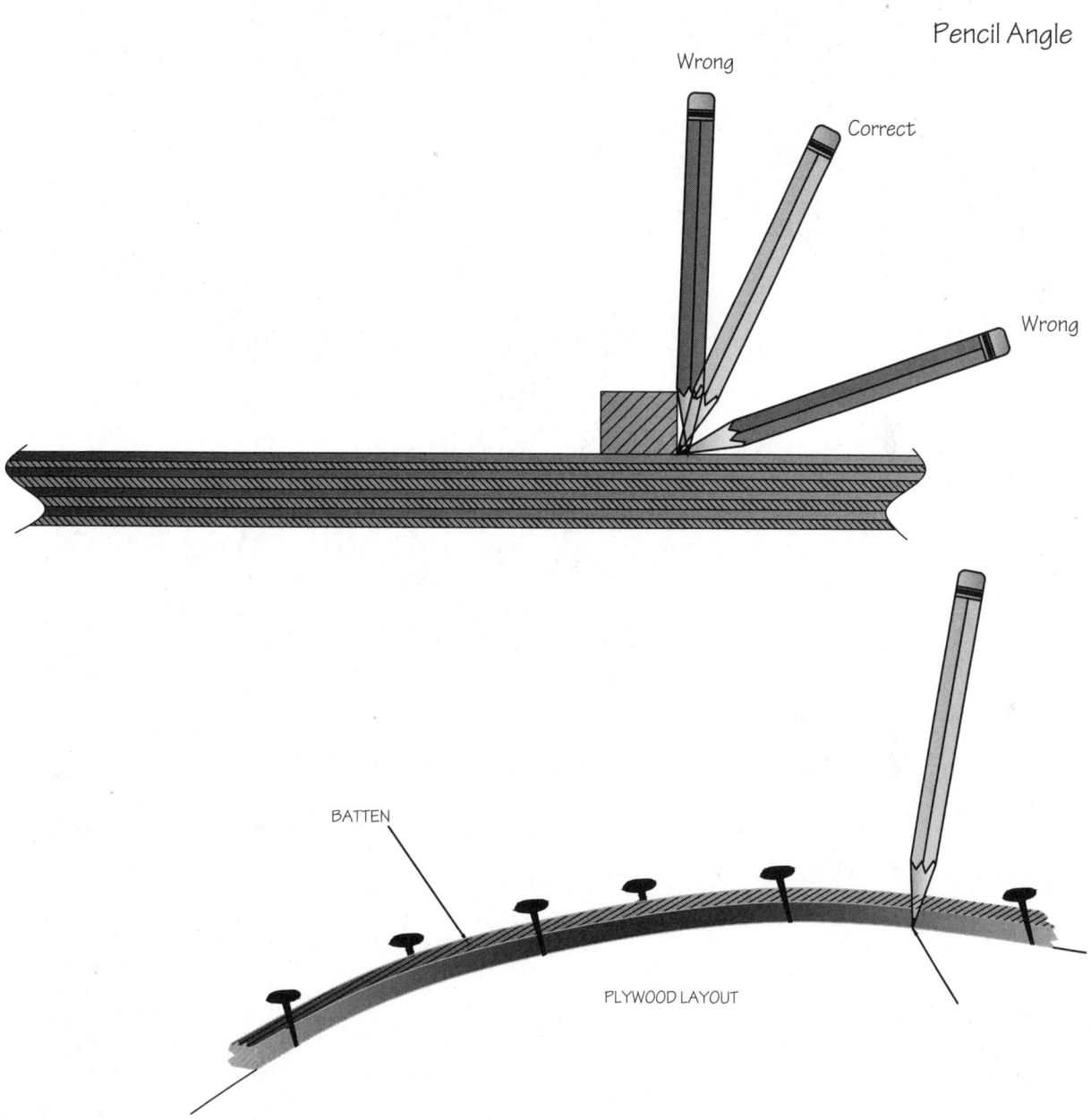

Figure 8-5. *An incorrect or varying pencil angle may introduce just enough imprecision to your lofting to cause annoying problems when you assemble the hull.*

Figure 8-6. *You can use a jigsaw to cut out your panels, but a small (5½-inch) circular saw may give you a cleaner line. Cut just outside your lofted line, and then you can easily refine it with a block plane.*

CHAPTER 9

Modeling

Building a scaled model of your Stitch-and-Glue boat before moving on to the full-sized job is not a trivial delay tactic for those apprehensive about the work ahead. Modeling can be a valuable learning tool, and it also can yield a lovely and lust-worthy mantel showpiece to help sustain momentum over the long run of the real boat construction. The basic assembly of a Stitch-and-Glue boat model is similar to the full-sized version, and it will help you gain understanding and confidence in the process. You can experiment with color schemes and contemplated modifications. At the advanced level, a floating and scale-weighted tow model can even be used to help establish a new design's hydrostatics, power requirements, and speed potential if you decide to try your hand at boat design.

I recommend working to the largest scale practical and one that can be conveniently read off the plans with an architect's scale rule. Try to end up with a model between 20 and 30 inches long. The scale you use may range from ¾ inch to 3 inches to the foot. The larger the scale, the more accurate your model.

You can build either a traditional half model, which essentially splits the hull in half down the centerline and mounts on a wooden backing plate (easier to hang on the wall for display), or a full model that can include as many details as you like, such as a deck and cabin (to be displayed on a table or mantel). Making the full model provides the most useful introduction to Stitch-and-Glue construction, so it's what I recommend if you are a first-time builder.

There is also a new modeling option made possible by CAD (computer-aided design) and CNC (computer numerically controlled) routers. CNC machines are growing in popularity, and among their many advantages is allowing very accurate cutting of the parts of a Stitch-and-Glue design at any scale. For any of our designs done fully in CAD format, we can create a model kit of however many parts and details the builder desires.

For the stock to build your own model, you can use thin balsa or basswood planks, which are available at hobby shops and many hardware stores. But for the most accurate modeling, I like to use modeler's 1/32-inch three-ply birch model aircraft plywood because it bends for the scale model very much as full-sized plywood does in the full-sized boat.

Building the model is much like building the full-sized boat, so the first step would be to decide whether you're going to build it right-side up or upside down. Reading Chapter 13 ahead now will explain the two processes. Either way, you'll begin by lofting all the bulkheads and setting them up on a workbench with their scaled spacing just like you would do on a full-sized boat (Figure 9-4). Lofting the longitudinals and setting them into position on the bulkheads follow. Then loft the bottom panels onto the modeling stock as outlined in Chapter 8, and cut them out with a small bandsaw or razor knife. A block plane can be useful for fairing the cuts, or you can make a miniature rasp by wrapping 150- or 220-grit sandpaper around a small wooden block. Next, lay one panel atop the other

Figure 9-1. *A carved half hull model is a fun and useful exercise in your boatbuilding journey. This one was crafted by the famed Canadian boat designer Bill Garden and is a work of art in itself.*

Figure 9-2. *A Stitch-and-Glue model is built much like the full-sized boat would be.*

as in a mirror image. Fasten the panels along the keel line with 1½-inch-long strips of strapping tape (the reinforced packaging tape typically sold for wrapping packages), then spread them out to form the entire hull bottom. If you're building right-side up and you have trouble holding the panels apart, cut a small stick to use as a bottom spreader, and tape that into place; it will hold the bottom panels apart until you can attach the sides to the model. If upside down, the bulkheads should mold the panels to the correct "V" form of the boat's bottom.

Now cut out the side panels, fair them to the precise shapes, and tape them to the bottom panels, starting at the bow end and working your way aft, completing one side before moving on to the other. We are using small tabs of the packaging tape in place of the wire stitches that we would use in the full-sized boat. Alternatively, I will sometimes wire together a model using light-gauge steel wire. As

Figure 9-3. *The stern view shows the twin rudder arrangement, unusual on a small boat. This is my Lit'l Coot 18 motorsailer.*

Figure 9-4. *This model was lofted from the full-size boat plans to a smaller scale.*

you will see, if you make a measuring error at this scale boatbuilding project, you waste a few dollars' worth of wood. The same error, scaled up to a scarfed 4 × 16-foot sheet of marine plywood, could cost a couple of hundred dollars.

After you are satisfied with the dry fit of the bulkheads, longitudinals, transom, and hull panels, you are ready to glue together the keel lines and chine lines with cyanoacrylate (CA) glue, an advanced cousin of the instant glues found in your local hardware and drug stores. Some of the gluing can be done from the exterior as the thin, water consistency of CA glue will capillary into the joint. I like a brand of CA glue called Zap-A-Gap, which works nicely with wood and fills small gaps between panels much as thickened epoxy will do in the full-sized boat. A companion product called

Zip Kicker will accelerate the curing time. Used together, these products will set up almost instantly, helping to assure perfect alignment. If you tend to work a bit sloppily, a third product to have on hand is Z-7 Debonder, which will help unglue mistakes. After the bottom and sides are fully bonded, glue in the bulkheads and transom from the inside or continue working on the outside until you have the stem, keel, or any other exterior structures added to the model.

If you had issues fitting any of the parts or find that you need some adjustment to the information in your building plans, make notations on the plans in colored pencil. These notes may provide very important information later. This is probably the greatest benefit of the whole modeling exercise—it gives you a scale preview of the full-sized building process, including warnings of possible trouble ahead.

How far you want to go on the scale model is up to you, but the more detail you build in, the more useful information you will carry forward to the full-sized boat project. Perhaps you are building a boat with a cabin and are considering altering the designer's configuration of its portlights, or windows (not recommended but sometimes undertaken by a zealous amateur builder). Even a small change here can have a profound effect on the boat's aesthetics, for better or worse. It takes

Figure 9-5. *Stitch the model's hull panels just as you will the full-sized boat, only using small-gauge wire.*

Figure 9-6. *This model is being stitched up without a building mold, a more difficult building method but one that works.*

Figure 9-7. *This model of the Pelicano 18 has been rolled right side up and has all the bulkheads installed.*

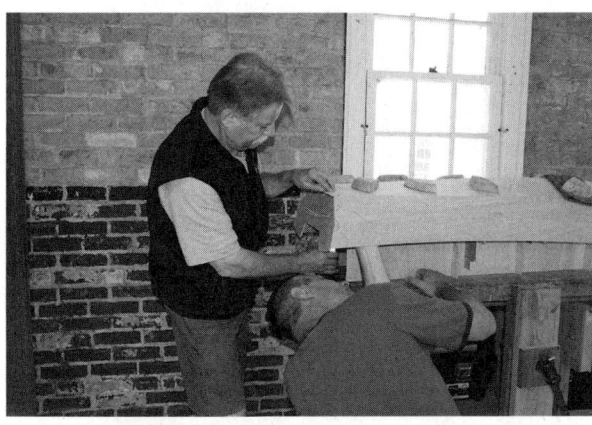

Figure 9-8. *These builders are tabbing the underside of the model to hold the hull panels together after its stitches are extracted—just as they will eventually do on the real boat.*

relatively little time and material to make two or more variations of the cabin and to try them on the model before making the full-sized and permanent commitment. One amateur builder I know even took photos of his cabin variants and e-mailed them to all his friends who were boat savvy for a vote. There is an obvious reasonable limit to the model's detailing, unless you actually are using it for strategic procrastination. However, I think it is quite useful to paint it in your prospective color scheme and then give it an honored display perch on your mantel or near your workbench. I have often found a model helpful in keeping me inspired and motivated while working on the full-sized version. It's not hard to generate motivational dreams, both in full color and in three dimensions, if the model is right in front of you.

Figure 9-9. *A presentation half model of a prospective design can even show trim detail. This model has a polished copper sheet on the back and appears like a full model with its mirror image.*

CHAPTER 10

Scantlings

An 8-foot dinghy or an 80-foot motorsailer can be built with the Stitch-and-Glue method. The big boat will be subject to many more and varied stresses than the small one, and it will need additional structure to tie together the exterior shell and help keep it rigid. In naval architecture, the term *scantlings* refers to the thickness and dimensions of any part of the boat's structure or shell (Figure 10-1).

Two forms of interior bracing help with these loads: stringers and bulkheads.

Stringers are longitudinal beams, usually laminated in place, running from bow to stern in as uninterrupted a manner as possible (Figure 10-3). They lie just inside the planking of the vessel and are usually fastened to the shell or planking of the Stitch-and-Glue boat with glue and fasteners. Gunwales in small boats and sheer clamps in larger boats fall into the stringer category. Stringers usually are made from dimensional wood stock, though there are exceptions where parts, or indeed the entire stringer, might be laminated from layers of plywood. My Surf Scoter 22, a raised-deck design, has so much shape in the stern portion of its sheer clamp—compound bends vertically and horizontally—that it precludes the easy use of just using dimensional stock. Fabricating these parts from several layers of marine plywood and then gluing them together makes the task approachable and easy.

Stringers are always glued, held into position and reinforced with mechanical fasteners until the glue sets solid. The fasteners are almost always left in place. In many instances, I've used stringers to land other structures, which effectively helps spread loads. An example might be a berth flat, settee top, or even a counter in the galley. If it can be placed over the stringer and then glued and screwed, you have a very rigid structure.

Bulkheads, the other basic structures in a Stitch-and-Glue boat, are cut from large marine plywood sheets and are set either athwartships, longitudinally, or horizontally (such as berth flats, counters, cabin soles, floorboards). Think of the Stitch-and-Glue boat as an egg crate or ice cube tray with an outside box and partitions erected at 90-degree angles to each other to hold the eggs and prevent the crate itself from being crushed. The Stitch-and-Glue boat works in much the same manner, redistributing stress and loads through a grid of bulkheads and stringers while protecting cargo and occupants in each of its compartments. These stresses must be distributed throughout the structure so that no single area carries a disproportionate load or strain. They all contribute to the strength and completeness of the whole structure.

Bulkheads provide an architectural definition of space along with their structural contributions. A galley compartment, for example, can be created using structural bulkheads. Bulkheads can double as berth faces, berth flats, cabin soles, cockpit soles, cabin sides, decks, cabin tops, galley faces, galley flats, head bulkheads, engine beds, anchor locker bulkheads, and more. By bonding each piece of plywood and every stick of dimensional wood with epoxy and fiberglass composite joints, every component becomes an integral part of the structure.

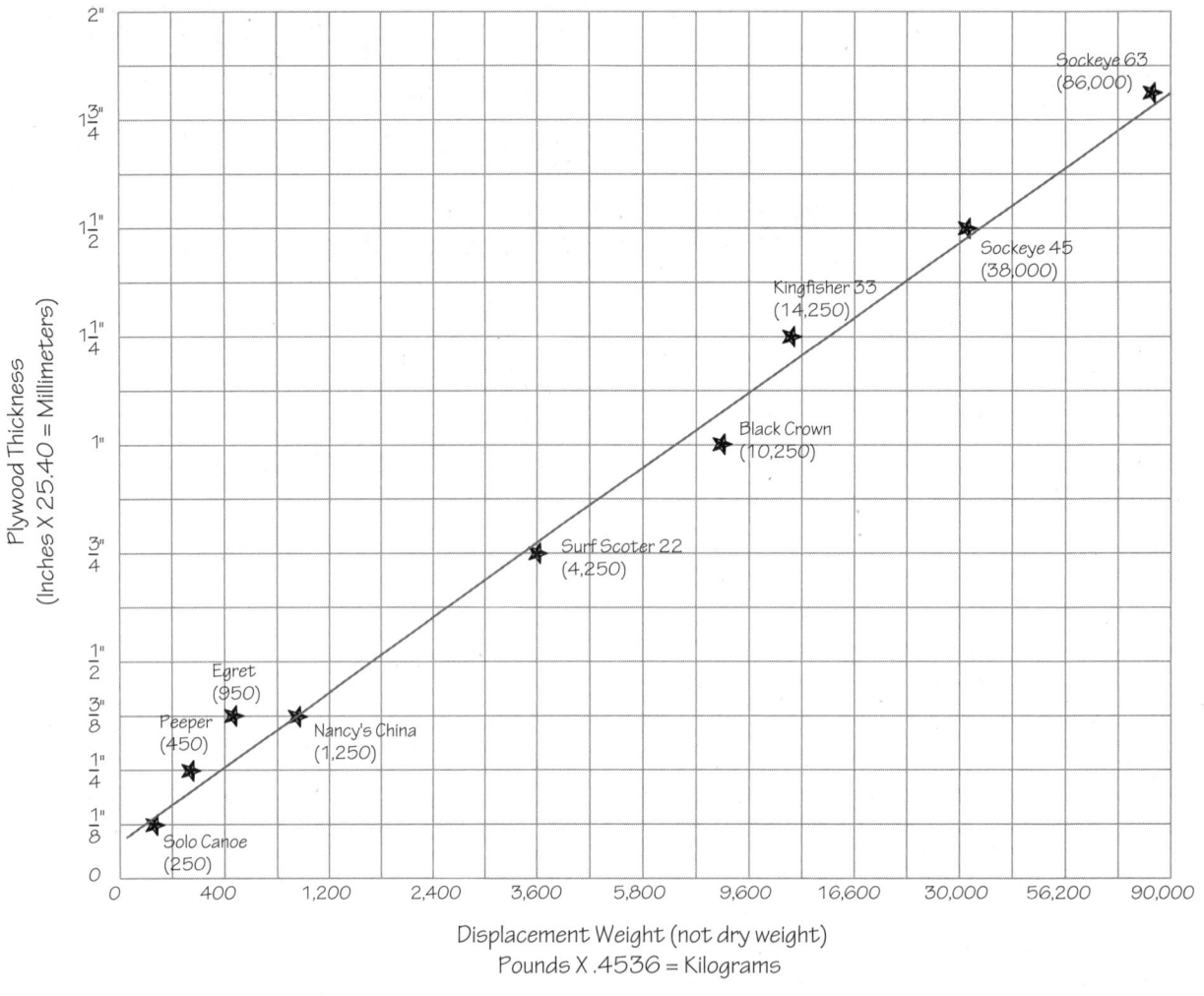

Figure 10-1. *This graph illustrates how hull thickness needs to increase with the size and weight of the boat.*

There is one drawback to this system: With all these pieces permanently locked in place, modifications become difficult (though not impossible). If bulkheads will define your battery compartment or the resting spot for your fuel tank, for example, be sure the compartment dimensions will accommodate the components you plan to use. If you need to cut holes for access, you can do so without difficulty and rebond the areas, but it is always smart to think well ahead to avoid unnecessary work.

Bulkheads must be stiff and beefy so that they will not deform under load, either during the boat's assembly or on a bad day at sea. For normal cruising and intermediate-performance designs, I use ¾-inch (18 mm) marine plywood for all major athwartships and longitudinal bulkheads and for any flat surface that may be jumped on or walked on or subjected to severe strain. For a component well supported by a primary framing structure, such as a foredeck, I use ½-inch plywood made from two layers of ¼-inch plywood, laminated together over its framing. For major bulkheads in really large Stitch-and-Glue boats, I will typically laminate them to a 1-inch thickness from two layers of ½-inch stock with staggered seams. This way a very large and strong bulkhead can be laid up with no need for scarfing.

Certainly, if I were constructing a multihull or racing monohull, overall weight would be a

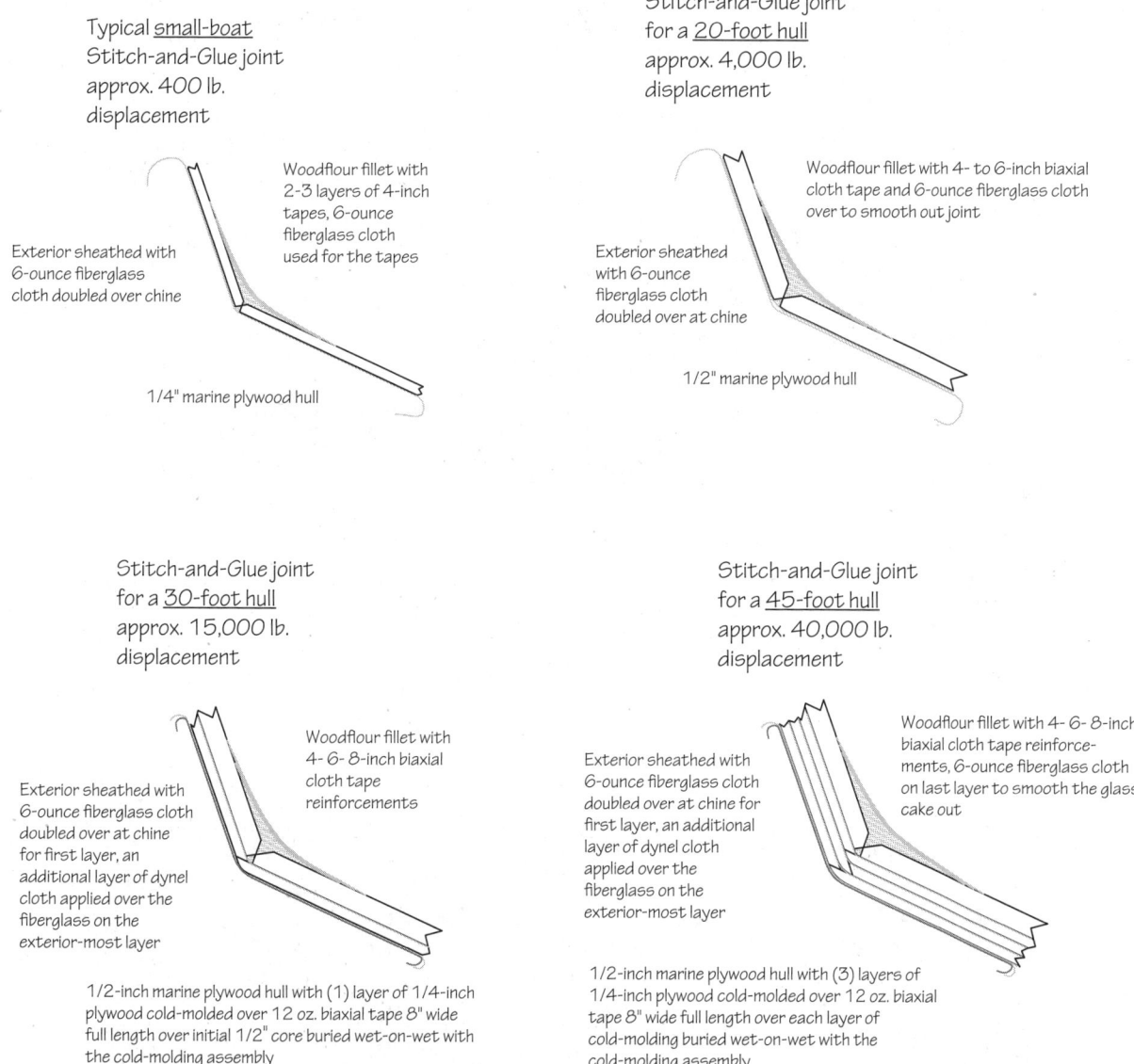

Typical <u>small-boat</u>
Stitch-and-Glue joint
approx. 400 lb.
displacement

Woodflour fillet with
2-3 layers of 4-inch
tapes, 6-ounce
fiberglass cloth
used for the tapes

Exterior sheathed with
6-ounce fiberglass
cloth doubled over chine

1/4" marine plywood hull

Stitch-and-Glue joint
for a <u>20-foot hull</u>
approx. 4,000 lb.
displacement

Woodflour fillet with 4- to 6-inch biaxial
cloth tape and 6-ounce fiberglass cloth
over to smooth out joint

Exterior sheathed
with 6-ounce
fiberglass cloth
doubled over at chine

1/2" marine plywood hull

Stitch-and-Glue joint
for a <u>30-foot hull</u>
approx. 15,000 lb.
displacement

Woodflour fillet with
4- 6- 8-inch biaxial
cloth tape
reinforcements

Exterior sheathed with
6-ounce fiberglass cloth
doubled over at chine
for first layer, an
additional layer of dynel
cloth applied over the
fiberglass on the
exterior-most layer

1/2-inch marine plywood hull with (1) layer of 1/4-inch
plywood cold-molded over 12 oz. biaxial tape 8" wide
full length over initial 1/2" core buried wet-on-wet with
the cold-molding assembly

Stitch-and-Glue joint
for a <u>45-foot hull</u>
approx. 40,000 lb.
displacement

Woodflour fillet with 4- 6- 8-inch
biaxial cloth tape reinforcements, 6-ounce fiberglass cloth
on last layer to smooth the glass
cake out

Exterior sheathed with
6-ounce fiberglass cloth
doubled over at chine for
first layer, an additional
layer of dynel cloth
applied over the
fiberglass on the
exterior-most layer

1/2-inch marine plywood hull with (3) layers of
1/4-inch plywood cold-molded over 12 oz. biaxial
tape 8" wide full length over each layer of
cold-molding buried wet-on-wet with the
cold-molding assembly

Figure 10-2. *These are typical hull scantlings for Stitch-and-Glue designs showing plywood and sheathing at the chines.*

concern. If keeping weight to a minimum is important, you might fabricate bulkheads made from one of the balsa-foam or honeycomb-core materials. These can be made considerably lighter than solid plywood, but they will cost much more, and you need to be extremely careful when you are laminating them into position.

Bulkhead placement is important, too. I like major athwartships bulkheads to be spaced no more than 60 inches apart, and I attempt to keep the maximum unreinforced area of skin between bulkheads, stringers, and other primary structures to no more than 12 square feet (that would be a maximum space of about 3 by 4 feet without support).

We can also increase the strength of our Stitch-and-Glue hull with additional layers of plywood laminated over the original stitched-up skin. This is usually not necessary in boats under 20 feet, but

Figure 10-3. *The Moon River 48 shown here is ready for planking with its bulkheads, longitudinals, and intermediate and sheer clamps all in place.*

Figure 10-4. *The Surf Scoter 23 under construction shows the lower side panel already attached to the intermediate clamp and the sheer clamp in position awaiting the upper panel.*

Figure 10-5A. *The deeply scalloped sheer line of the Banjo 20 requires a sheer clamp sawn to shape rather than the usual laminated and bent type.*

Figure 10-5B. *A view from behind the Banjo 20's cockpit shows the sheer clamps and carlins, both sawn to shape.*

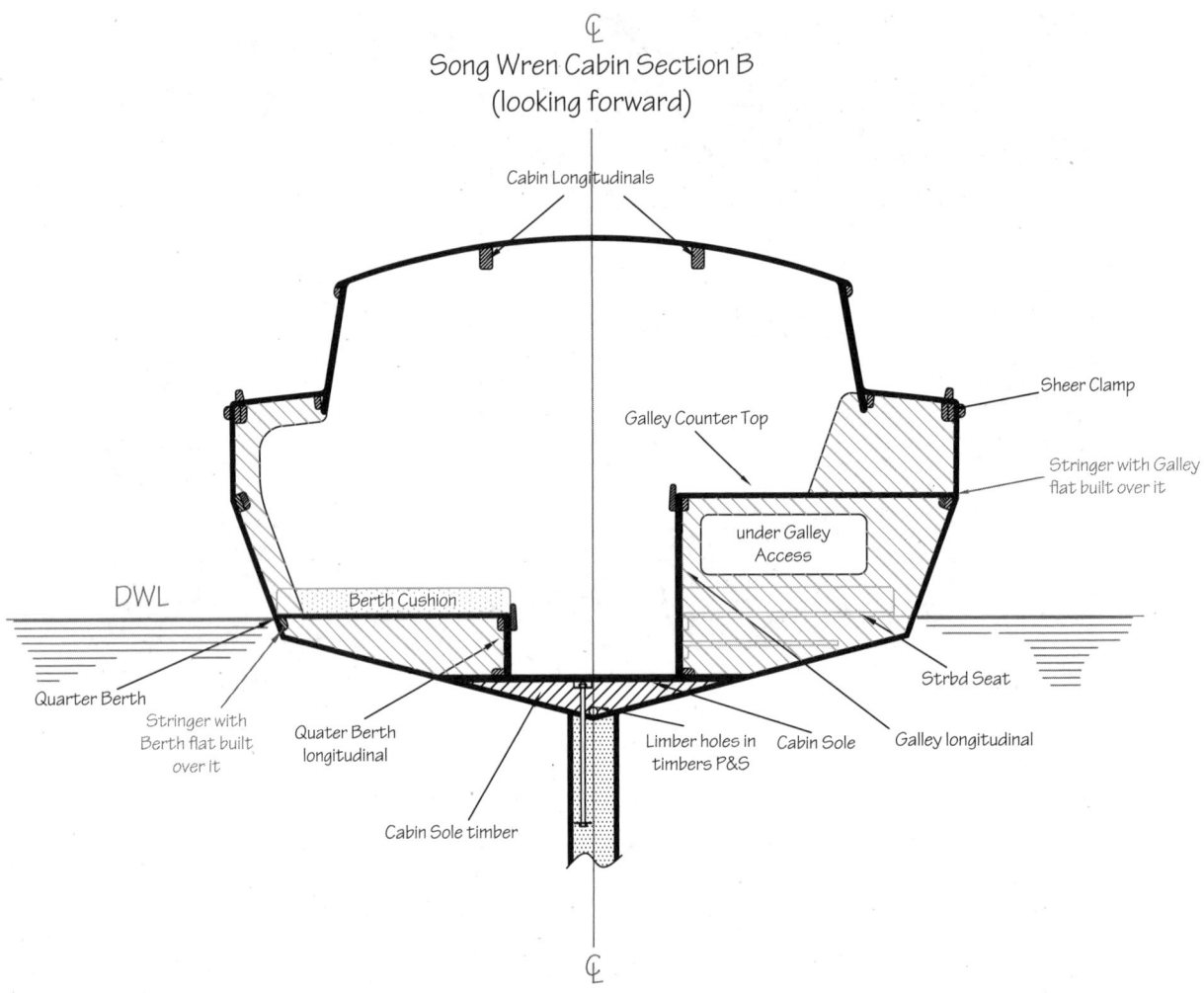

Figure 10-6. *This midsection of the Song Wren 21 sailboat shows how berth and galley flats land on intermediate chine stringers and become part of the boat's structure.*

larger boats may need a thicker skin, or at least a more durable bottom, than what a single layer of up to ½-inch plywood can provide. Thicker plywood fiercely resists bending into boatlike shapes, so we laminate layers of thinner sheets. For a complete discussion, see Chapter 16.

In my designs, I use several types of rigid-clamp support systems at the sheer. These are really stringers, but they are set at the edges of the plywood panels and are called sheer clamps, shelf clamps, beam shelves, or inwales, depending on what country you're from and what type of boat you're building. They provide a strong, stiff, and fair curve at the sheer (or upper) edge of the hull. In

the initial building stages of Stitch-and-Glue construction, the sheerline is defined only by the pressure of spreaders or by bulkheads against the skin of the hull planking, and it can look a bit scalloped. A laminated sheer clamp stiffens and helps hold the span between the bulkheads smooth and fair. Later, the sheer clamp will also serve nicely as the landing and fastening point for athwartships deck beams and ultimately for the deck. A more thorough discussion of this process is presented in Chapter 22.

Depending on the design and on what I am trying to accomplish from a structural and aesthetic sense, I sometimes eliminate athwartships deck or cabin top beams and replace them with longitudinal

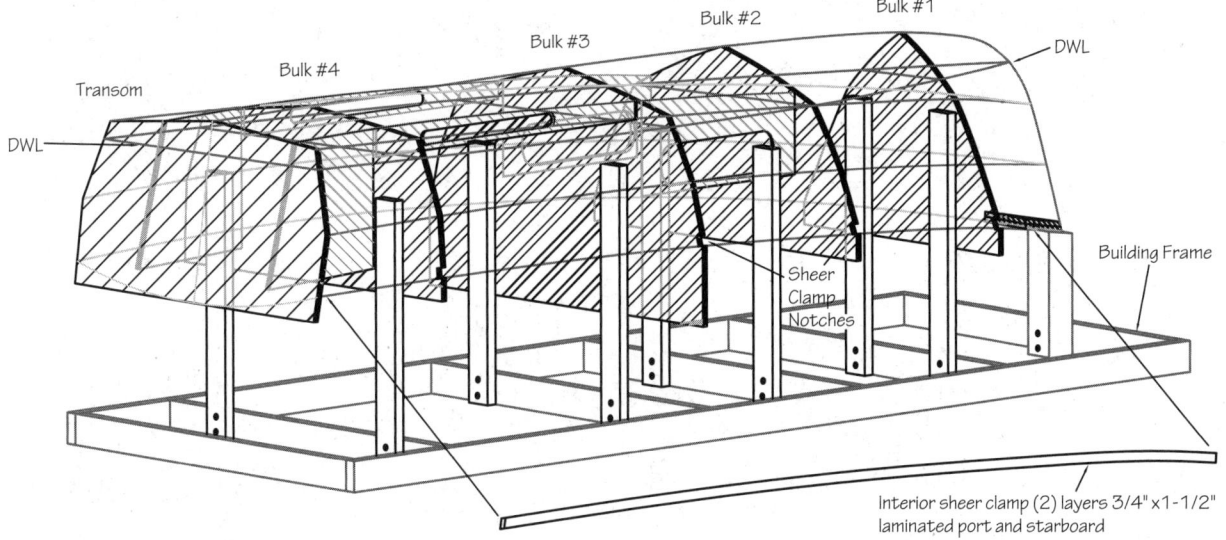

Figure labels: Bulk #1, Bulk #2, Bulk #3, Bulk #4, Transom, DWL, DWL, Sheer Clamp Notches, Building Frame, Interior sheer clamp (2) layers 3/4" x1-1/2" laminated port and starboard

Figure 10-7. *Bulkheads and longitudinals are set up on the building frame in preparation for upside-down hull construction. Note the pre-cut notches in the sides of the bulkheads that will receive the sheer clamps.*

beams on port and starboard (left and right) stretching fore and aft. This makes sense, since the middle of the belowdecks area, especially in a small boat, gets the heaviest fore-and-aft traffic flow. If headroom is already scant, athwartships beams mean more head-knockers. A couple of larger scantling, stiff, longitudinal beams allow you to leave the center 20 to 28 inches uncluttered for maximum headroom. If carefully planned and crafted, longitudinal beams look just as proper in their supporting role as would athwartships beams, and with the addition of a laminated topping of two or more layers of marine plywood, provide for a very strong deck. Furthermore, fairing the deck or cabin top is made easier since there are fewer parts to fit.

In either construction, the sheer clamps/stringers serve as strong landing points for the deck. This is a critical joint. Dock rash, bumps, scrapes, gouges, bashes, and miscellaneous insults all seem to focus on the sheer and hull-to-deck joint—in my experience, 70 to 90 percent of a boat's wear and tear. And a sailboat with a deck-stepped mast adds to all this the load transfer from deck to hull. A strong sheer clamp heavily backs up, reinforces, and protects this joint.

Decks and cabin tops are also subject to heavy stress, so don't be tempted to make the decks too light. Just imagine a clumsy, overweight oaf jumping from a high fuel dock onto your deck. Laminate the decks over the deck framing with two or three layers of thin plywood. Laminated decks also help tie the deck framing to the sheer clamp. A laminated deck acts as a huge breasthook or knee. I've found that a ½- to ¾-inch laminate

Figure 10-8A. *As with the deck, longitudinal and athwartships beams support the cabin top.*

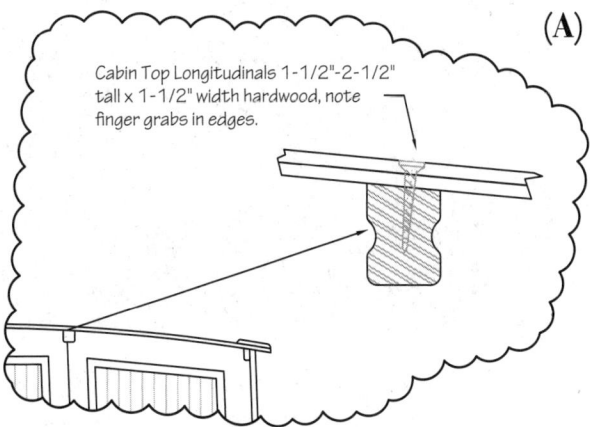

(A)

Cabin Top Longitudinals 1-1/2"-2-1/2" tall x 1-1/2" width hardwood, note finger grabs in edges.

Figure 10-8B. *The pilothouse top of the Red Salmon 33 is being laminated onto the longitudinal beams.*

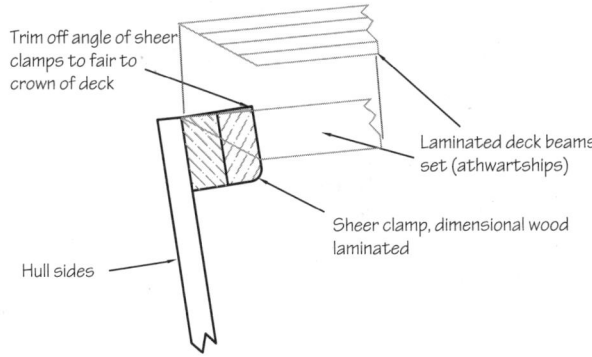

Figure 10-9. *Athwartships deck beams can be let into the sheer clamps with simple notching.*

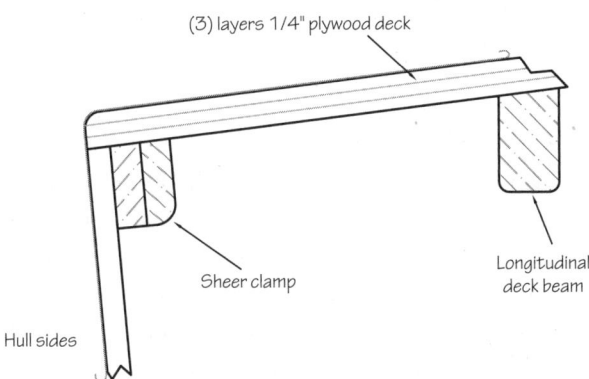

Figure 10-10. *Laminating a crowned deck from two or more layers of ¼" marine plywood will help make it stiff and strong.*

usually is a reasonable compromise between adequate stiffness and excessive weight.

Stitch-and-Glue construction is different from traditional plywood construction in the bilge or keel areas. Traditional plywood construction requires a rabbet joint in the keel to attach the plywood skin

to the boat's backbone and framework. This is a tricky joint to make and offers major potential for leaks into the hull itself at the junctions of the hull planking to the keel. In a Stitch-and-Glue boat, keels, skegs, and all other appendages are bonded into place over the one-piece glass/epoxy sheathed hull, assuring the hull's watertight integrity.

While it is tempting to laminate these structures using large pieces of dimensional wood, it is difficult, if not impossible, to encapsulate and seal those same heavy, thick chunks of wood effectively with epoxy. If moisture is allowed into the wood in any way, the wood's dimensions will change as it swells and contracts with varying temperatures and with changes in the moisture content of the wood. This dimensional instability results in an incredible strain in the epoxy joints, which ultimately can lead to failure. The larger the piece of wood, the greater the dimensional change it can and will undergo.

In general, I'd say that no piece of wood thicker than 18 mm or ¾ inch can be adequately sealed with epoxy. If you have a thick keel to build, you are better off laminating it from many layers of thin stock rather than using fewer pieces of thicker stock. (We will undertake a thorough discussion of building keels and skegs in Chapter 17.) As time has shown me with my experience of building, I mostly assemble these stem and keel structures from marine plywood these days. Glassed with many layers of cloth set in epoxy, they seem to be the best combination of strength and utility. I build a hull so that its structural integrity is not compromised by problems with the stem or keel, as explained above. I consider these parts replaceable, but even so, they must be copiously sealed with epoxy for the longest possible life and reliability. Replacing boat parts below the waterline is nobody's idea of a fun summer.

CHAPTER 11

Building Cradles

Most forms of boatbuilding, ranging from traditional plank-on-frame to modern fiberglass and aluminum, require some type of backbone-supported framework or molds around which the hull is shaped. Stitch-and-Glue designs actually don't need it, though for boats of more than 12 to 15 feet it is much easier to first set up the bulkheads of the vessel on a backbone and stitch the plywood panels around them. We will go through this process in detail in Chapter 13.

For smaller boats that can be built right-side up (those small enough that you can reach comfortably into the hull at least to the centerline for the epoxy/glass fusing of the hull panels), the builder must provide a cradle or other support system at the outset to hold the boat upright and level both athwartships and longitudinally. This is to help keep the hull level and the panels in proximate relationship once they are stitched together. It also assists with the positioning and setting of interior parts, helps to support the hull if the builder must climb in to work on the inside, and keeps the panels from flexing out of their true relationship with one another. For a very small boat, these cradles might be nothing more than a couple of sawhorses to which you clamp some simple cut-to-shape wedges to support the hull. A more elegant option and one more suited to a slightly larger boat is a built-to-shape wooden support cradle. Or you might use four or more boatyard-type jackstands (screw jacks), which are not inexpensive but will last you for a lifetime of boatbuilding. The basic rule is that if you cannot

reach the centerline of your boat without climbing into it, you will certainly need to build or provide a cradle to work in the interior and firmly hold the boat in a level position. And if you build your hull upside down, after rolling it right-side up you will still need a cradle or jackstands to level and hold it firmly for all your later work inside and on top.

A cradle's most important purpose is to keep the boat level and square—that is, in simple words, not twisted. Once cradled, it is vastly easier to level the boat longitudinally and athwartships, and if it is a good cradle, she will stay in proper position throughout the entire building process. During every stage of building, it is important to keep checking the levelness of the hull structure; failure to keep it level and not twisted will quickly lead to the misplacement, misalignment, or even twisting of interior structures—and a ruined hull. Misalignment is not as great a potential problem if you build upside down over bulkheads, as the hull is fused on her interior seams and joints while upside down and before rollover, and once she is fused together, there is not much that can distort the form. But you will still want a stable work platform as you will be entering and leaving that boat many, many times while constructing it when she is right-side up.

If you want to build the cradle, I suggest a design that uses longitudinal beams as skids for the base, built up with cross members that touch and support the hulls at approximately stations 2, 5, and 8, based on a 10-station plan. If your plan does not specify the cradle dimension details, you can

Figure 11-1. *Scrap wood braces screwed into sawhorses will help to stabilize the hull while you're working on it.*

Figure 11-2. *Foam wedges can easily be cut to shape to cradle the hull.*

Figure 11-3A. *Jackstands provide very good and adaptable support for larger boats such as the Storm Petrel 33 here.*

Figure 11-3B. *Note that jackstands must be securely tied together in pairs with chain or rope to prevent the boat's load from pushing them out of place—with disastrous consequences.*

Figure 11-4. *This small boat has been stitched right-side up on its cradle.*

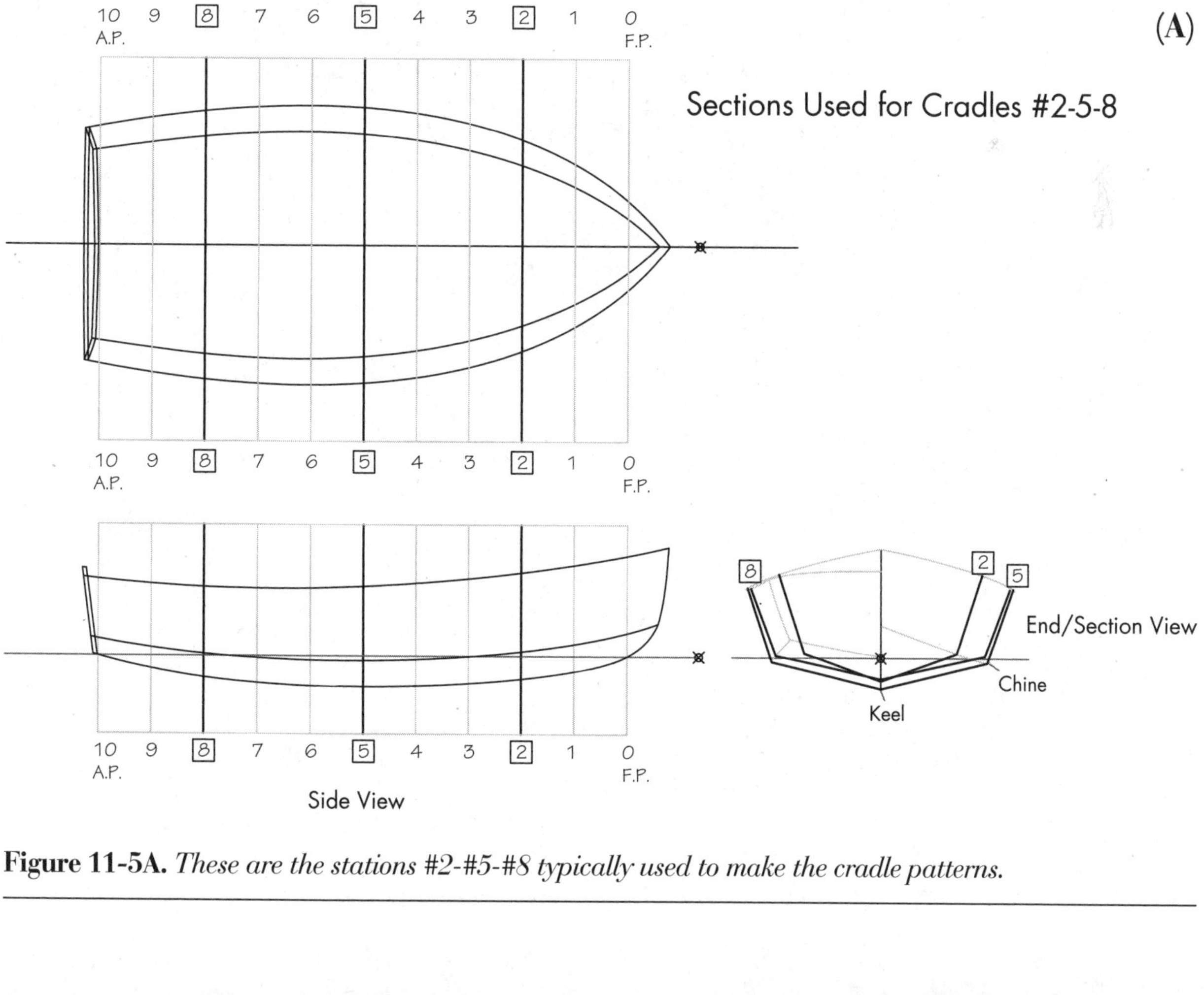

(A)

Sections Used for Cradles #2-5-8

10 9 8 7 6 5 4 3 2 1 0
A.P.   F.P.

10 9 8 7 6 5 4 3 2 1 0
A.P.   F.P.

10 9 8 7 6 5 4 3 2 1 0
A.P.   F.P.

Side View

End/Section View

8   2 5

Chine

Keel

Figure 11-5A. *These are the stations #2-#5-#8 typically used to make the cradle patterns.*

Figure 11-5B. *Spray insulation foam for weatherproofing windows will hold cradles or mold floors in position.*

easily figure out the cradle points from the station or bulkhead lines of your design. You could also use the inverted hull as a form for building part of the right-side up cradle on top of it. This would help you perfectly set the angles of the cradle members making contact with the hull.

To build a cradle for a boat under 24 feet long, use 2 × 6-inch skids and 2 × 6-inch cross members and uprights. When making cradles for larger boats (up to 32 feet), I use 4 × 8-inch skids and 2 × 8-inch cross members and uprights. Any boat larger than this would require jackstands and a more elaborate support scheme. You will be climbing inside the hull, so the cradle must be strong enough to support your weight (and that of helpers,

Figure 11-6. *A well-constructed and padded cradle for the right-side up work awaits the upside-down hull next to it.*

if you'll need them) without flexing or any possibility of tipping. This is not a place to scrimp on materials. Align the skid members first. Thinking ahead to your launching, if you'll be using a truck or flatbed trailer to move and launch the boat, its width will determine the width spacing of your skids. My shop flatbed is 5 feet 8 inches between the rails, so the skids on many of my cradles are just a bit narrower than that. This gives me a few inches of adjustment when loading or unloading. Build up the cross rails at the #2-#5-#8 station intervals and at appropriate angles and lengths. The metal joist brackets available at lumberyards are perfect for attaching and stabilizing the cross members, or you can make scrap plywood gussets to reinforce the cradle.

Place the uprights on top of the cross rails. If you are building a keel (power or sail) boat, build the uprights in two sections so that they support the hull port and starboard, leaving sufficient room for the keel. On larger keel boats, you might want to build two sets of uprights: one for the hull in a low (keel-less) position, and one for the hull with the keel in place. The lower uprights will make interior work easier by eliminating the additional climbing that would otherwise be necessary until you set the ballast keel on her toward the end of the build.

Reinforce all the potentially moveable joints in the cradle structure with plywood gussets. Protect the cradle surfaces that will come into contact with the hull with layers of scrap carpet padding,

nailing or stapling the padding along the sides of or underneath the uprights to prevent nail heads from damaging the boat. Use a good exterior-grade house paint to protect all exposed cradle surfaces from weathering during outdoor storage.

Figure 11-8. *Details of cradle support system. Triangular gussets or buttresses are solid insurance that the cradle won't move or collapse.*

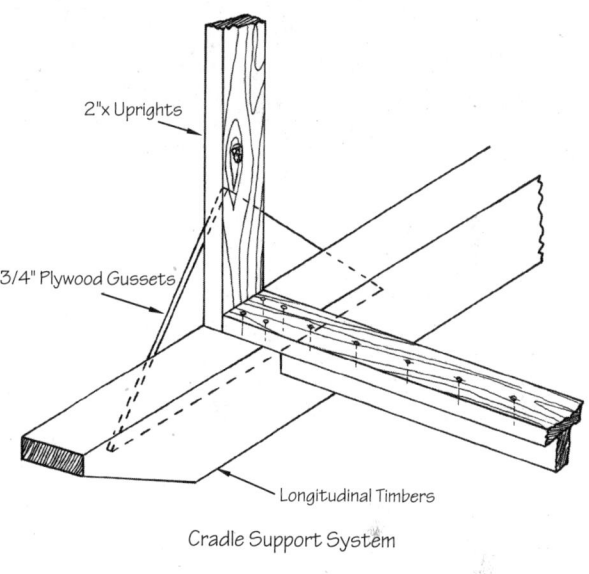

2"x Uprights

3/4" Plywood Gussets

Longitudinal Timbers

Cradle Support System

Figure 11-9. *Details of cradle support system. Triangular gussets or buttresses are solid insurance that the cradle won't move or collapse.*

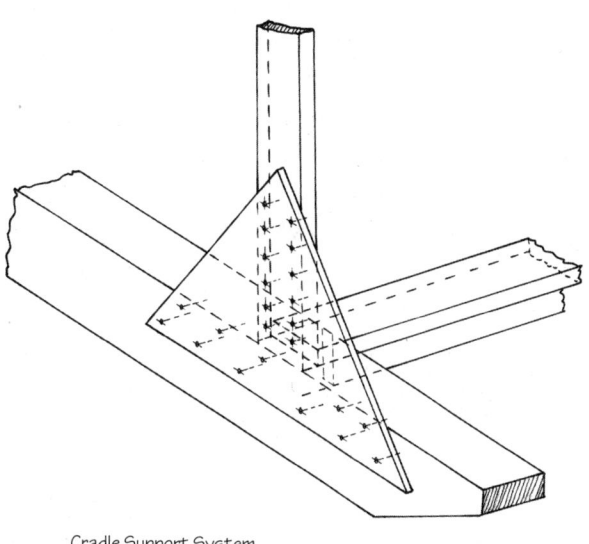

Cradle Support System
Reverse View

Figure 11-7. *A typical shop-made wooden cradle.*

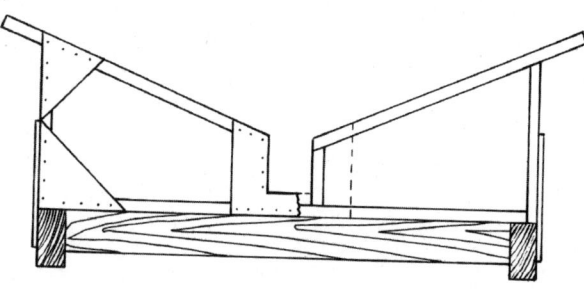

Building Cradle Powerboat
Type "End View"

Figure 11-10. *The building jig uses foam and wedges to lock its geometry in place.*

Built this way, your cradle can live a long and useful life.

In many cases, it pays to make the boat under construction mobile from the start. As in every phase of the boatbuilding process, playing the construction movie forward in your mind will avert much future grief and wasted time. If the boat is going to be too large to be turned over inside the workspace, heavy-duty casters bolted to the cradle skids will make it easy to roll outside. If the hull is constructed upside down, you may want to attach casters to your building jig *and* the right-side up cradle that will await it at the eventual turnover party outside. Online shopping makes heavy-duty casters easy to access these days; long gone are the hours I spent at Boeing Surplus looking for casters that could be bought somewhat affordably.

CHAPTER 12

Stapling Your Stitch-and-Glue Boat

Shall we stitch or staple? I had been thinking of stapling as a potential substitute or augmentation for wire stitching over a couple of decades, but every time I was stitching up a new hull, the process worked so well and quickly that I stalled on trying something new. I guess the adage *if it ain't broke, don't fix it* applied. But one day I found myself in a real predicament, having to build a small boat in a matter of days; every minute I might conserve in the process could help save me from working between midnight and 6 a.m., when I really do appreciate some sleep. My younger son Mackenzie was helping me work at our home shop, not the normally busier main boat shop, and with no one else looking over our shoulders, I decided, in some desperation, to try stapling.

The bottom line: It worked spectacularly. And very quickly.

How quickly? A few years later I was teaching a class at the Wooden Boat School in Brooklin, Maine. We were building three of my Bella 10 skiffs in the class. I demonstrated the stapling method on the first boat, talking through the process with my students as I was working, and in about 20 minutes we had the first boat assembled and ready for tabbing the interior seams. I had the students tackle the second boat on their own with just outside coaching from me. I timed the third boat while the students

did the work, and they were able to assemble the 10-foot hull in just 10 minutes and 23 seconds. This was practically breathtaking, and the stapling performed as well as on any fully stitched boat. It just took much less time to complete.

Today in our shop we use a combination of stapling and stitching, generally resorting to the stitches only where we need greater holding strength, such as pulling the panels together at the bow (Figure 12-1). You will need to initially

---

Figure 12-1. *An upholsterer's pneumatic stapler and lineman's pliers form the basic tool kit for stapling your Stitch-and-Glue boat.*

assemble the boat upside down, but that is not a problem if the boat is small, as you can easily roll her over to work the inside later. And a larger boat would be built upside down anyway.

A professional shop will likely use a pneumatic upholstery staple gun (these cost in the $85–$110 range), but for the home builder one of the inexpensive ($20–$50), heavy-duty, spring-powered guns such as those by Arrow and DeWalt will do. Also coming on the market now are cordless staplers by many of the cordless tool manufacturers, and I believe these will do a fine job. We use stainless steel Fine Wire Staples #EE710SS that measure a ⅜-inch wide crown and are available in ¼-inch, ⅜-inch, and ½-inch leg lengths in our own upholstery staple gun. The staple wire is fine enough to avoid blowing out the grain of the plywood if the legs poke all the way through. Another advantage is that the wire is so smooth that stray epoxy does not seem to adhere to it quite as tenaciously as it might to thicker galvanized or coated wire staples. All metal staples must be removed before fiberglassing, as they might start moving later in the boat's life and cause a breach in the integrity of the joints. I reviewed a composite staple at a recent Wooden Boat Show display, and these polymer composites might work very well as the staple legs could simply be sanded off flush. The technology in the tool and fastenings realms is changing very rapidly, and I encourage you to explore your own options.

Still another advantage of stapling is that if you are planning to bright-finish (varnish) your hull, you would not have the unsightly filled ⅛-inch holes that the wire stitches leave. The tracks of the fine wire staples are literally invisible.

For larger boats in sizes up to 48 feet, I have had very good results with a Senco air stapler using "M" wire staples. These also have a ⅜-inch crown but are a much heavier gauge of wire and are available in ½-inch to 1¼-inch sizes.

The stapling process is much simpler than drilling holes and twisting wires. Just set the stapler to straddle the two panels and pull the trigger (Figure 12-2). It probably takes a slightly more experienced hand to hold the panels in position, as the staples don't quite have the initial strength of clamping that the wire stitches do. You may also

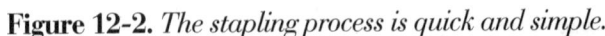

Figure 12-2. *The stapling process is quick and simple.*

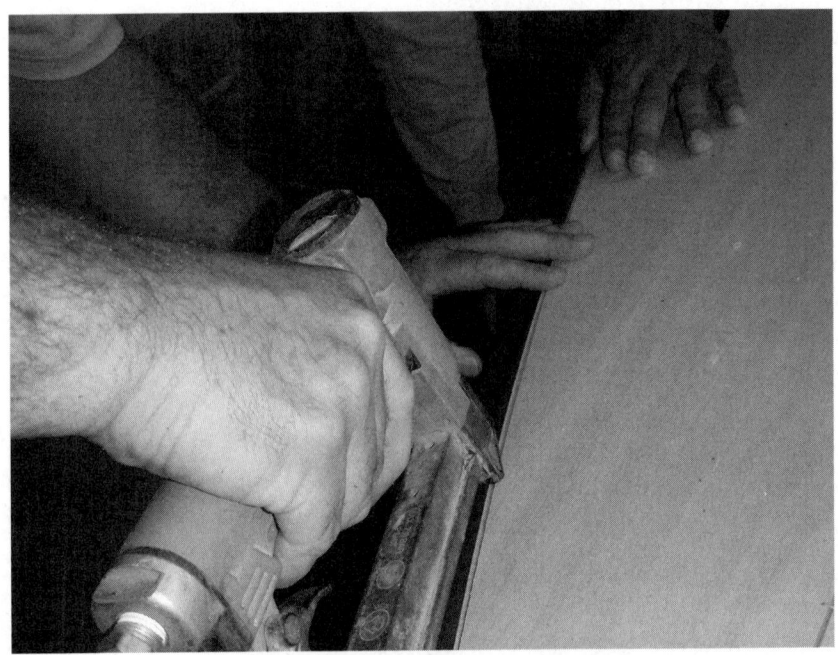

need to stitch the bow and transom ends because the greater bending of the plywood at the hull's extremes requires more force. I actually recommend using a combination of stitches and staples, as we do in our shop (Figures 12-3, 12-4, 12-5, and 12-6). The stitching process will be explained in the next chapter. Combining the two may require a bit of experimentation and experience, but it

Figure 12-3. *Many boats will need a combination of conventional stitching and staples, with the stronger stitches used to pull the panels together at the bow and stern.*

Figure 12-4. *Stapling is very easy at the intermediate chine.*

represents a significant labor-saving improvement in the Stitch-and-Glue method.

Removing the staples is easy with an awl (Figures 12-7, 12-8, and 12-9). Professional upholsterers use a special trident-shaped remover (available from an upholstery supplier). I have also had very good service from an old flat-blade screwdriver that had its end heated and bent to an angle in the metal vise. A bit of filing results in a great little tool that will make the process simple and easy.

We have of late been experimenting with another stapling variation on smaller boats that can easily be rolled by one or two people, where we do the minimum amount of normal wire stitching and fasten the rest of the exterior with the staples as shown and described previously. But then we roll the boat over and after installing the spreaders and confirming our happiness with the boat shape,

Figure 12-5. *Combination stitching and stapling has been completed on this hull.*

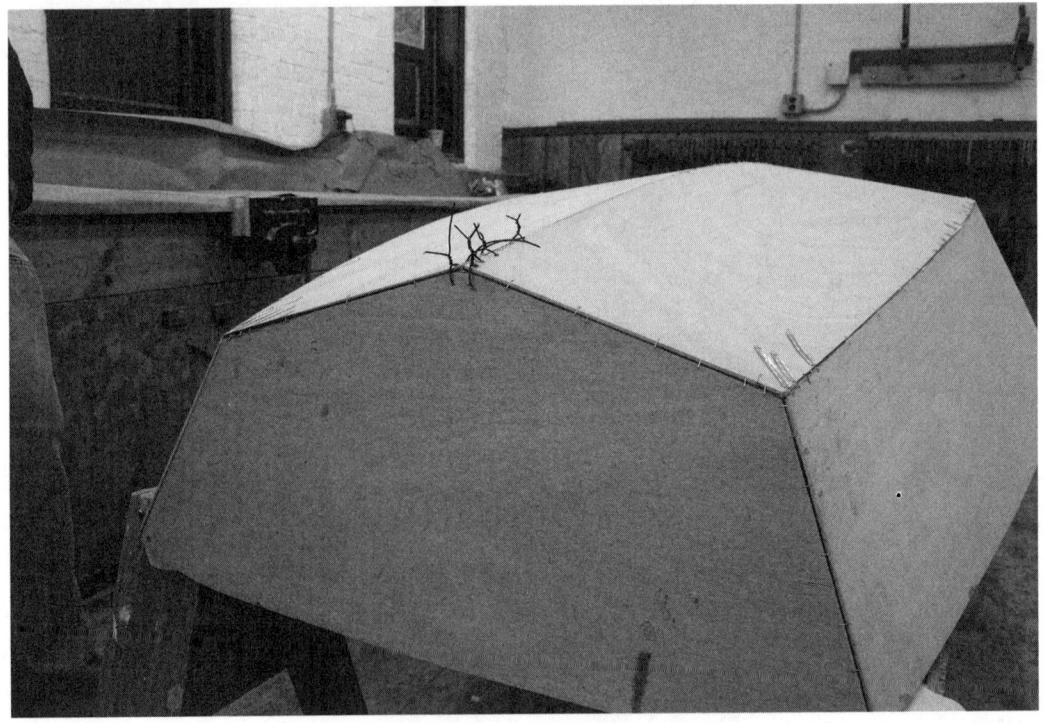

Figure 12-6. *Heavy-duty stitches are usually needed to convince the panels to join up at the stem.*

Figure 12-8. *Bending the awl's tip makes it more effective in prying out the staples.*

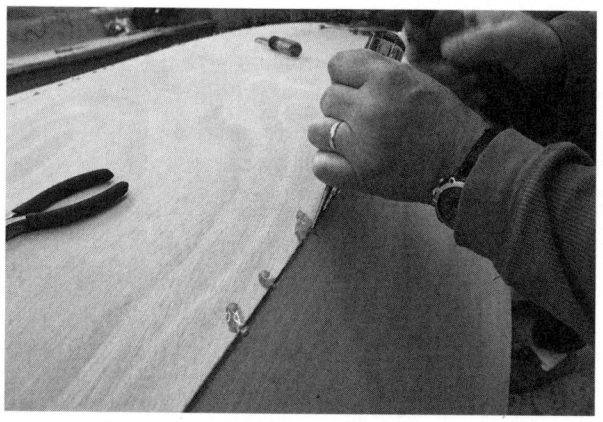

Figure 12-7. *Use an awl to remove staples after tabbing inside.*

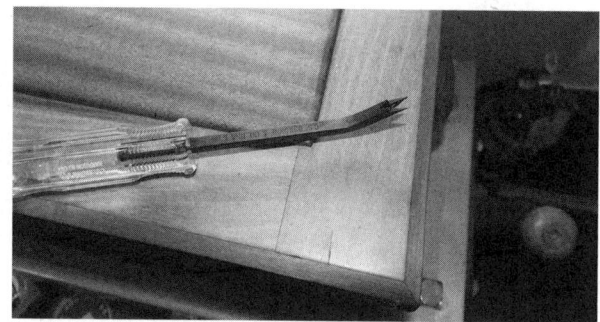

Figure 12-9. *A commercial staple-removing tool is available from upholstery suppliers.*

we proceed to also staple the interior seams of the boat. Holding the staple gun at an angle prevents the wire staples from going completely through the plywood planking. Rolling the boat again to the upside-down configuration we can then pull out all the exterior wire stitches before rolling it once again right-side up and fully taping the interior. Why would we want to do this? Simply to allow us to save another day in the building process and skip the intermediate step of tabbing. This allows us to literally bury those interior staples with the wood flour fillets and epoxy taping of the interior seams, and skip the time-consuming removal of glued wire stitches. Experiment with stapling yourself; it's a great technique to add to your Stitch-and-Glue toolbox.

CHAPTER 13

Upside Down or Right-Side Up

We are now ready to leave the preliminaries behind us, roll up our sleeves, and build a boat. Although this method will take some explaining and preparation, you may be pleasantly surprised at how simple and quick the Stitch-and-Glue process is. One of its distinct advantages is that you will find yourself looking at a boat-like shape very early in the building process, which will pay immense dividends in encouraging and keeping those boatbuilding energies lit and bright.

We used to build all our boats right-side up in our shop, literally crawling inside the open hull after the stitching to undertake the tabbing process. We built boats up to 38 feet this way successfully and accurately. But you might already suspect the complication: As we rambled around in the open hull, our shifting weights would sometimes flex or wrench the panels out of position. We needed a very strong and complex building cradle to nest the hull, holding things in fair and proper relationships until we had the chance to fuse the plywood hull seams together. There was another disadvantage to building larger boats right-side up. We would stitch the hulls together, climb inside, and tab and install bulkheads and other structures before we dared pull the wire stitches, and then we would roll the boat upside down to cold-mold the hull panels, attach the stem and keel structures, and sheathe the

hull with cloth and epoxy. Then we would again roll it right-side up so that we could finish construction. We therefore gained considerable savings in labor when we abandoned right-side up building in favor of the upside-down method for all boats that were too large for us to reach inside (from outside the hull) to the centerline for the tabbing and interior glassing. This effectively eliminated building anything much larger than 12–15 feet in length right-side up. We have not looked back wistfully.

In the upside-down method, we cut out the bulkheads first, position them upside down on a simple backbone or ladder framework, and continue assembling the hull by draping and stitching the hull panels around them. The extra time needed to set up the bulkheads is more than reclaimed in the ease of stapling and stitching the panels together and positioning them on this form with no further worries about the proper hull shape, and in rolling over just once.

Now that most of our designs are available in CNC-cut kit form, the upside-down method makes even more sense, and I recommend it for all sizes of Stitch-and-Glue boats. I looked at my building list from the last 15 years and found that we built almost every boat in our shop with a precut plywood CNC kit. Building upside down with perfectly accurate parts has eliminated much labor and has resulted in our building even better boats than in the old days of right-side-up building with hand lofting and cutting-out. For professionals and amateurs alike, it's as near foolproof as boatbuilding can get. The kit very often comes with a building setup jig that includes precut slots for the bulkheads to slide into, and this serves as a perfectly aligned form for you to wrap the hull panels around and stitch or staple them together. No large gaps will appear between hull and bulkheads, and no twisted-banana asymmetry will occur because of mistakes in alignment. It's really much easier to mold a hull around its bulkheads than to later fit bulkheads into a stitched hull as would happen with right-side up building.

I'm going to describe the upside-down process first and then follow with the right-side-up procedure. If you choose to read both descriptions, you'll easily see why we now prefer the former. But if I've already convinced you to build upside down, you really don't need to risk confusing yourself with the alternative method.

UPSIDE DOWN

Make a straight, rigid building jig or framework, or assemble the components of the kit setup jig with the supplied instructions (Figures 13-1A through 13-1E). If you're building from scratch, mark the locations of the permanent bulkheads or temporary molds on the jig's top surface, taking the intervals from the plans. Draw a centerline to keep them aligned.

Cut out or assemble your bulkhead puzzle pieces and prepare them for inserting in the jig. Like the floor timbers (which we will install later), the bulkheads must have limber holes to allow water to drain to a central low point in the bilge where it can be pumped out. Think hard about when and where to create them. Drilling or cutting them before the hull wraps around them at first seems easy. However, my shop prefers to do all the interior glassing of the bulkheads first without worrying about trying to glass around limber holes. After the epoxy sets, I then arm myself with a hole saw on a cordless

drill and cut the limbers. At this point, the hull is right-side up and leveled fore and aft and sideways. It is pretty easy to see the low points in all the bulkheads and longitudinals and then cut out the limbers that ultimately will lead any water down to low-point collection areas (called bilge sumps). If you are going to install multiple bilge pumps, the limbers can lead to those deepest sumps. However you cut them, limbers must have their edges sealed very thoroughly with epoxy; you never want to give water a migratory path into a bulkhead's inner plies with porous sealing.

With small boats (say, under 22 feet in length), I resist the temptation to compartmentalize any one area of the boat and give bilge water an easy path to the lowest point. The main exception is the engine box (if the boat has an inboard engine), which must be a separate bilge since it is neither legal nor environmentally sound to pump engine-related petroleum products overboard with normal bilge water. Isolating the engine compartment will help to contain any fuel oil or lubricant spills until they can be cleaned up with oil soak rags or sponges and disposed of properly. It may also be very smart in the long run to set up a separate bilge sump in the fuel tank area. Any leaks years down the line would then be contained and not have any possibility of fouling the boat's main bilges. In a large hull, there might be as many as three or four bilge sump areas: cockpit, main cabin, fo'c'sle, and engine bilge.

Now you can set up your bulkheads on the jig (Figure 13-1F). If need be, brace them to set vertically (unless your plans call for some to be set at angles) from the jig floor. All the waterlines should be at exactly the same measurement from the top surface of the jig, and the centerlines of all the molds must be in alignment. The more careful you are at this setup stage, the easier your building project will be. Most of your plans will have a mouse hole shown on each of the bulkheads, which will help with alignment of these structures. Thoroughly seal all the bulkhead edges if these are permanent marine plywood bulkheads; some designs use temporary bulkheads, and they would not need to be epoxy sealed as they are only there for structural

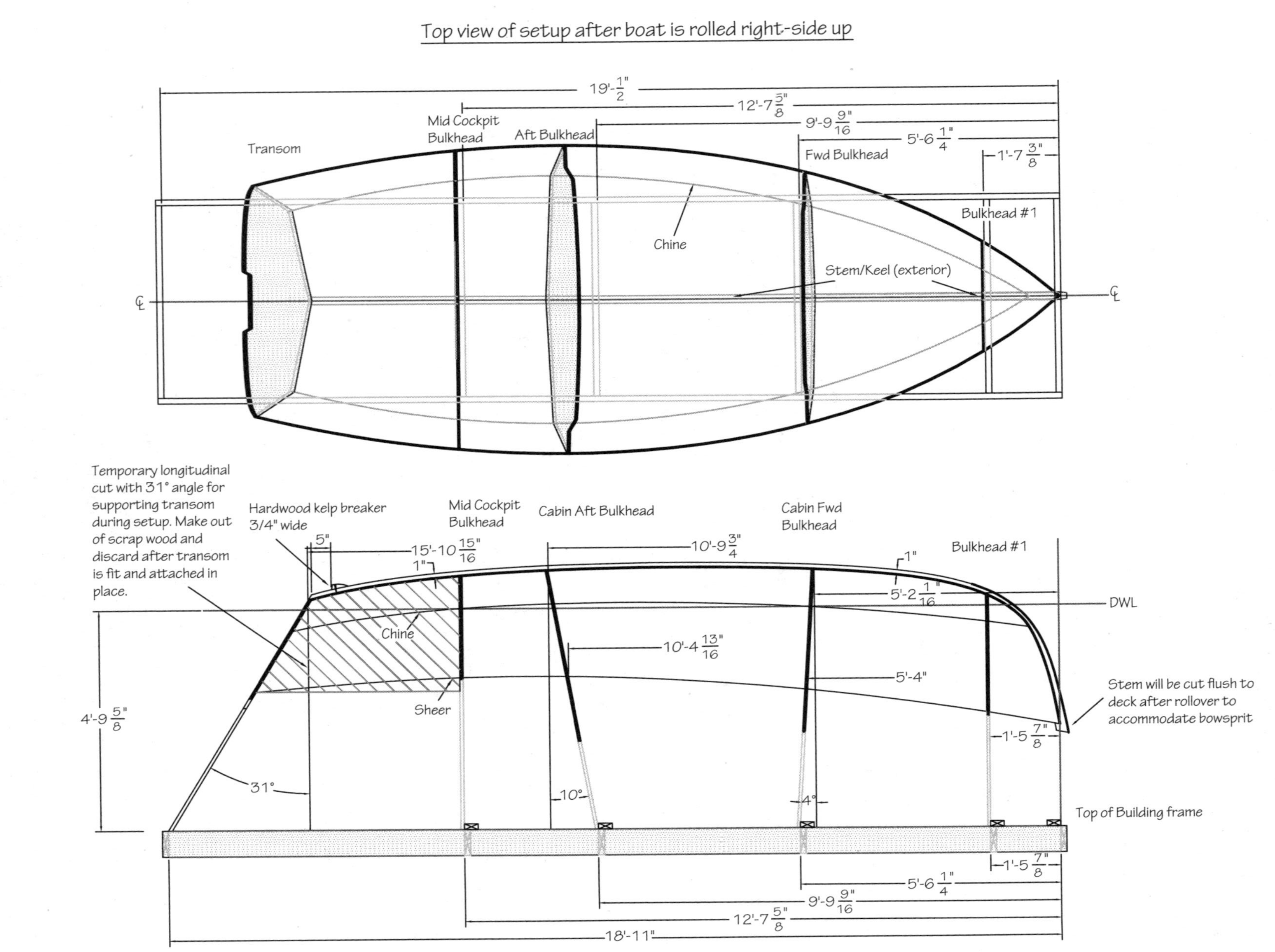

Top view of setup after boat is rolled right-side up

Top of Building frame

Bulkhead #1

Stem/Keel (exterior)

Fwd Bulkhead

Chine

Mid Cockpit Bulkhead

Aft Bulkhead

Transom

1'-7 3/8"

5'-6 1/4"

9'-9 9/16"

12'-7 5/8"

19'-1 1/2"

Mid Cockpit Bulkhead

Cabin Aft Bulkhead

Cabin Fwd Bulkhead

Bulkhead #1

DWL

Stem will be cut flush to deck after rollover to accommodate bowsprit

Top of Building frame

Hardwood kelp breaker 3/4" wide

Sheer

Chine

Temporary longitudinal cut with 31° angle for supporting transom during setup. Make out of scrap wood and discard after transom is fit and attached in place.

1"

5"

15'-10 15/16"

10'-9 3/4"

10'-4 13/16"

5'-2 1/16"

1'-5 7/8"

1"

5'-4"

4"

10"

31°

4'-9 5/8"

1'-5 7/8"

5'-6 1/4"

9'-9 9/16"

12'-7 5/8"

18'-11"

Figure 13. *The building begins with a setup frame on the floor.*

shaping of the hull panels. As always, prepare at least two epoxy coats: the first to soak into the exposed plywood end grain, and the second to form a solid plastic barrier against water intrusion. The bulkhead edges that will be on top after the boat is rolled over can be sealed later; these edges may first need some trimming to fit as deck landings.

If the design you are building uses interior sheer clamps, you will see notches at the corners of the bulkheads to accommodate them. Now is the best time to fabricate and install the sheer clamps; otherwise you will have to be reaching under the hull's side panels to work on them. Look ahead to Chapter 22 for a discussion of sheer clamps, and

Figure 13-1A. *The upside-down building sequence begins with the frame drawing showing where the bulkheads will be placed on the building jig.*

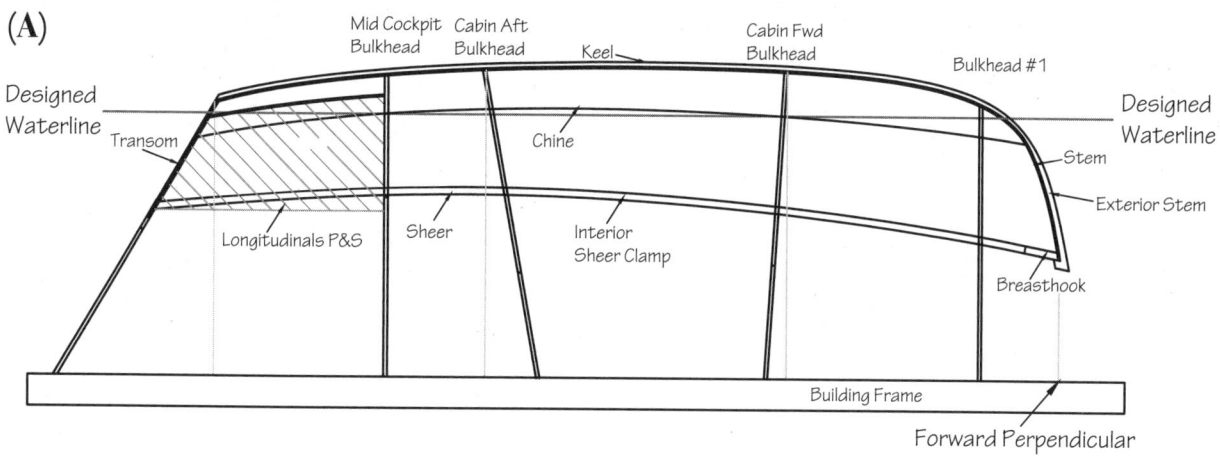

Figure 13-1B. *Bulkheads and longitudinals are set up on the building jig.*

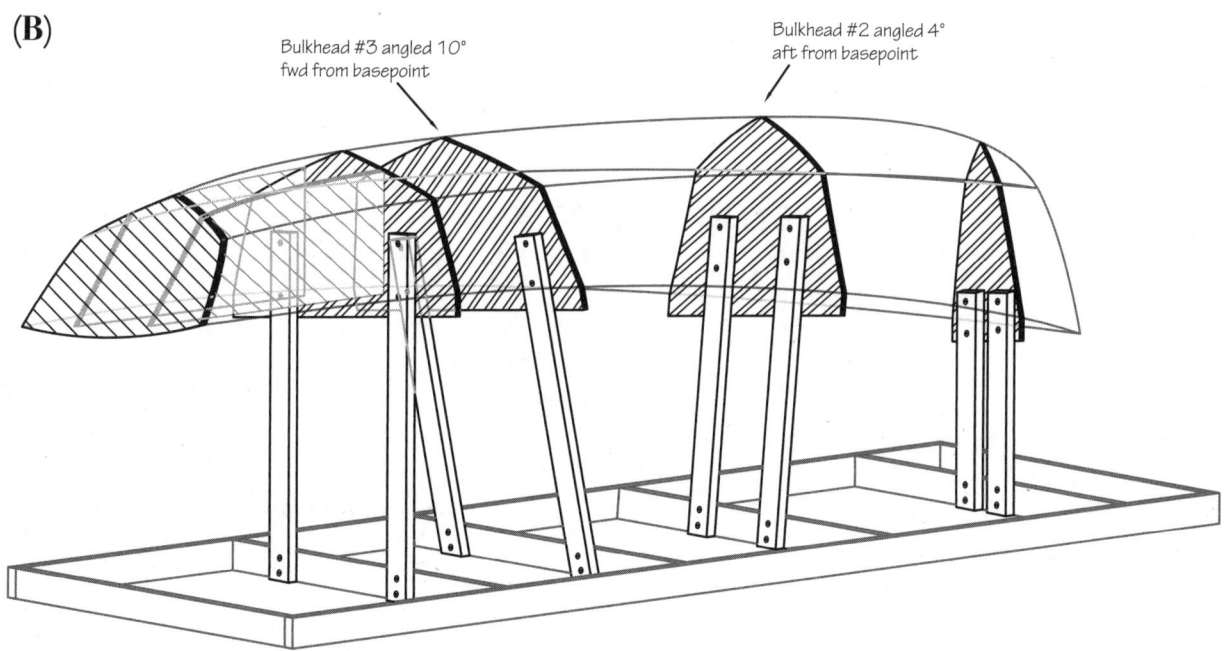

(C)

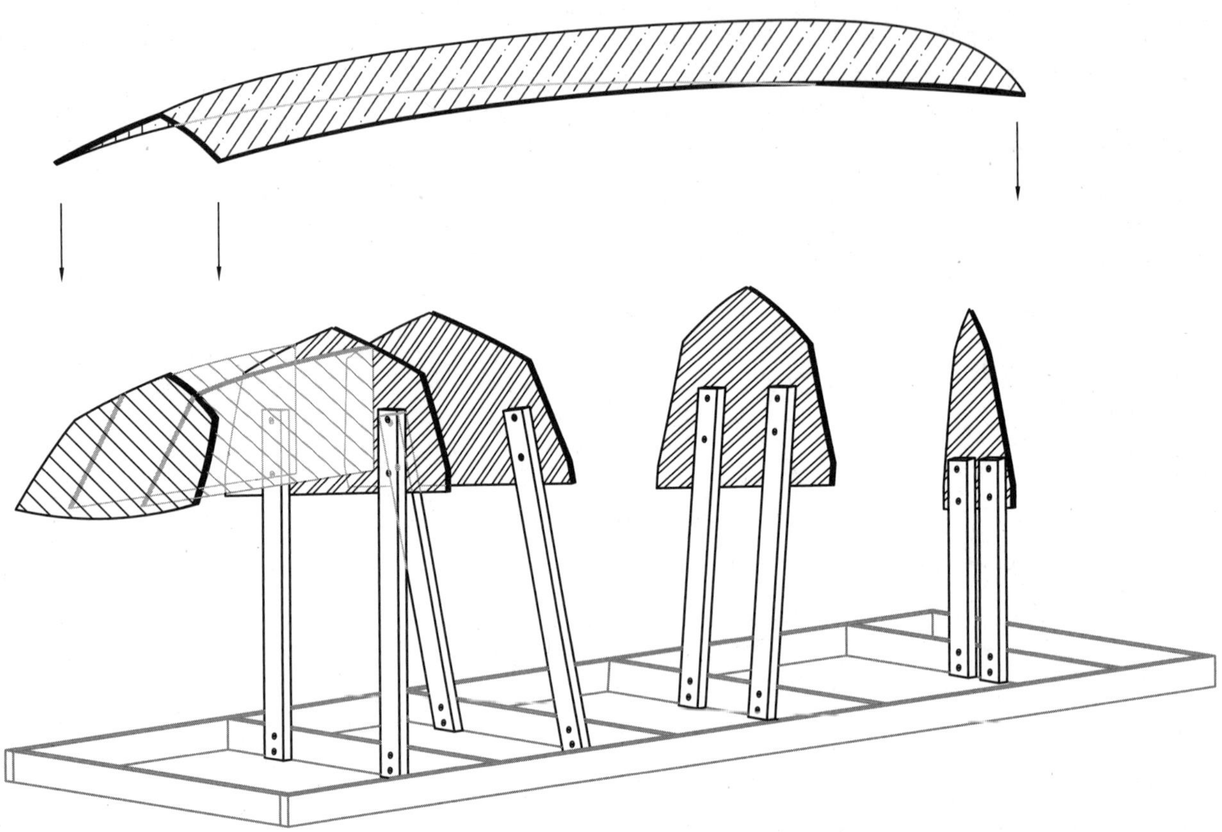

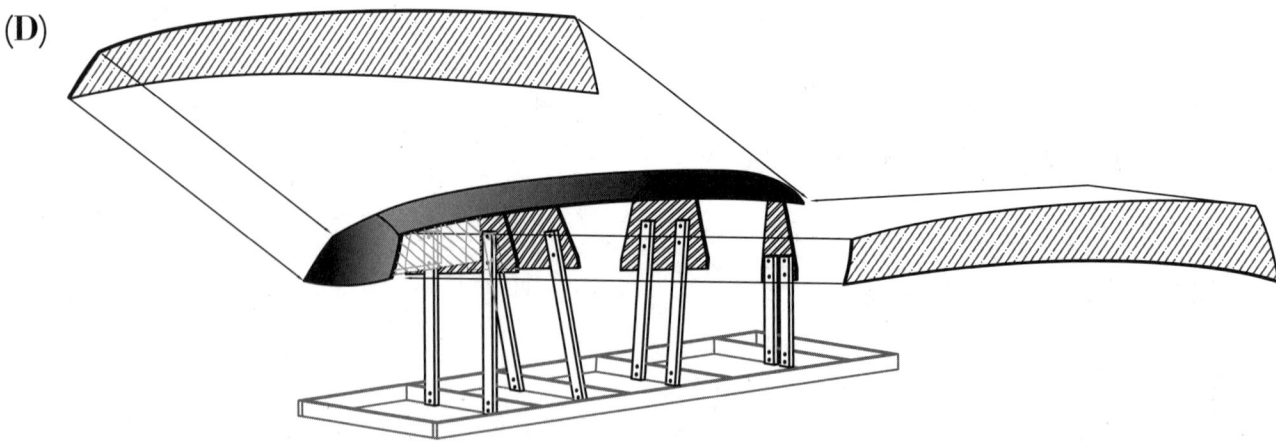

Figure 13-1C. *With the bulkheads in place, the bottom panels can be stitched together face-to-face and then opened over them like a book upside down.*

(D)

Figure 13-1D. *Now the sides can be stitched to the bottom panels starting at the bow and working aft to the stern of the boat, one side at a time.*

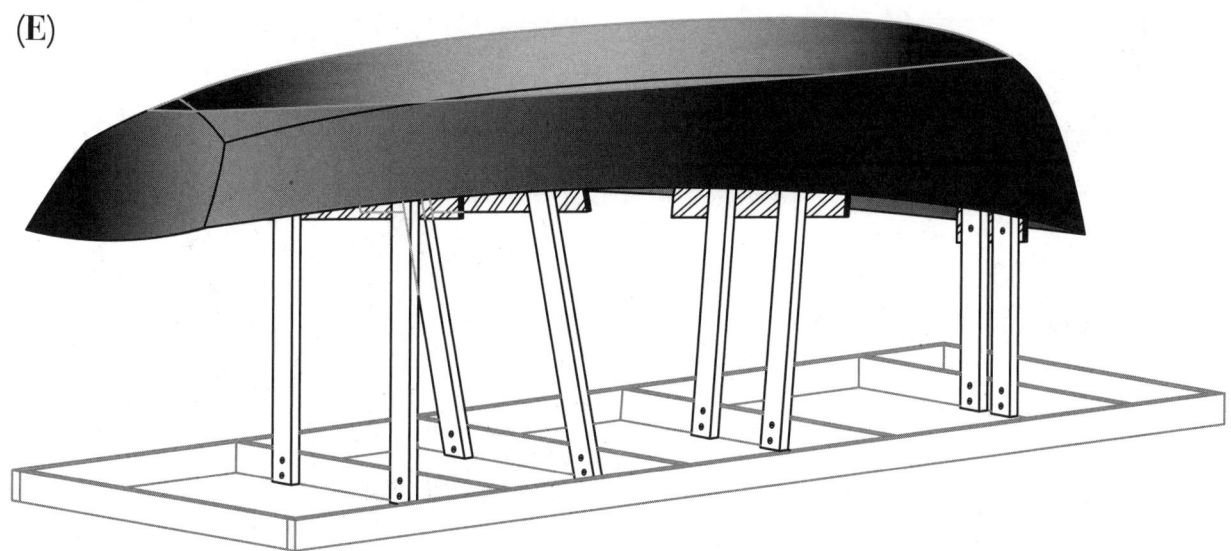

(E)

Figure 13-1E. *With the bottoms and sides now stitched together, carefully and thoroughly check the hull for symmetry and fairness. Then duck under the hull and tack-weld the panels together with small tabs of wood/flour fillets.*

refer to Figures 4-5 and 4-6. There are two good reasons for having these key stringers permanently in place before you stitch the hull panels over the form: One, it will help you align the panels and provide landings for them that already have a fair curve. Two, the stiffening clamps will fortify the hull for that momentous day when you roll it right-side up to begin your work on the interior.

We have discussed stapling previously, and you should use it as you wish, but now we will also discuss assembling our boat with conventional wire stitches. Make up a stitching kit in a box or bucket consisting of segments of steel wire precut to ready-to-use length, about 4 to 6 inches depending on the plywood thickness. Galvanized electric fence wire works best—17 gauge for boats up to about 20 feet and 14 gauge for larger boats. I buy it at local farm supply and feed stores. I use a pair of heavy-duty lineman's pliers with integrated side cutters for both the cutting and twisting.

Lay out the two bottom panels, interior face to interior face, on the floor or on sawhorses and clamp them together. You will now prepare stitching holes to help pull them together. Starting at the bow or stern end, scribe stitch lines that parallel their keel edges but are set back by the thickness of one of the plywood sheets plus ⅛ inch. If the plywood is ½ inch thick, the stitch lines will be scribed ⅝ inch in from the keel edges of the bottom panels. Drill a series of stitching holes about 2 inches apart through both panels along this line within 1 or 2 feet of each end of the panels, wherever you're seeing the sharpest curves. Make sure these holes are perpendicular to the plywood faces. Once you get into the straighter parts of the bottom panels you can stretch out the stitching hole interval to 6 or 8 inches. At the other end, shorten up the hole interval like you did for the starting end (Figures 13-2C, 13-2D, 13-2E, and 13-2F).

Now separate the two bottom panels and bevel their inside **keel** and **chine** edges to about 45 degrees with a block plane, or use a 45-degree angle bit on a small router. Bevel only to the halfway point of the plywood's thickness (Figure 13-3A). If you were to attempt to align the panels without this bevel, you'd find that the keel and chine edges would want to override, or one panel might slip forward or aft along the other. The bevels create friction along the joint, allowing us to hold the proper alignment more easily.

Figure 13-1F. *Many boat kits include an easily assembled precut building jig that will precisely locate the bulkheads or molds for the hull panels.*

Figure 13-2A. *Close-up photo of a limber hole drilled and sealed after glassing the bulkheads.*

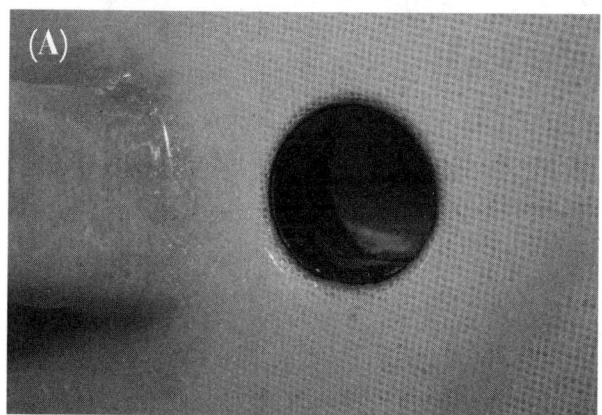

After beveling, bring the two panels back over each other, inside beveled face to inside beveled face, and prestitch the two panels together with your wire stitches (Figures 13-3B and 13-3C). Don't twist them up too tightly yet. You can slip a pencil or a ¼-inch dowel under the twist and tighten until the wires are moderately tight on it, but you can still remove the dowel (Figure 13-4). Bring the bottom panel pair over to the bulkhead setup. Spread the panels apart and drape them over the bulkheads, like placing a book upside down and opening the book up so that the pages are equally spread out. If you are building a really large boat, an overhead gantry and crane or block and tackle

Figure 13-2B. *This Candlefish 13 shows the limber holes in the bottom of the bulkheads.*

Figure 13-2C. *Panels are scarfed to length and ready to stitch over the jig. The flakeboard uprights in the jig are not bulkheads, but only temporary molds that will spread the hull to its proper shape.*

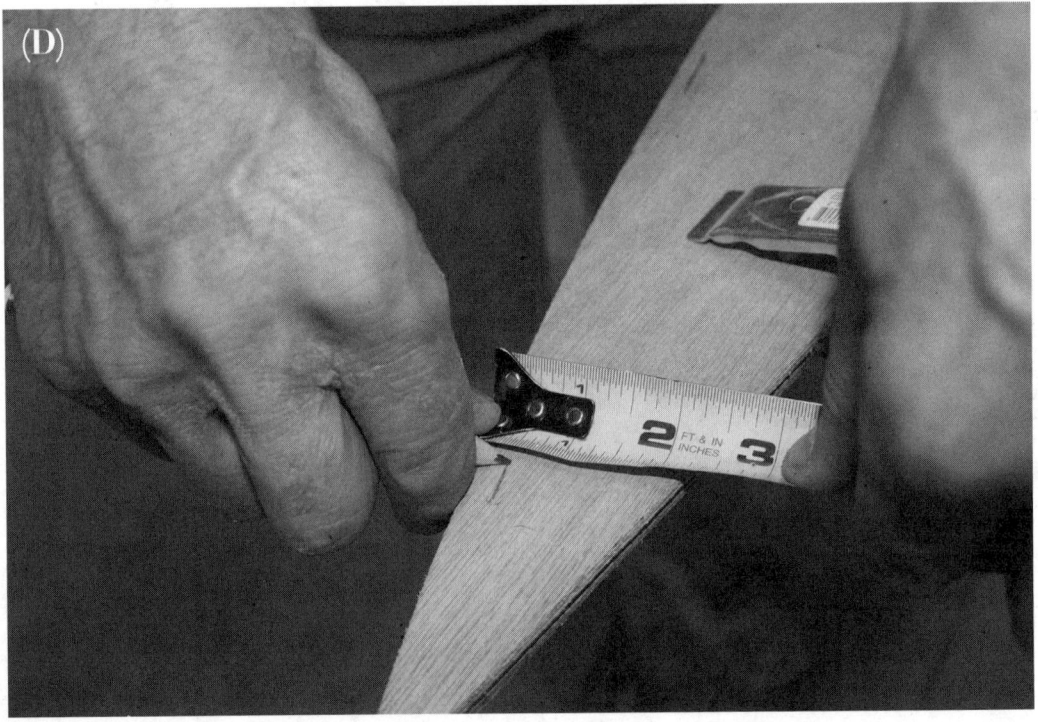

Figure 13-2D. *Scribing the stitch line on the bottom panels in preparation for stitching up.*

Figure 13-2E. *Planing the 45-degree bevels on the inside faces of the bottom panels.*

Figure 13-2F. *Drilling the stitching holes in preparation for the wires.*

bulkhead setup. The reference bulkhead's position with regard to the bottom hull panels is shown on the plans, or else small marks are made by the CNC cutter on the panels. Once you are sure of the positioning both fore and aft and side to side, drive a couple of screws through the panels and into the reference bulkhead to pin them in proper position very near the keel edge of the bottom panels. These screws are simply holding that fore and aft orientation to the middle of the boat. Sometimes, instead of the screws, we wire the panels to the reference bulkhead, as the wires give a slight bit of adjustment and flexibility to help with the next step.

mounted on a sturdy beam will help considerably. With simple lifting straps, you can lift up very long panels and set them in place safely. You will need to check where the bottom panels contact the middle (or reference) bulkhead to make sure that you have not slid them too far forward or aft on the

With the bottom panels spread out evenly over the hull, you can twist each wire medium-tight

along the backbone of the boat from bow to stern. Watch your eyes, hands, and clothing when stitching: The wire ends are sharp! (See Figures 13-5, 13-6, 13-7A, 13-7B, and 13-7C.)

Go to your side panels and bevel the inside edges of the chines and the stem end to 45 degrees as you did on the bottom panels. Don't bevel the sheer edges as they will receive the sheer clamps. Mark and predrill stitching holes on the edges where you will need them to fit along the chine. You can align the side panels together on your workbench and drill through both together, saving time and assuring symmetry. Now lift the side panels into place, drill whatever stitch holes are necessary in the bottom panels, and do a careful dry-fit with a few stitches or staples to make sure the side panels conform reasonably closely to the sheer clamps. Confirm the symmetry of the two sides. It's not a problem if the panel edges overlap or fall short of the sheer clamps by a small measure—less than ¼ inch, I would hope—because you'll plane the sheers into fairness after the hull is right-side up. If there are larger discrepancies, this is the time to

figure out why and correct them. Once you've committed to epoxy, you are quite literally stuck with whatever you've done.

If all looks good, you're ready to go. You want to start stitching at the bow to make sure that each

Figure 13-3B. *Both panels are laid over each other and wire stitches are inserted and twisted semi-tight. If you are planning to staple the seam you only need to wire the first few holes at the bow end.*

(B)   Bevel inside edges of panels to 45 degrees, and stitch panels together

Figure 13-3A. *These are the 45-degree bevels that need to be planed or routed on the inside faces of the keel and chine edges of the bottom panels.*

(A) Bevel inside edges of panels to 45 degrees, and drill stitch lines along the keel edges

router or block plane the edge to 45 degree; bevel half the thickness of the plywood

Bottom Panel

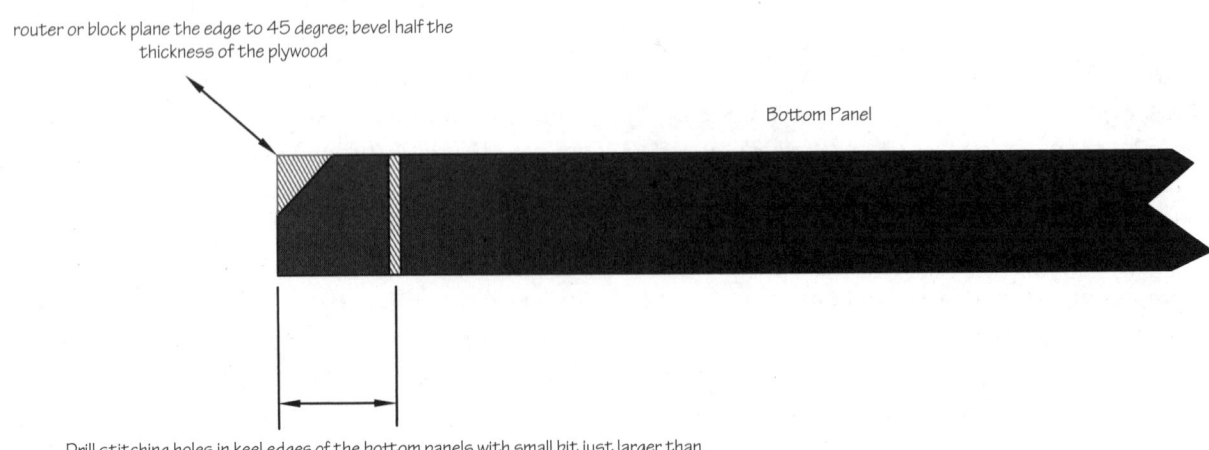

Drill stitching holes in keel edges of the bottom panels with small bit just larger than the wire stitches; plywood thickness +1/8"-2 mm

Figure 13-3C. *These panels are ready to open up and place on the building jig.*

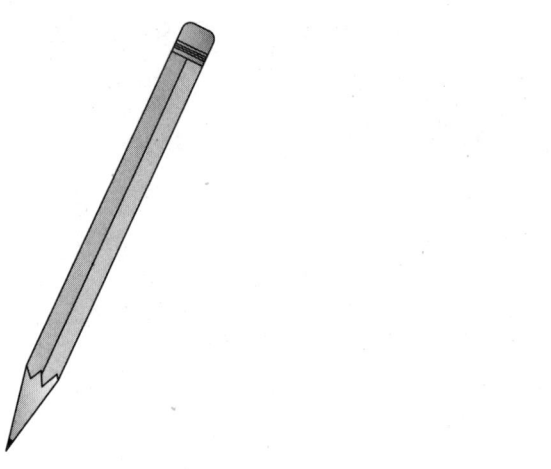

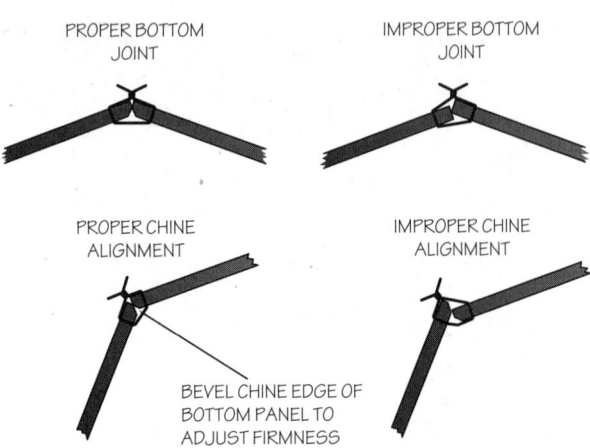

PROPER BOTTOM JOINT IMPROPER BOTTOM JOINT

PROPER CHINE ALIGNMENT IMPROPER CHINE ALIGNMENT

BEVEL CHINE EDGE OF BOTTOM PANEL TO ADJUST FIRMNESS

Figure 13-6. *You will easily be able to see whether the hull joints are in proper alignment. Inspect all the joints along their full lengths.*

Figure 13-4. *You can use a pencil in the stitch to help predict how tightly to twist it. Once you get the hang of the tension, tuck your pencil back into your work apron.*

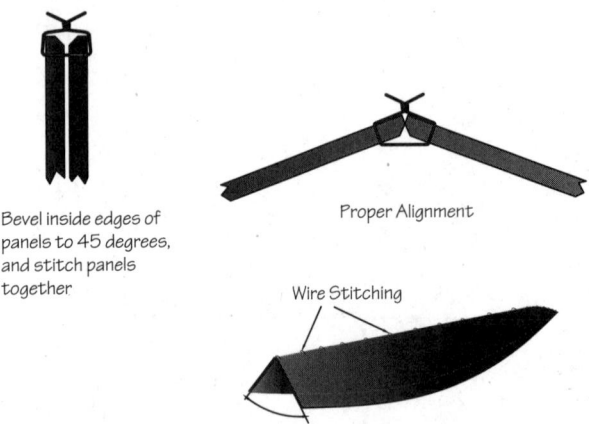

Bevel inside edges of panels to 45 degrees, and stitch panels together

Proper Alignment

Wire Stitching

Figure 13-5. *With the bottom panels spread, this is the proper orientation of the joint before draping over the jig.*

Continue with the next panels up if the vessel is multi-chined (more than a single side panel per side) until the entire boat is stitched from stem to stern over those bulkheads. Do not yet glue the hull panels to the bulkheads. But you will spring each hull panel away from its sheer clamp using small scrap wood wedges, brush in thickened epoxy, and clamp the pieces together with stainless steel screws about 8 inches apart. (If it is a multi-chined boat you will glue the first side panels to the intermediate stringers before continuing on to the next pair.) Be sure to clean up any excess glue that gets squeezed out of the area and wipe down a couple of times with clean rags and denatured alcohol to avoid much laborious sanding later of glue globs. These screws will be permanent, so before you begin painting the hull you might want to mark their locations so you don't later drive screws from the deck right into them. If you have some slow hardener on hand, you could brush epoxy onto the full length of the plywood faying surfaces and sheer clamps before beginning the stitching, as the slow curing would give you an hour or two to work before the glue would set. Wedging after stitching the panels is always a sure way to avoid glue setting prematurely and gives you as much working time as you need.

If the boat has a transom, it could have been part of the setup of the bulkheads and should

panel extends evenly out from the mating edges of the bottom panels in a nice, fair line (Figure 13-8). Whatever you do on the first side panel, mirror the same stitching and position on the opposite side.

Figure 13-7A. *The hull bottoms are now spread and fit over the molds on the jig.*

Figure 13-7B. *On a 48-foot boat we need a forklift and a crew to set the hull bottom in place.*

Figure 13-7C. *Another boat, another view of a hull bottom awaiting the side panels.*

already be in place. If it isn't, you can install it now. Use staples, stitches, or screws (or all three), firmly attaching each panel onto the mating face of the transom. It is important to work symmetrically side to side on the transom attachment, avoiding the inclination to do one entire side before moving on to the other side.

Inspect the panels to be sure all the lines are fair (smooth) and don't show any bumps or unsightliness. It is pretty easy with the upside-down hull to sight along the panels of the boat, but if you need extra help, a batten might be used to reveal any bumps or depressions. At this time avoid fastening the panels into the bulkheads. For now we want the hull skins to float a bit above the faying or landing surfaces of the bulkheads. On larger boats the skins can be quite stiff, and you would be surprised at how much they will ease themselves into place if left alone for a few hours. The weight of the panels will help shape the boat. If small gaps remain, we can easily close them with wood flour/epoxy fillets and fiberglass tape. Any boat that uses flat stock for planking has had that planking conically developed

to predict the shape of the panel; modern naval architecture software has algorithms that help the designer do that task. But since different types of plywood vary in stiffness, no computer program can perfectly predict the panel shapes and how they will intersect the bulkheads inside the vessel. With long experience designing Stitch-and-Glue boats, I purposely don't loft those curvatures and I leave

Figure 13-8A. *This drawing shows how the side panels should join the bottom panels and each other. Both sides should run past the bottoms in equal measure.*

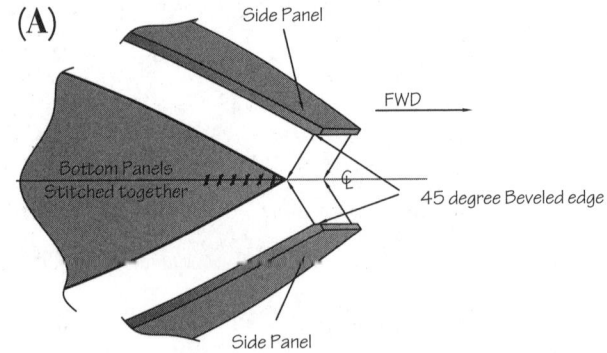

(B)

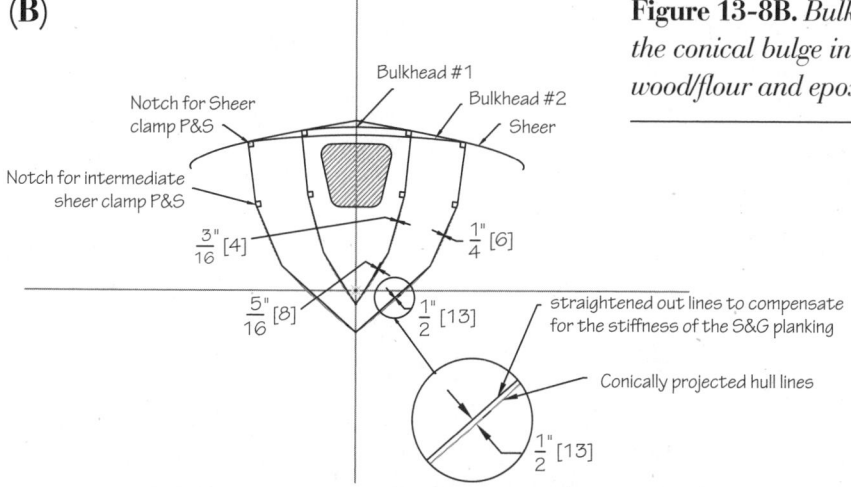

Bulkhead #1
Bulkhead #2

Notch for Sheer
clamp P&S

Sheer

Notch for intermediate
sheer clamp P&S

$\frac{3}{16}$" [4] $\frac{1}{4}$" [6]

$\frac{5}{16}$" [8] $\frac{1}{2}$" [13]

straightened out lines to compensate
for the stiffness of the S&G planking

Conically projected hull lines

$\frac{1}{2}$" [13]

Figure 13-8B. *Bulkheads are lofted and cut without the conical bulge in there sections. Fill the gaps with wood/flour and epoxy mixture.*

Figure 13-9. *All the panels have been installed on the Duckling build, and the hull looks essentially complete though still upside down.*

those lines straight in section on the bulkhead cut-out patterns. That means your hull panels may not lay exactly on the mating surfaces of the bottom side of the bulkheads. This is most noticeable on bulkheads no. 1 and 2.

Now is the time to begin using your boatbuilding eye to work with the wood, not against it. You might add small wedges between bulkheads and panels on the underside of the upside-down hull to help keep the hull fair and smoothly shaped, or you might even add judicious tightening screws from the exterior for the same purpose—though it is far more important to let the plywood take its natural shape than to force it to conform. The more time you spend fairing and smoothing out all the junctions and unions between the panels, the easier the time you will have later in smoothing the hull to final shape.

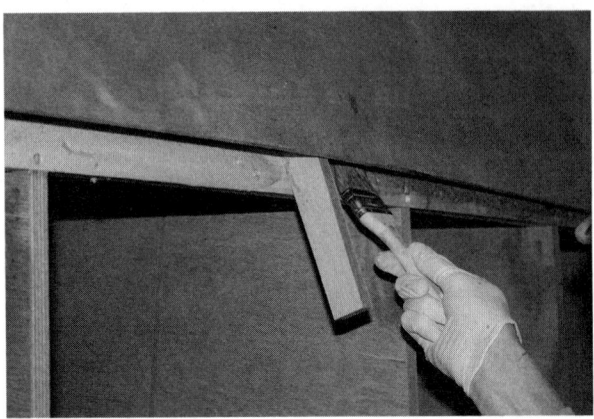

Figure 13-10. *Using wedges to pry the side panels slightly off the intermediate clamps and sheer clamps, spread thickened epoxy on the faying surfaces for permanent gluing.*

In multi-chined designs, you might find adding some stealers is an easy way to help hold the panel edges in proper relationship. These are scrap blocks of marine plywood or even dimensional wood, and with a couple of decking-type screws added, the panels can be forced into close alignment. Figure 13-14 shows a detail.

Builders have been known to succumb to bursts of jubilation and enthusiasm at this point, rushing on to the next step without taking enough time here to accurately adjust the hull panels for fairness and symmetry. Don't succumb. It is easy to adjust and solve problems at this stage. Once epoxy and fiberglass tape have replaced staples and stitches, however, the fixes will lie somewhere between difficult and impossible.

When you are confident in the final shape of the hull, it is finally time to mix some epoxy and wood flour and tab the hull together from the inside. Tabbing is a lot like tack welding in metal construction, consisting of small fillets of thickened epoxy spanning the spaces between the wire stitches. Tab only the spans between the stitches or staples because after the epoxy cures we are going to remove the steel fasteners, fair the exterior seams of the hull, and finally sheathe the exterior before rolling it right-side up or in larger boats add extra layers of plywood to build up our skin thicknesses

Figure 13-11. *"Stealers" are scraps of plywood that can be screwed to help hold both hull panels in the same plane, making the later fairing work easier. Don't worry about all the holes you're drilling in the hull; it's quick and easy to fill them with thickened epoxy.*

before sheathing. Tab the inner bow area generously so that the stressed plywood won't pull apart after removing your stitching. In fact, I recommend reinforcing the junction with a few wood screws running from one panel to the other, which you'll eventually remove, to guard against the panels dramatically springing open.

The steps described above will take you to the stage where the hull is a complete unit. What follows after tabbing the interior and removing the wire stitches or staples on the exterior is to climb under the hull and tape all the bulkhead seams and chines that are easy to reach, mainly the vertical sides of the bulkheads. I don't try to fully glass seams overhead as it produces a terrific epoxy mess,

Figure 13-12. *Staples and stealers work together to hold these hull pieces in position.*

Figure 13-13. *These stealers have aligned the side panels to perfection.*

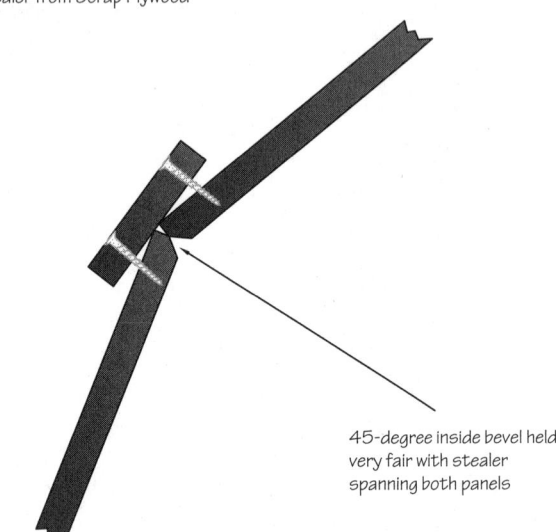

45-degree inside bevel held very fair with stealer spanning both panels

Figure 13-14. *Drawing of a stealer and orientation to panels.*

and gravity is working against us. These are much more easily done after rolling the boat right-side up. Allow those joints that you can tape to cure completely, but don't undertake further finishing or sealing on the interior until the boat is right-side up. On larger boats, however, it may be advantageous to fillet and glass the bulkheads to the longitudinals while the hull is still upside down and before adding the planking.

It will be helpful to review Chapter 6 and look ahead to Chapter 14 now for a thorough explanation of epoxy–fiberglass reinforcements.

When building upside down, you will go on to sheathe and finish the exterior of the hull at least to a primer paint stage before you roll it right-side up. I even like to scribe the bottom paint waterline on the hull before rolling her right-side up.

Work on those critical underwater portions of the hull will never be easier than it is now. The next

several chapters will cover these steps. If you're building a larger boat that requires cold-molding of the hull panels (adding additional layers of plywood to build up the skin thickness), after tabbing you would pull the stitches and immediately move into the cold-molding process. This process is described in Chapter 17.

RIGHT-SIDE UP (SMALLER STITCH-AND-GLUE BOATS)

Lay out the two bottom panels, interior face to interior face, on the floor or on sawhorses and clamp them together. Scribe full-length stitch lines that parallel their keel edges but are set back by the thickness of one of the plywood sheets plus ⅛ inch. If the plywood is ¼-inch thick, the stitch lines will be scribed ⅜ inch in from the edges. Drill a series of stitching holes about 6 inches apart through both panels along this line. Within 1 or 2 feet of each end of the panels, where greater stresses will occur on the stitching, tighten up the spacing to about 2 inches. Make sure these holes you are drilling through both panels are perpendicular to the plywood faces. Now separate the panels and bevel their inside keel and chine edges to 45 degrees, using a block plane as described in the preceding section (when the plywood is this thin, it becomes more difficult to use a router for the beveling). Make up your stitching kit, also as described above, but use the thinner 17-gauge wire.

Insert the wires along the keel edge of the panels so that the wire ends will be on what is about to become the outside of the hull. Twist each wire medium-tight before you go on to the next. The tension on the wire should be tight enough to take the pressure when the panels are spread apart.

Once all the stitches are in place and tensioned, open the panels just like you were opening up a book. Align the keel seam for evenness or run-by (extra material that sticks out past the other panel). Light taps with a small hammer on one of the panels or the other will help to align them. Use finesse; a heavy hand is not necessary here. Check all the stitches for uniform tightness. The 45-degree bevels should provide enough friction to the joint to allow the panels to stay locked together smoothly and fairly. If you need further adjustment or tightening of the wire stitches, now is the time.

You will now need to make a very simple keel-spreader to assist in holding the two panels open (Figure 13-15). This functions like an extra pair of hands. Cut a stick of roughly 1½ × ¾-inch scrap wood to a length about 6 inches less than the total width of the two opened bottom panels. Place a length of the electric fence wire around the spreader, then loop it through the top of one of the keel stitches. Twist it back onto itself so that you have a loop of wire that secures the spreader to the stitch in the hull panels. Now twist the wire (like a Spanish windlass) to tension up until the spread angle of the bottom is about 160 degrees. Don't worry about precise measurement; this improvised clamp just serves as an extra pair of hands to hold these two panels in approximate place while you stitch up the sides. You will be pleasantly surprised at how easy this step is.

Now prepare the side panels by cutting 45-degree bevels on the edges where they will join

Figure 13-15. *This photo shows a keel spreader in place on an upright stitching of a smaller boat.*

the bottom and, if it is a multi-chine boat, the mating sides of the chines and bow end of the panels, just as you did with the bottom panels at the keel and chine lines. Bevel the bottom panel chine edges similarly if they weren't done previously. If you have seen either the first edition of this book or our boat-building video, you may remember a discussion of something called the "transition joint," where the chine angle had to shift from an overlap to a butt joint, near the bow. Thanks to the double 45-degree bevels, this annoying complication is now unnecessary. Scribe the stitch lines along the side panels at the chine edges and drill holes to receive the stitches just as you had done in the bottom panels. You will want the chines to look just like the keel line, with two opposing 45-degree bevels facing each other on the inside of the boat.

Lay one of the side panels on top of the other, inside face to inside face, and scribe the stitch line along the chine edge of these panels at the same distance from the edge as the thickness of the panel plus ⅛ inch (for example, if you are using ¼-inch plywood, drill the stitch line ⅜ inch from the chine edge). Drill both of the panels together just as you had drilled the two bottom panels. This will give you one side of the stitching line predrilled, saving you work later.

Next, starting at the bow end, take one of the side panels and incrementally drill and stitch your way aft along the chine one side at a time. By running the forward end of each side panel slightly beyond the front of the bottom panel, you will end up with a very nice alignment and no open gap at the bow. Align one of the side panel chine edges with the bottom panel chine edges at the bow. Drill a couple of holes in the bottom panel to line up exactly opposite and approximately the same distance from the edge opposite the first two predrilled side-panel holes. Insert the stitches from the inside out so that the twists will again end up on the outside of the hull and tighten them. You will be able to easily reach into the boat for drilling these holes and inserting stitches. Continue drilling and stitching down the length of the panel from bow to stern, and then do the same on the opposite side. Caution: You will have a lot of sharp wires bristling at you, so safety glasses and care are necessary to avoid nasty pokes and cuts.

If it isn't already obvious, a helper is really beneficial at this phase. He or she can hold the side panel in position at the transom end and control the twist and the relationship of the side panel to the bottom while you drill and stitch. Simple directions telling the holder to lift or drop the panel or move it laterally in or out will help keep it aligned to the bottom panels and allow you to stitch up easily.

After both sides are stitched in place and the bow is stitched up to the sheer, spread the side panels to allow the transom panel to fit (Figure 13-16). I often cut the same 45-degree bevels on the transom as I did with the other joints, providing these joints with an extra bit of friction. Don't worry about the small angular crevices this will leave in the joint; they will fill with thickened epoxy for a very strong, secure junction. Wire the transom in place using wire stitches about 4 inches on center. If the side panels exhibit some run-by (extra plywood that sticks past the transom), it can be trimmed later. But the amount that it runs by should be the same on both sides of the boat. If any run-by or overhangs exist on one side, you must replicate them on the other. If you don't, you are about to build an asymmetric boat! Correct the problem while you still can.

Figure 13-16. *Stitching up the transom.*

Once the transom is in place, you are ready to fit some spreaders into the sheer, opening up the hull to its designed and more appealing final shape. In an upside-down build, this would have been taken care of by the width of the bulkheads, but in a right-side-up build, you need to install temporary spreaders that will open the sheer to the designed width. Usually, you will need more than one spreader. I most often do the sheer spreaders in two phases, first setting some simple temporary spreaders at the sheer and opening up the boat to its approximate shape, using the friction of the two hull side panels to hold the spreaders in place (Figures 13-17A and 13-17B). On some of the smaller pointed-bow boats such as the Polliwog or Guppy dinghy, it is very common for amateur builders to end up with a powder-horned sheer that is not very attractive. This can be averted by placing an extra spreader far enough forward to push out the sides of the hull to the proper width that the bow should exhibit. Look carefully at the drawings of the boat and then at the hull from the side view, crouching low enough to really see the true and smooth curves of the sheer. Ask yourself: Is more shaping needed from another spreader, or do I need to spread the sheer more to make it fair and true? This is the beginning of learning to trust your own vision and to make aesthetic judgments.

Once you get the sheer spreaders in place, the temporary keel spreader has finished its work, and so you may remove it. The hull structure may be a bit wobbly at this point, but that is to be expected. Now is the time for you to support the hull in a right-side-up level securely before you start on your epoxy work. You can screw some cleats or even Styrofoam blocks onto the sawhorses or bench top to secure the hull.

Next you will fasten the true spreaders in place that fix the shape of the boat in its final form. Take measurements from the plan at each of the bulkhead locations (or other spreader measurement points amidships if there are no bulkheads). Cut the spreaders to these lengths. If there are no inwales or sheer clamps in the plans, you can now insert the spreaders from sheer to sheer. Small screws or nails set through the sides of the boat and into the ends of the spreaders will hold them in place while you work on the interior. For symmetry's sake, these spreaders must be absolutely perpendicular to the centerline; measure carefully and frequently as you

work. If inwales or sheer clamps are soon to follow, set the spreaders a couple of inches below the sheer to allow working space, slightly shortening them if necessary to establish the correct sheer-to-sheer measurements. A little forethought here will help the project flow smoothly.

Take a pause and step back and look at what you have brought to life. This is a boat now, and you should start to have a sense of this new life you are bringing into the world. This will be one of those moments that you will reflect back upon many times in this boatbuilding journey, and it will help you immensely on the darker days when you need the encouragement and enthusiasm. Don't hesitate to spend a few moments now in celebration.

For most boat designs there will be some bulkheads or interior structures to be bonded to the inside of the hull. Cut these out now and temporarily fit them with small screws or wire brad nails to

Figure 13-17A. *Position spreaders to keep the boat from looking powderhorned from the side. This is especially important for small boats that have a lot of shape in them.*

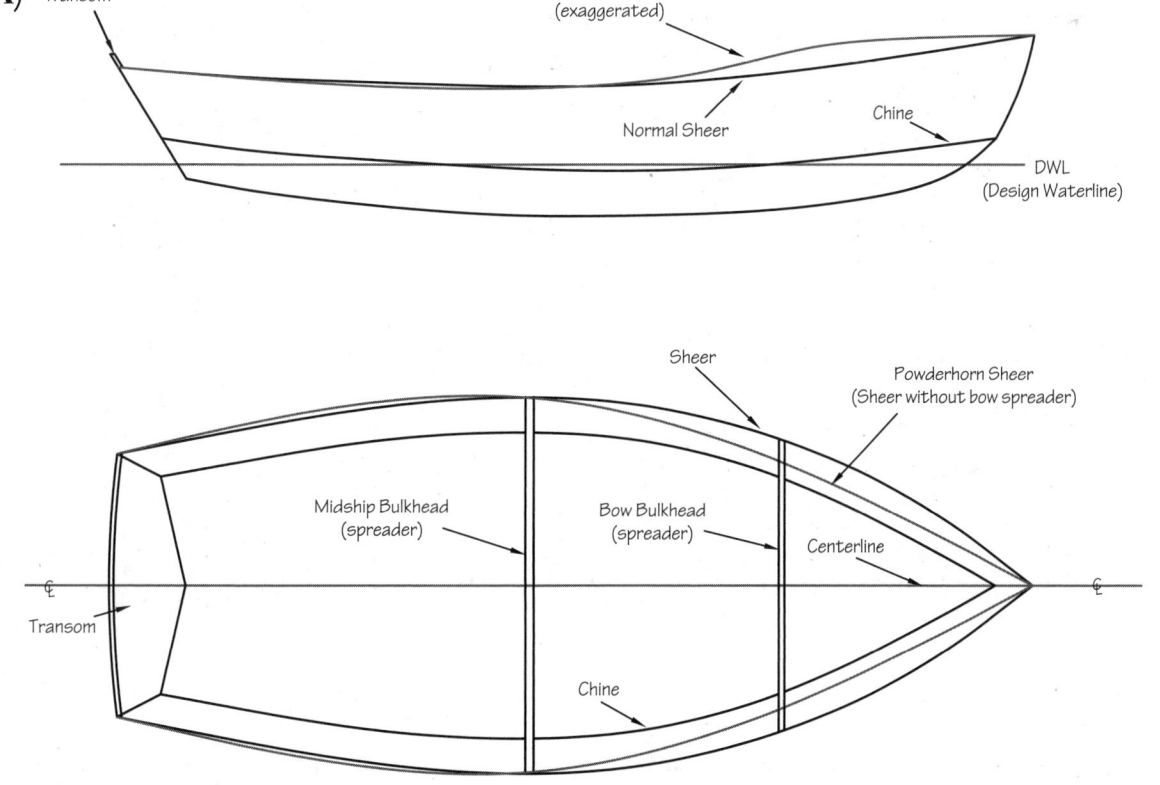

135

Figure 13-17B. *Final spreaders are in place and we are ready to true the hull to square and to the water-line. These steps are unnecessary in the upside-down method as the building jig does these steps for us automatically.*

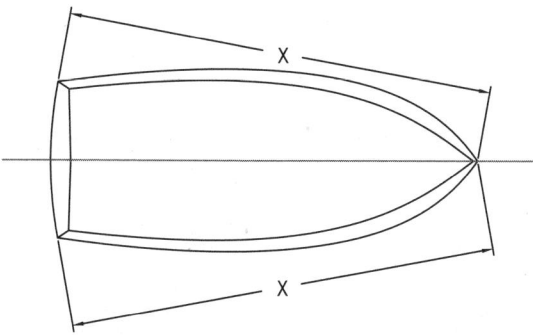

Measure diagonally from stem to each corner of the transom; for the boat to not be twisted, the measurements will be exactly the same port to starboard (top view)

Figure 13-18. *Avoid the dreaded Twisted Hull Syndrome by measuring the diagonals from stem to both corners of the transom.*

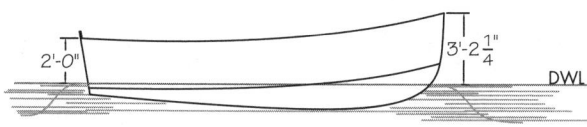

The difference between this example boat's bow and stern freeboard is 14-1/4 inches. Measuring from the transom edges and the bow to your shop floor or some other level surface should produce the same difference: 38.25" minus 24" equals 14.25"

Figure 13-19. *Level the stitched boat to the waterline as accurately as you can.*

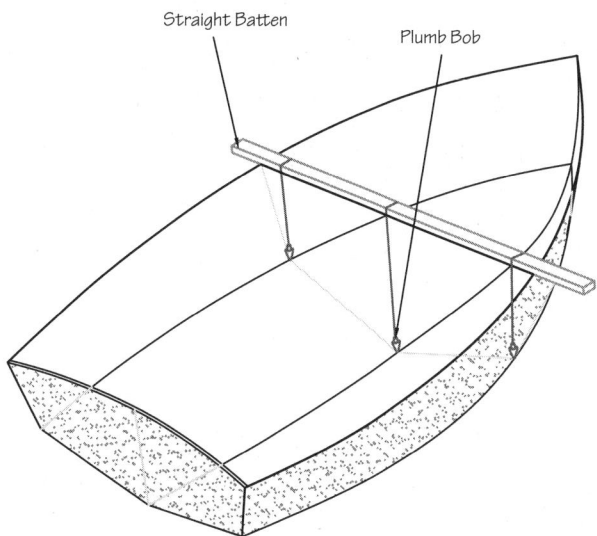

Figure 13-20. *Mark where the bulkheads will be placed on the sheer of the boat and at the chines and keel intersections.*

pin them in place. Fitting them to the crisp angles in the stitched hull is much easier than it will be later after the interior hull joints have been glued and coved with epoxy fillets. As you proceed, remember that the boat must be leveled along the waterline fore and aft and sideways before starting to make these structures. And check the truing of the hull, measuring diagonally from the stem to each side of the transom. These measurements must be exactly the same from side to side to make sure that the hull is not twisted or otherwise untrue.

The method I call "proofing the bulkheads" checks that the parts that you will be lofting and cutting out will fit exactly into the interior of the hull. Wood is a dynamic medium, and while there is reasonable consistency in marine plywood from one sheet to another, small differences in the thickness or density of the individual plies can translate into slightly different shaping for each individual hull that gets built. You, as the builder, will adapt to reality rather than the designer's ideal. Your building plans will call out measurements for these parts, which you should now compare with the realities of your own unique hull with this simple proofing procedure.

Place a straightedge across the sheer in each of the exact positions of the bulkheads as measured aft from the stem. Next take several measurements and mark them in red on the plans for each of these locations. These would be the width of the boat at the sheer or gunwale, the width at the chine or chines in the same plane (you can use a plumb bob to help with this measurement, dropping it alongside the straightedge and marking its intersection with the chine of the boat on both sides), the depth of the chine from the straightedge to the same marks, and the depth of the keel as measured from the straightedge. With these measurements in hand, you can determine how these realities compare to those dimensions on the plan, and you can now loft, cut, and fit bulkheads without any waste. Remember to keep checking for symmetry, using both your

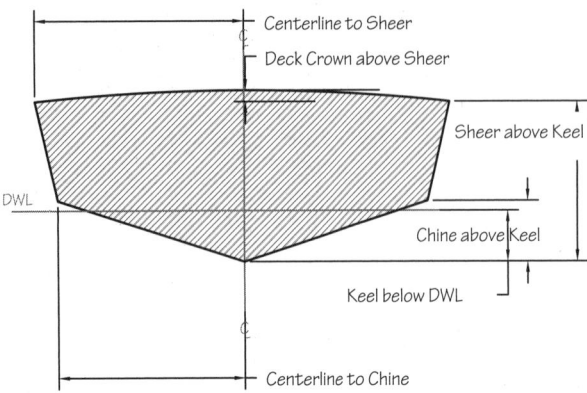

Figure 13-21. *Proofing the bulkheads is a vitally important step to check the accuracy of your build so far.*

measuring tools and your increasingly aware boat-building eye. You are approaching the point where you will have to trust your judgment in making a final call on whether the hull shape is good enough to fuse together with permanent epoxied seams.

After dry-fitting each bulkhead, you will remove them and you should paint the faying areas (the contact areas of the parts that were pre-fit into the boat) with epoxy resin before you reinstall these components. Now you can tab the rest of the seams or joints of the boat in between the stitches or staples. Once the epoxy has set solid for the tabbing and these parts are pinned into place, you will have a very rigid hull structure, with each step adding to the rigidity of the hull. We will drill down on this procedure in the next chapter.

STEP-BY-STEP SUMMARY OF THE BULKHEAD FITTING PROCEDURE (RIGHT-SIDE UP STITCH-AND-GLUE CONSTRUCTION)

Level the stitched hull with the design waterline fore and aft.

Level the hull athwartships to the design waterline.

Measure diagonally from the stem to the corners of the transom. These measurements should be the same if the hull is level and true.

Mark the major bulkhead locations on the top of the sheer on both sides, port and starboard.

Lay a straightedge sideways across the boat, at the intersecting marks on sheer.

Drop a plumb bob to align vertical bulkheads and mark the bulkhead position at the chine or chines and at the keel centerline.

Measure the vertical distance (heights) from the straightedge to the chine and to the keel centerline at the locations marked in preceding step.

Measure the sheer and chine beams (widths) to the inside of planking.

Check these measurements against your plans, and correct the bulkhead dimensions as necessary on your plans.

When you are satisfied with your actual dimensions, lay out the plywood stock for the bulkheads.

For each bulkhead, draw in the centerline.

Mark the sheer height on the bulkhead centerline.

Mark the chine heights on the centerline.

Draw horizontal lines through the height marks, and along these mark off the half-beams (half widths) at the sheer and chines.

Draw the deck camber as dimensioned in the plans.

Draw in any other dimensions or details indicated on the plans.

Connect the dots marking the perimeter and details of the bulkhead, and cut out the bulkhead.

Position the bulkheads in the boat and pencil-scribe and trim to adjust into place as necessary.

Coat the edges of the bulkheads and the hull where the bulkheads will land with epoxy, and drop the bulkheads back into the hull, making sure the marks at the sheer, chines, and keel align to the proper side of the bulkhead (usually the forward side).

Pin each bulkhead into place with a minimum of two or three screws or small brad nails per hull panel. Screw holes can be pre-bored from inside the hull before final placement of the bulkhead to aid the positioning of the screws.

After all major bulkhead pieces are in place, finish tabbing the interior. After the epoxy has cured, remove the hull panel stitches and any temporary screws or brads. Fillet and glass the inside hull-to-bulkhead joints.

CHAPTER 14

Fusing the Stitch-and-Glue Hull

Fillets (coved epoxy paste thickened with fillers) and fiberglass tape reinforcement are the tendons and sinews of Stitch-and-Glue construction. Not only do these epoxied joints fuse the plywood panels together, they also effectively transfer structural loads from one surface to another while dispersing stress concentrations. This is the essence of what makes a well-constructed Stitch-and-Glue boat such a strong and durable vessel. These joints can

be executed easily and nicely by even the beginning boatbuilder.

Before you begin the fillet-and-fiberglass welding process, it is important to understand its physics. The load on every seam or joint of your boat must be distributed over a wide area on the panels on either side of the seam itself. This is the reason for the cove-shaped fillet. And the fillet's load-spreading ability will then be reinforced and extended by layers of glass cloth or tape of varied widths.

As a rule of thumb, I strive for a joint that is as strong as the pieces it bonds—but no stronger. Meeting that specification every time would be perfect engineering. Some say that L. Francis Herreshoff's design for a turnbuckle was so perfect that when strength-tested, no one part failed before the other. Failure occurred virtually simultaneously in every part. That is my ideal for a Stitch-and-Glue boat design, tailoring fillet thicknesses and weights and widths of glass cloth reinforcing to achieve uniform strength throughout the hull.

A fillet's minimum depth should be approximately equal to the thickness of the plywood being bonded. (If pieces are laminated, use their *total* thickness to figure your fillet depth.) For example, a dinghy or small rowing skiff constructed of ¼-inch plywood skin should have a fillet about

Figure 14-1. *Your filleting mixture should be the consistency of thick, slightly chunky peanut butter. If you hold your stir stick up and any slides off the stick it needs a bit more wood flour added.*

¼ inch deep. The runout of the fillet (the B dimension in Figure 14-2) should be between 1.5× and 2× the thickness of the plywood joined, or ½ inch from the joint's center in this instance, using ¼-inch-thick plywood. This fillet helps to smoothly transfer the strength from one panel to another without an abrupt interface. Layers of fiberglass tape over the fillet should be tapered in a stairstep manner, allowing the fabric to strengthen the joint while gradually distributing loads and avoiding stress concentrations where reinforcement comes to an abrupt end. In a small skiff, 3- and 4-inch-wide, 6-ounce cloth layers would be more than adequate. We will be filleting and taping the interior seams of the boat, the stem, chines, keel line, and any intersections of bulkheads, to weld the boat together. If you have only 4-inch tape, another way to taper the joint would be to overlap the layers slightly. The more gradual the transition from one panel to the next without excessively strong or weak spots, the better.

On smaller boats, Volan-finished, 6-ounce fiberglass cloth that you cut into tapes makes strong, light joints. If you cut slightly diagonally across the width of the cloth, you will have a tape without a long yarn wanting to unravel from the edge. On larger boats, I use biaxial fiberglass cloth tapes for joint reinforcement, usually of 12-ounce weight. I don't like to use triaxial cloth because the mat layers do not saturate very well with epoxy. By planning ahead and knowing which areas on your vessel's interior will be finished, you can use woven cloth tapes in multiple layers for the visible areas rather than the more opaque laminate of biaxials. Use biaxial cloth for the major hull joints, however, since they are significantly stronger and are usually hidden by location or under paint in the finished boat.

A typical fillet in a 30-foot boat with a skin thickness of ¾ inch would be ¾-inch deep, on top of which you would align 4-, 6-, and 8-inch-wide laminations of biaxial cloth tapes. On top of that, a 10-inch-wide layer of 6-ounce glass cloth will help level out the roughness of the biaxial tape and allow you to smooth and sand without cutting into the rougher biaxial tape structure. You can also use a layer of peel ply on top of the laminates to help smooth out the texture and help to avoid air bubbles forming in the tape layers.

You will be applying your fillets and glass cloth with the "wet-on-wet" method, meaning the glass cloth and epoxy saturation will go over the non-cured fillets before their thickened epoxy sets up. This is the most efficient way to proceed, and it does not require frantic speed. Just work patiently

Figure 14-2. *Figuring the fillet scantlings for small- to medium-sized Stitch-and-Glue boats.*

Stitch-and-Glue joint for small- to medium-sized Stitch-and-Glue designs.

"B" fillet runout = 1.5 to 2 times the thickness of the plywood in the finished joint

B

A

"A" fillet depth at joint is at least 75% of the thickness of the plywood in the finished joint. Apply three tapes of 6 oz. fiberglass tape in 4, 6, and 8" widths over the interior joints; alternatively, you could use 4" tapes for all joints but stagger back the layers to allow a tapered joint.

For small- to medium-sized Stitch-and-Glue boats, a layer of 6 oz. woven fiberglass cloth sheathing is applied. It is also smart to apply a single layer of 6 oz. fiberglass tape on exterior seams before starting with the sheathing layer, or make sure to double over all exterior seams with the cloth sheathing layers, minimum two layers at all seams.

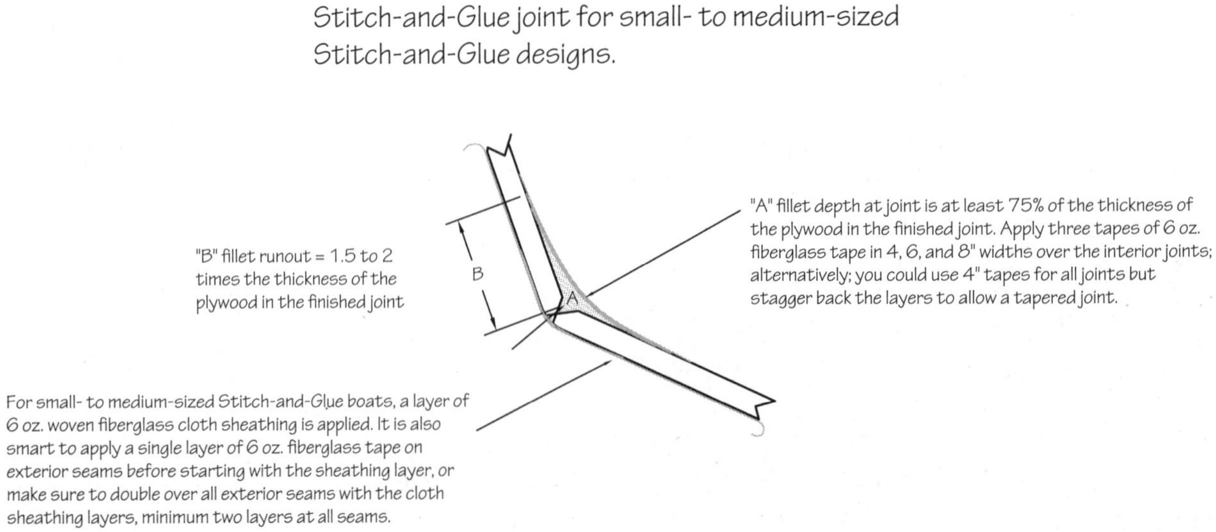

Stitch-and-Glue joint for larger cold-molded Stitch-and-Glue designs.

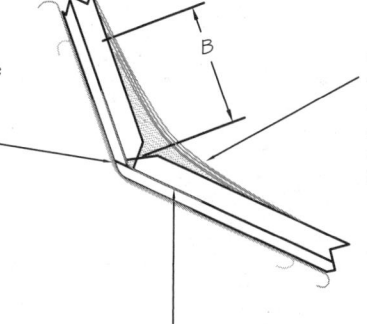

"B" fillet runout = 1.5 to 2 times the thickness of the plywood in the finished joint

For cold-molded Stitch-and-Glue boats, a layer of 6 oz. woven fiberglass cloth sheathing is applied, then an additional and final layer of dynel or xynol polyester cloth is sheathed over the entire hull. It is also smart to apply a single layer of 12 oz. biaxial tape on exterior seams before starting with the sheathing layers.

"A" fillet depth at joint is at least 75% of the thickness of the plywood in the finished joint. Apply three tapes of 12 oz. biaxial in 4, 6, and 8" widths over the interior joints. A layer of 6 oz. glass cloth tape over them will help to smooth out the joint.

For cold-molded Stitch-and-Glue boats, a single layer of 12 oz. biaxial tape 6 or 8" width is applied interlaminate wet-on-wet with the cold-molding process.

Figure 14-3. *Here are the fillet scantlings for larger cold-molded Stitch-and-Glue hulls.*

and methodically. I also suggest that you work symmetrically, completing a section of the seams and joints on one side of the hull and then moving to the corresponding section on the other side. If you are building right-side up, keep a tape measure at hand and recheck the symmetry of the hull at every step, measuring from the stem to both corners of the transom and corresponding corners of each bulkhead that you are welding in place. If you are building a 12-foot dinghy, a ⅛-inch discrepancy is not too alarming, but larger errors may be. I still remember the small duck-hunting boat we once built that never posed a threat to any duck. After the hull seams were glassed, something about the hull seemed odd, although I couldn't put my finger on it. A couple of days later we discovered that it was ¾ inch out of square, a banana-like twist permanently locked into it thanks to the immutable miracle of epoxy. We stashed that worthless hull in the rafters of the shop for years, a silent reminder to measure early and often.

With interior glassing your first step is to clean out the inside of the hull with a vacuum and/or clean cloth wipedown, and for heaven's sake don't introduce moisture to any of the bare wood areas. Then measure all the sections to be glassed and precut the glass tapes you will need. Your plan should include notes or illustrations for what we call the lamination schedule. For example, the interior chine fillets might ask for overlapping layers of glass cloth 4, 6, and 8 inches wide. Cut to length, arrange these

Figure 14-4A. *Small squares of biaxial tape are precut for tabbing inside the hull.*

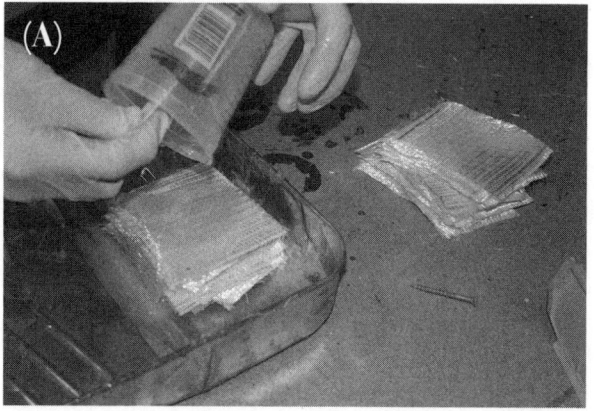

(A)

Figure 14-4B. *Tabbing is easier when you presaturate the small cloth squares before crawling under the hull.*

Figure 14-4C. *The sanded tabs are now awaiting new full-length fillets to be applied, and following those the glass taping of the full interior seams.*

into sets, and label them with masking tape or a permanent marker on the final layer. I like to array them near the area of the hull that will be glassed. The more organized you are, the easier the job will be (Figures 14-5A through 14-5D).

You will find it convenient if you take a few minutes to make a glassing box out of scrap wood. Take a piece of plywood 12 to 14 inches wide and perhaps 6 feet long, and nail side and end walls about an inch high around the perimeter. After you measure and precut your glass tape pieces, you can saturate them with epoxy in this box and even carry them to the boat with minimal handling and mess.

Next, make a set of rounded squeegees for the fillet material. Refer to the explanation of fillet depth and width above to figure out the radii you will need. Make some cardboard templates, then cut

Figure 14-4D. *For convenience, a smaller boat can use exterior tabbing between staples or stitches. Squeegee thickened epoxy into the slight gaps between the panels, then tab as usual on the interior with small fiberglass squares.*

Figure 14-4E. *Exterior tabbing is just about completed.*

Figure 14-4F. *Squeegeeing off excess tabbing epoxy while it's wet is far easier than sanding it after it's cured.*

Figure 14-5A. *Precut and dry-fit all the glass taping you'll need before mixing the epoxy.*

a selection of spreaders from yellow plastic squeegees (typically available where you buy your epoxy) with a band saw or jigsaw. Smooth out the working edges with a sharp block plane or fine sandpaper.

For the strongest possible hull joints, you can roughly outline the borders for the widest of the

tapes and then distress the surface of the plywood with a stiff phenolic grinding pad and 36-grit paper. This cuts into the plywood surface veneers to give the epoxy a better "bite." Coat the joints and adjoining surfaces with unthickened epoxy, brushing out a little beyond where the widest cloth will land to ensure

Figure 14-5B. *Keep the tape strips organized and close to their destinations in the boat.*

Figure 14-5C. *A neat and clean cutting table helps the work go quickly.*

Figure 14-5D. *These are the precut tape layers for different parts of the boat, folded over and with scrap wood blocks labeling where they will go on the boat.*

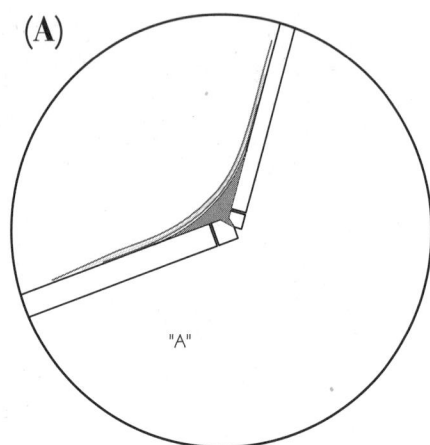

4"-6"-8" staggered glass tapes applied over fillet

Figure 14-6A. *This is a typical three-layer chine reinforcement of 4-inch, 6-inch, and 8-inch overlapping tapes. The narrowest goes on over the fillet first.*

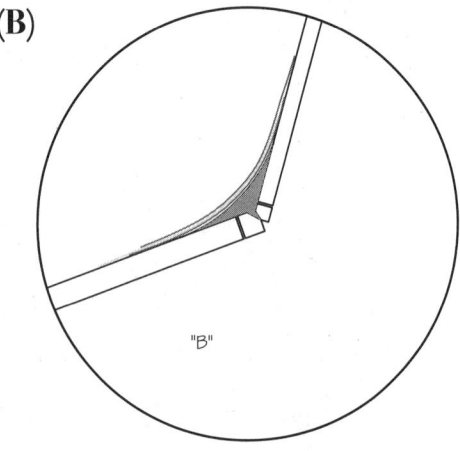

Figure 14-6B. *If you only have one width of tape, this scheme will still allow you to overlap three layers.*

that no part of the system will be epoxy-starved. This first epoxy painting of the joint area allows the epoxy to be slightly absorbed into the bare wood.

Once the joint areas have been coated, mix some wood flour and silica thickener such as CAB-O-SIL into the epoxy to create the filleting material. I like a ratio of 75 percent wood flour to 25 percent silica; this is a silky, easy-to-work-with blend. It's easiest to mix the dry fillers together before getting a cup of sticky epoxy in hand. Now measure out your ratios of resin to hardener, stirring and mixing very well before adding the wood flour thickener. The fillet mixture should be neither syrupy nor too stiff to work. To test for the proper consistency, hold the stirring stick vertically with a golf ball–sized glob on

the end of it; if any part of the mixture slides, it is still too thin. When the wood flour is too stiff, the mixture will appear a bit dry and will result in fillets peppered with hard-to-remove air holes. We are looking for the consistency of a moderately stiff peanut butter.

Apply the thickened mixture to the joints using your custom squeegees. Be sure to prod the mixture neatly into the joint and smooth out the surface and edges as nicely as possible. We want a smooth, curved transition from one panel to the next.

Those of you with an inventive bend might want to make up plastic or freezer paper cones (sort of like cake icing cones) to squeeze out a very even and neat bead of wood flour/epoxy mix.

With the fillets sculpted in the section of the hull that you are going to glass, the next step is to apply the glass tape reinforcements.

I prefer that the laminate be built up with the narrower cloth tapes placed over the fillets first, followed by incrementally wider layers. Each subsequent layer helps to smooth out the edge of the narrower tape below it.

Figure 14-7. *A collection of radiused squeegees will suit various fillet sizes you'll need to make.*

Figure 14-8. *This epoxy/wood flour mixture was a bit on the dry or stiff side, still usable but needing more work to smooth.*

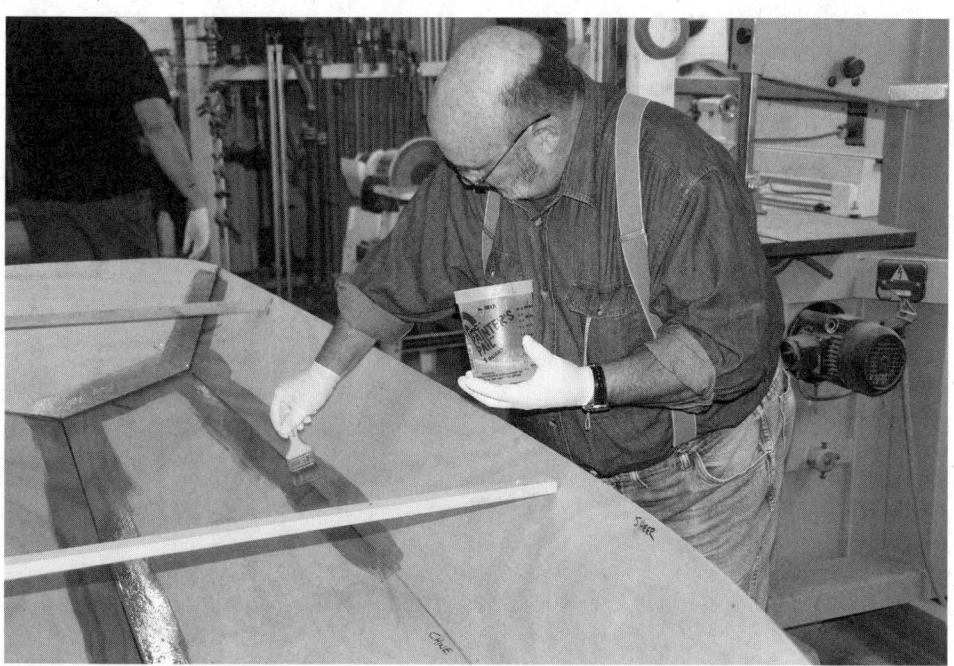

Figure 14-9. *Paint out the bare plywood at the interior seams with unthickened epoxy before adding the fillets and tape layers.*

Lay the first layer of fiberglass tape over the fillet joint and lightly smooth it into place with your gloved hand or a dry bristled 2-inch chip brush. Take care not to dimple or shift the underlying fillet material. Keep a sharp eye out for air bubbles, as they weaken the laminate. A small, toothed fiberglass tooling roller will work well to smooth out bubbles. Brush additional unthickened epoxy over the fiberglass tape to complete the saturation. When the cloth is sufficiently saturated, its appearance will change from dry and silver-white to translucent. Air bubbles should be rolled or squeegeed out before applying the next layer of tape. Repeat the process for each layer. Try to avoid applying additional epoxy as an overcoat to the final layer; it usually absorbs the excess epoxy from previous layers, creating a better resin-to-cloth ratio. After several minutes, if it is obvious that there are going to be some dry spots that won't saturate properly, carefully brush on a bit of epoxy as needed to complete the saturation.

For first-class results that require much less sanding for a smooth seam, apply a layer of peel ply on top of your taped seams. Peel ply is a finely woven polyester cloth (available from your epoxy or glass cloth supplier), and epoxy will saturate it but won't adhere well to it. Just lay or lightly press a slightly wider strip of it down over the top of the fresh joint, and allow the uncured excess wet epoxy of the joint to soak into it. Use a plastic squeegee to smooth the strip over the layers of wet-out fiberglass tape, thus smoothing out and removing excess

Figure 14-10. *Epoxy thickened with wood flour is now sculpted into fillets.*

Figure 14-11. *A Candy Stripe or other type of paint roller aids in smoothing the glass taping and helps to keep air bubbles out of the laminate.*

Figure 14-12. *We like to use a glassing box to confine the epoxy and tape while saturating, then lift all the layers of saturated glass out as one piece for installation.*

Figure 14-13. *Peel ply applied over the wet epoxy will greatly reduce the amount of sanding you have to do later.*

epoxy from the laminate. After the epoxy has cured solid, simply peel off the strip, leaving a smooth joint surface. Peel ply is a wonderful time-saver, almost eliminating the need to sand the surface once it is removed. Remember this: The care and time you expend in making these glassed fillets neat and clean will be repaid about fourfold later.

Some small fillets that are not structural hull joints will not require fiberglass tape overlays. An example might be a partition for a storage bin or locker. For these joints, mix the same wood flour fillet material and use the rounded end of a tongue depressor to cove the fillet surface. Let the fillet cure until it reaches the consistency of stiff modeling clay. At this point, use a piece of tightly woven cotton cloth—soaked in either denatured alcohol or, less desirably, in lacquer thinner—over the fingertip of a solvent-proof glove to smooth the surface. If you avoid denting or moving the semi-stiff fillet material, the surface will become

Figure 14-14. *In our shop, we have always let the new guys strip the peel ply off the hulls. It's a fun job with a high level of satisfaction.*

so smooth that there will be little need for sanding later. We very often do these wipeups as a two-session process, catching the initial cleanup on the first pass and then the final smoothing in a second pass half an hour later, when the epoxy is stiffer. On a bright-finished surface, the cured, wood-colored fillets will blend very well with the natural wood color of the plywood panels. A boat constructed with these nicely sculpted fillets is not only beautiful but also practical: there are few if any nasty corners to trap dirt, debris, or moisture.

Figure 14-15. *We made this very low-tech machine to saturate multiple layers of tape at once, and somehow it works great.*

Figure 14-16. *These are well-executed nonstructural fillets where the skeg meets the hull bottom and transom, smoothed before setting hard by wiping with a rag wet out in denatured alcohol. Some boat plans may call for larger structural fillets.*

CHAPTER 15

Removing Stitches

Wire stitches (even if they are cut off flush to the plywood) cannot be left in the hull because the normal heating and cooling cycles of the boat's plywood structure are very different than the expansion and heating factor of metal and may cause them to migrate to the surface, marring the finish or even compromising the structural integrity of the hull. I feel so strongly about this that at times we count the pulled wires to make sure every suture comes out.

You may recall that in Chapter 12 we discussed stapling the interior seams to save time, and I said these staples can be left in and buried deep in the epoxy-wood flour fillets. The distinction is between interior and exterior stapling. Staples driven in from the inside are unlikely to go anywhere, and as long as their tips aren't poking through the panels to the outside, they may be left in place and filleted over. Staples on the outside and *all* wire stitches *must* be removed.

In a small boat, you can tab with fillet material the interior seams between the wire stitches, allow these fillets to cure, then pull out the wires before glassing the joints inside and out. But you may, for a variety of reasons, find that it may be better to complete all the interior hull seams with the full epoxy fillet/fiberglass cloth composite joints, allow them to cure, then remove the stitches later. This sounds more difficult than it is; once you begin you'll be happily surprised at how quickly the work proceeds.

The wire stitches are left in place only until the filleted, taped seams have fully cured. Applying heat to the wire ends softens the cured epoxy enough to allow the stitches to be pulled with a pair of pliers. Some builders prefer to use a fast epoxy and remove the wires immediately, while the epoxy is still setting, but I like to wait overnight and remove the wires when the epoxy is more fully cured. Exterior staples often can be removed with an awl without applying heat, as little if any of the staple would come into contact with epoxy. But if heat is needed on staples, the soldering-gun technique is the most advisable.

Don't bother with the dinky 15- or 30-watt soldering irons intended for delicate electrical circuit boards; use a quick-heating 100-watt gun. And don't be tempted to try a heat gun or propane torch. The heat gun will soften the epoxy in the whole neighborhood of your target wire. The torch will set fire to your boat.

To remove stitches, begin by untwisting the wire suture two or three turns, then snip the wire on both sides about a quarter-inch out from the hull with a pair of wire cutters. Apply heat to one end, and when that end is glowing red-hot, allow half a minute or so for the heat to transfer throughout the wire. Then use pliers to pull the wire toward you with a levering motion against a scrap wood block. The wire should ease out nicely. If the wire breaks inside the suture, simply heat the other end and repeat the removal process. A dull pair of parrot-beaked wire-cutting pliers works well for this purpose as long as they're dull enough to not cut the wire but grip it enough to lever backwards and pry the stitch from the joint.

Be careful to not apply heat for too long because it can flame the epoxy and even burn the plywood panel. Begin by experimenting with brief heat applications of a few seconds, and increase only as much as necessary to pull out the wires easily. For efficiency, I usually heat one or two wires ahead of myself, and then I return to pull the wires once the heat has softened their bond to the epoxy.

Be sure to do a thorough job removing all bits of metal. If a small piece breaks off inside the plywood, don't be tempted to leave it—pry it, dig it, or drill it out. You can easily fill the hole with thickened epoxy, and you won't be leaving a potential breach in the hull integrity.

Many of the kayaks that are sold as kits are supplied with copper wire for the sutures and the kits advise the builder that the copper wire can be trimmed off flush and glassed over. My tests and experience have shown me that, given time, the epoxy's hold on the metal can release, and the metal can start to work its way out of the joint. It is not much of a chore to remove the wire, and you will have a much better project for the effort. I am also not a fan of using nylon zip cable ties for a Stitch-and-Glue project. They can function as pretty good stitches, but the holes drilled in the plywood are larger than those needed for the wire stitches, and again, leaving anything in the joint is not advised.

Figure 15-1. *A pair of lineman's pliers makes a great tool for removing stitches. If epoxy is gripping the stitch, heating the wire with a soldering gun will allow easy extraction.*

You now have a constellation of holes in your hull, which you will need to deal with. But don't bother with filling them now. You'll fill them with resin, easy to do during the exterior sheathing process—which isn't far off now.

CHAPTER 16

Cold-Molding the Stitch-and-Glue Hull

When I first began building boats with the Stitch-and-Glue method, I was working with small craft that could rely on a single-thickness plywood skin for their hull integrity. But before long I realized that our boat designs would grow beyond what we could build with a single layer of plywood. The shapes and curves in my designs seemed to enforce a limit of ½-inch or 12 mm marine plywood (excepting the topside skins in certain narrow and fine-lined vessels). I did not want to moderate my hull shapes and risk polluting the water with clunky, misshapen boats built from overly stiff marine plywood panels, so I needed to find a way to build in the strength and rigidity that larger boats would require while still working within the versatile Stitch-and-Glue method and within the bending capability of thinner initial skins.

Other construction methods could attack the problem by constructing a thicket of longitudinal hull stringers or athwartships frames close together and fastening the plywood skin onto those stringers. But in the Stitch-and-Glue method we are literally starting construction from the outside, so the prospect of trying to add thick and stiff stringers full length into the complex bends of the hull, already planked up with our skin, was not at all practical. Another solution occurred to

me: adapting the technique of cold-molding to the Stitch-and-Glue method.

A number of professional wooden boatbuilders today use cold-molding, another post–World War II development that owes its existence to the superb gluing qualities of epoxies. It is an intimidating challenge for most amateur builders, however. It involves building up a hull with several very thin veneer-like layers of wood over temporary molds, then swaddling a giant vacuum bag over the entire structure to apply uniform pressure while the epoxy cures. Essentially, you're creating curved plywood on a complex boat-shaped form.

Stitch-and-Glue cold-molding, I suspected, would be a lot simpler. Why not use the initial skin of the Stitch-and-Glue hull as the form, then simply add more layers of marine plywood until we reach the thickness needed for the desired strength and stiffness? A ¾-inch hull, for example, could be built from a core layer of ½-inch plywood plus an outer layer of ¼-inch plywood. So with the theory in hand and an order for a 25-foot motor-sailer on the books, I tried this new form of cold-molding. It worked easily, quickly, and reliably. It has allowed our shop, as well as other professionals and many amateur builders, to construct larger boats that wouldn't otherwise be possible with the

Stitch-and-Glue method. In our Stitch-and-Glue catalog of designs, many boats now use the cold-molding method to reinforce their hulls and make them immensely strong and capable.

If you have contemplated cold-molding as it applies to round-bottom boats, you learned that you would have to learn to spile (individually scribe and cut to shape) each veneer of thin wood in each of the layers that would have to fit to its neighbor precisely and wrap around the hull's compound curves. This is a long and exacting process. In Stitch-and-Glue construction, the hull is made entirely of developed shape marine plywood panels, so you can avoid spiling entirely. Plywood planked the boat initially, and plywood will plank subsequent layers. The only shaping you need to do is rough-cut your cold-molding panels to place them on the boat. You can use up to full-sized sheets of plywood for most of these extra layers. And there are labor-saving shortcuts. One we use when we know that we will be cold-molding extra layers of plywood onto the hull is to lay out under our full-length scarfed first-layer plywood the required number of secondary layers without scarfing, but with the plywood sheets' butt joints staggered so as to not weaken the eventual stack of the hull panels. Then, using a circular saw, we cut the entire stack of first skins and the cold-molding layers in one pass. We know the outer layers will have to be a bit wider because they will have to span a larger radius at the joints of the chines and keel, so we cut one long side of each set of panels evenly, then offset the stack to allow for the extra widths as we cut the opposite side. We then label each panel with a permanent marker and stack them against a wall until needed.

The rule of successful plywood cold-molding is for the outer layer or layers each to be no more

Figure 16-1. *Sand the bottom of the hull panels to remove any excess epoxy and to smooth them out, then prepare the thinner panels of plywood that will be used for the cold-molding layers.*

than half the thickness of the innermost or initial skin. The shape of a Stitch-and-Glue hull is deceptively complex, and I have found that if I use ½-inch or 12 mm plywood for the initial stitch-up of the bottom of a particular hull, I cannot conform another layer of ½-inch ply to it without using a blizzard of screws or fastenings. But one or several outer skins of thinner and more flexible ¼-inch ply can be held to it with much simpler fastening until the epoxy cures.

Although this cold-molding procedure is simple, you must prepare well. You have many large panels to assemble, and as you know by now, once the epoxy is mixed you have a limited window of time to work. You must also work very cleanly, as sawdust or any miscellaneous shop grit between layers would adversely affect the integrity of the hull laminate.

Whether you have cut them out in advance or one by one, taking the patterns off the stitched hull now, take your precut panels for the first outer layer and dry-fit them in position on the hull. You can use staples or small screws to hold each panel until you have completed the dry test-fitting of the entire boat, all layers in place. You can mark the panels with notes to help with alignment. This dry-fitting will reveal whether there are any issues with the total task at hand, which is essential before you get the sticky stuff on your hands and tools. You will find that some boat designs have so much shape that even ¼-inch plywood in large sheets won't accept the bends. These problems will usually occur in the bow or in an unusually complex stern shape. In such cases you will cut 4-inch-wide strips from the ¼-inch cold-molding panels and dry-fit them diagonally to the hull. You can then label and number each panel with its position, taking care to indicate which side is "up." Remove all the dry-fitted pieces from the hull and stack in an order that allows you to proceed methodically with the gluing.

One thing we will want to do is to drill 3 mm or ⅛-inch-diameter holes about 8 to 10 inches on center through each cold-molding panel. These small holes help to prevent air entrapment between layers. Improbable as it seems, it is very common when laminating curved plywood panels into a hull form or a crowned deck to trap large bubbles of air between layers, the epoxy acting like a large gasket. These voids have the potential to cause future trouble. The small holes allow air to escape and also will verify that you have good contact between layers when you see small beads or drools of epoxy resin ooze out of each hole. The squeezed epoxy also forms, in effect, plastic epoxy "nails" that will help bond the layers together.

When everything is ready, and you have fit and marked all the panels to be applied, mix epoxy *without thickeners* and roll a generous amount on the face-down sides of the panels to be cold-molded. We call these faces that take the glue the "faying" surfaces. Set these aside while rolling an equally generous coat onto corresponding areas of the upside-down hull. If you notice gouges or holes in the hull or the panels, fill them flush with lightly thickened epoxy before starting the uniform coating of epoxy. Make sure every surface and edge of the hull and the cold-molding stock is coated with epoxy before positioning and fastening the plywood cold-molding layers to the hull.

I usually use pneumatic staplers with stainless steel or Monel bronze wire staples in a variety of sizes and lengths to fasten the panels to the hull. If you don't own a power stapler, consider renting one. Cordless or hand staplers, bronze ring-shank boat nails, and small screws will also work, though they all require more effort and often create more mess. Place a fastener at least every 6 to 8 inches to guarantee contact between pieces until the epoxy cures. You should have a good ooze out of the small holes; entrapped air produces a hollow sound when you run your hand over the panels, and it's a good idea to sound out the whole surface. If you encounter an area that does not show the epoxy ooze out of the predrilled holes, add more fasteners to ensure good contact between the sheets. If the staples aren't pulling the sheets down properly, add some short screws. We would typically use #6 × ½-inch flat head screws.

If cold-molding more than one outer layer, it's best to wait until the first layer cures before proceeding to the next one. Pull all the mechanical fasteners within 4 inches of the chines, stem, keel, and other edges, since those areas must be faired smooth before adding the next layer; the balance of the other fasteners can be left in place interlaminate.

If you are concerned about building a hull that will stand up to the worst abuse imaginable, cold-molding offers opportunities to build in even more strength and puncture resistance than a single layer of plywood with its glass sheathing. You can and should lay down a strip of biaxial fiberglass tape on the major hull seams under each of the cold-molded layers, either allowing it to cure and slightly sanding it smooth, or applying the next plywood layer wet-on-wet (the preferred method in my shop). I use 12-ounce biaxial tape in about 8-inch widths running the length of the

chines for these interlaminate layers, which makes for immensely strong hull joints. You can also sandwich a sheathing of Kevlar cloth set in epoxy between the plywood cold-molding layers, making your boat literally bulletproof. For most boats destined for reasonable use, interlaminate Kevlar is an unnecessary expense, but for work vessels or those expecting extreme demands, it can be a real benefit.

As I noted in Chapter 6, Kevlar is a nightmare for sanding and finishing, but if you bury it between plywood cold-molded sheets and work wet-on-wet, you will not face that issue.

Apply as many cold-molded layers as your design calls for. You may decide to apply more if needed to prepare for hazards such as ice or deadheads (partially waterlogged logs floating vertically, which is fairly common in Pacific Northwest and

Figure 16-2A. *On any design needing cold-molding layers, we first add longitudinal strips of biaxial cloth on the chines. Then either finish the cold-molding wet-on-wet or let the epoxy cure on the tape and sand smooth before adding the plywood lamination.*

Alaskan waters) or for prolonged dry storage on a trailer. Additional layers in a localized area of the hull or deck may be necessary to dissipate such stresses and strains as a side-mounted winch for oceanographic work or commercial fishing. You should always consult with your designer about changes like these but the capability of tailoring a Stitch-and-Glue boat to suit a specific use is remarkably easy. A little extra protection is easy to add now but a lot harder later. Cold-molding extra layers of plywood greatly enhances the strength and integrity of the hull.

After the epoxy has cured beneath the final layer, prepare your exterior surface for fiberglass or Dynel sheathing. Pull *all* fasteners from the outer layer. If you don't, these staples, screws, or nails may telegraph their forms through to the final finished surface of the hull (especially if you

Figure 16-2C. *This build was done with the wet-on-wet method, installing the cold-molded panels before the seam epoxy hardened. It's tricky to sand that rough edge of the glass taping, but not impossible.*

Figure 16-2B. *On this boat we allowed the glassed reinforcements to cure, sanded them smooth, then applied the cold-molding layers.*

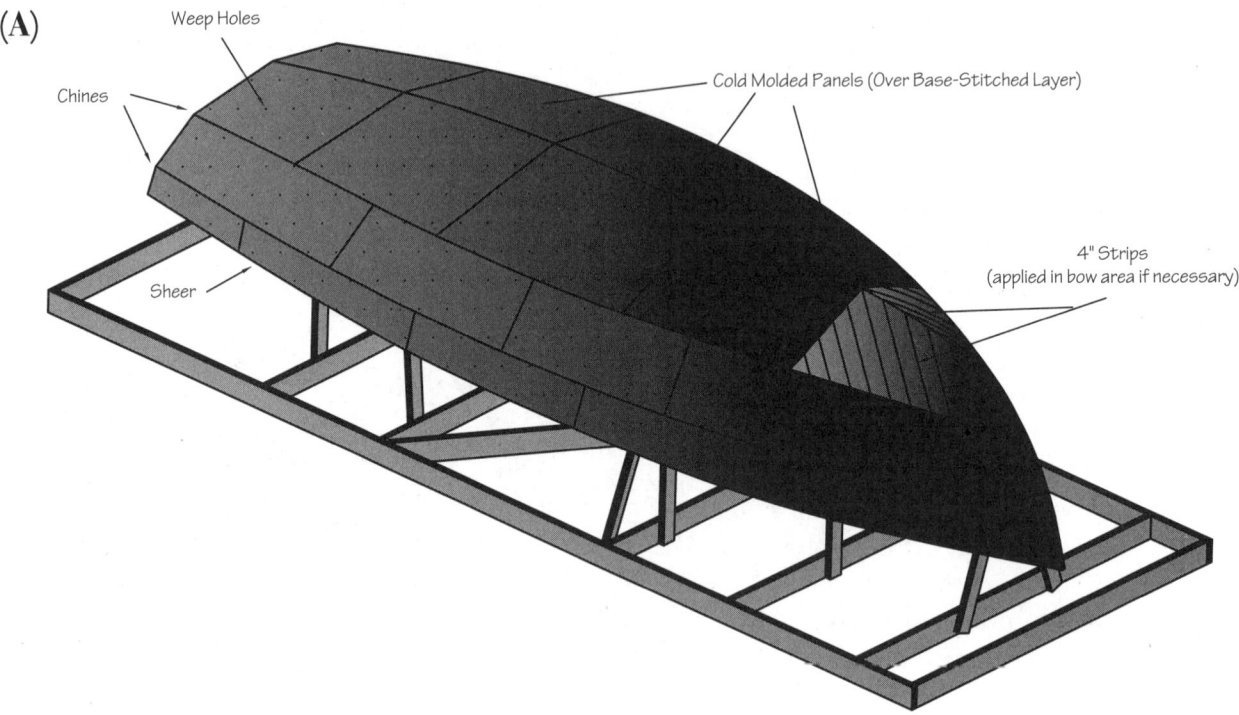

(A)

Weep Holes

Chines

Cold Molded Panels (Over Base-Stitched Layer)

Sheer

4" Strips
(applied in bow area if necessary)

Figure 16-3A. *On some boats it may be easier to add the first few feet of cold-molding as narrow, diagonal strips, then transition to large sheets a few feet aft of the bow.*

Figure 16-3B. *Note the use of vertical strips on the first few feet of the cold-molding layers to more easily conform to the shape of the hull.*

Figure 16-4A. *Small ⅛-inch holes about 10 inches on center allow air to escape and insure good wood-to-epoxy-to-wood contact.*

(A)

paint your hull a dark color). To make them easily removable, you must either do the final fastening through small blocks of scrap 6 mm plywood or use screws fastened through finish washers. If I am working on a really large hull with a final thickness of 1½ inch (38 mm) or greater, then I have enough thickness to allow me to use drywall screws through finish washers for my final fastening. After pulling these screws, I epoxy wooden matchsticks into the holes. This method has given me the best final finish on my hull even when it is painted dark.

You may be wondering: What about using structural foam in place of the inner laminations, as some fiberglass boatbuilders do? I see no real advantage. The current price of structural foam is more than twice the cost of premium marine plywood, and while the composite foam-plywood hull

Figure 16-4B. *Clean the epoxy that oozes out of the weep holes.*

Figure 16-5A. *Drywall screws with finish washers are a fine way to fasten cold-molded layers. After the epoxy cures, remove the screws and fill the holes.*

(A)

might be lighter in weight, I believe nothing beats the overall physical properties of marine plywood glued together with epoxy resins into one complex, strong, and tough shape.

Cold-molding invites us to do things with boat forms that once seemed unimaginable in plywood. An example is my TUGZILLA, an honest-to-God working tugboat that was featured on *Wooden-Boat*'s cover in 2020 and won a Classic Boat Award in Britain the next year. Like any self-respecting classic tug, TUGZILLA has an ellipse-shaped fantail stern. We built a pair of heavy plywood harpins, horizontal structural members that curled around the stern and served as a form for wrapping a plywood skin around them. Since ½-inch plywood

Figure 16-5B. *Another fastening technique is to sheathe scrap plywood strips in packaging tape, then screw the strips to the hull with drywall screws.*

(B)

Figure 16-6. *Teamwork is helpful in cold-molding, especially on large boats.*

Figure 16-7. *This team is completing the bottom panels approaching the bow.*

Figures 16-8A–C. *We have glued wooden matchsticks into the leftover screw holes to avoid print-through of the fastening scheme in dark-painted boats.*

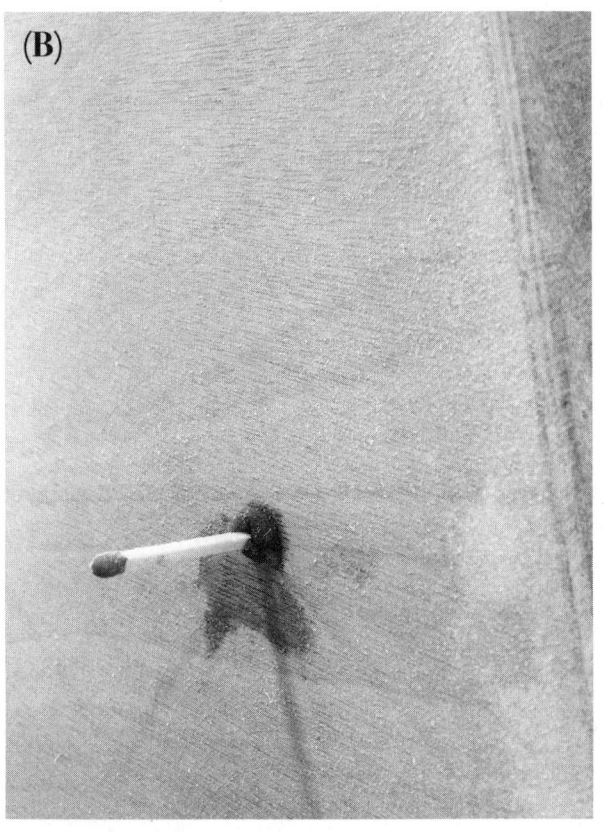

(B)

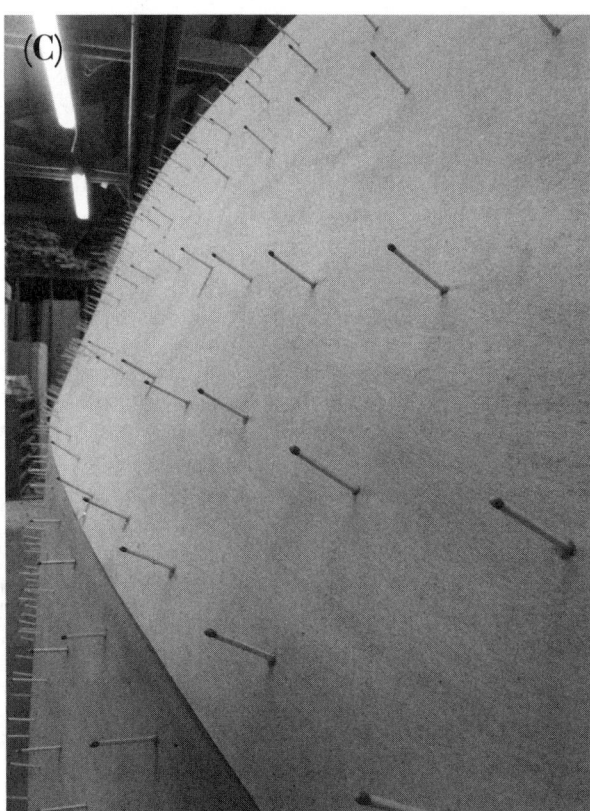

(C)

Figures 16-8A–C. (Continued)

Figure 16-9. *This photo of a transom corner of a Kingfisher 33 shows the very securely bonded cold-molded layers.*

Figure 16-10. *A cutout from the hull of one of our Sockeye 45s for the bow thruster tunnel shows its cold-molded layers.*

would not dream of taking a curve with a radius of just 4 feet, instead of ½ inch + ¼ inch, we cold-molded three layers of ¼ inch. If the boat had been smaller and its fantail even tighter, I would have set a small circular saw for a cut depth of slightly less than ⅛ inch and cut a series of parallel kerfs across the width of each of the plywood panels, allowing them to bend much more tightly. Epoxy used in laminating the panels would fill the kerfs, restoring the strength lost through the cuts.

While I don't recommend that a first-time boatbuilder tackle a fantail stern, the process isn't really that difficult, and several of my customers have executed very nice ones. Cold-molding opens up a whole new galaxy of possibilities for the Stitch-and-Glue builder, and it's nothing to be feared.

Figure 16-11. *A classic fantail stern can be cold-molded by wrapping layers of ¼-inch ply around framing members called harpins. This is my Camarone 39 motorsailer design.*

CHAPTER 17

Stems, Keels, Skegs, and Other Appendages

The hull of a Stitch-and-Glue boat wears its appendages, or external structural members, differently than boats built by most other methods. The stem, keel, and skeg are not integrated with the initial hull structure as it is being built, but rather are constructed separately and attached with mechanical fasteners (screws or bolts) or more commonly can be fused to the hull with epoxy and fiberglass cloth. Since keels and skegs are potentially high-wear or heavily loaded structures, the Stitch-and-Glue boatbuilder should take great care in securing them to the hull and protecting them from potential moisture penetration.

I learned the hard way how *not* to build and attach these appendages, though it took considerable time for the lesson to become apparent. In the early days of our boatbuilding business, we built the stems and keels from dimensional wood laminated from large pieces and mechanically fastened them to the hull. We then applied and ran the glass sheathing at the junctions up and onto the faces of the dimensional wood members. While this seemed to work fine in the short term, on inspection of boats that were 10 years old or older, we began to see cracks where the sheathing terminated and the plain epoxy-sealed wood began. Although these

were not structural cracks that would compromise the structure of the boat, such issues could lead to water intrusion into the wood of the stem-keel itself, and as a boatbuilder, I never want to tempt fate.

We have learned to avert this problem by constructing stems, keels, and skegs not with dimensional hardwoods but with plywood laminations, or a core of dimensional wood with plywood layers laminated to either side. This way, when we fuse appendages to the hull with epoxy/glass sheathing, we are glassing plywood to plywood and not to dimensional wood alone. This allows the joint to respond to temperature and possible moisture content fluctuations in the same way as the rest of the structure and eliminates any potential for even cosmetic cracks to appear.

To construct a laminated plywood stem, make a pattern out of scrap thin plywood or stiff poster board, taking care to refine it until the inner curve of your pattern fits the curvature of your hull very closely. Then transfer this pattern onto your marine plywood (the stuff with which you are ultimately going to build the stem), with as many sheets or layers clamped together as you need to build up to the required width. A typical 1½-inch-wide stem, for example, can be laminated from three pieces of ½-inch ply. The laminations run parallel to the centerline of your boat; you will not be bending plywood to fabricate the stem. Rather, you are cutting pieces that are fit to the shape of the hull and laminated together to their final desired thickness, and you want to be sure to stagger any joints in the laminates so that you end up with one complete structure ready to be mounted and fused to the hull planking. Cut out the pieces with a jigsaw or bandsaw and glue the laminations together with epoxy. The plan should illustrate the fastening method, which will typically use thickened epoxy and screws. Seal the exposed end grain of the plywood with several coats of epoxy, then glass the structure in place with fiberglass tape-reinforced seams similar to the glassing on our interior hull seams. Both stems and keels need protection from battering and grounding, so I recommend attaching a brass or stainless steel half-oval strip to the outermost edge. Use small screws of the same material as the chafe guard half-ovals and bed the strips in polyurethane adhesive bedding compound such as 3M 5200. For the skeg or bottom of the keel, which usually has little or no curvature to its exposed edge, a sacrificial "shoe" made of a very hard wood can substitute for the metal strip if you like. Ironbark is the species traditionally used, but I have had perfectly good results with the much less expensive and more available purpleheart.

If you have built a plywood stem, you will of course be painting it along with the hull. You would have a very difficult time bright-finishing (showing the wood/grain) this type of structure. Occasionally, there comes along a builder who insists on a bright-finished hardwood stem to define and highlight the profile of the bow. In this case, I would recommend laminating it in the same way as the plywood above, taking care with the appearance of the joined planks you will need to form a "sheet" large enough. Take care also to make a very flat, precise landing for this stem structure on the stitch-and-glued hull. You might decide this is entirely too much care to take and follow the path of lesser

resistance with the plywood bonded-on stem. But if you insist, continue on. After bedding and bolting the hardwood stem to the hull, stabilize it with epoxy thinned by about 15 percent denatured alcohol. Remember that any exposed epoxy-coated surface needs some form of UV protection, as the epoxy itself has none, so you will be either painting or varnishing your stem. This type of stem would be fastened onto the flat landing on the sheathed hull with wood screws plus 3M 5200 or Sikaflex to help bond and seal it.

Rudders, centerboards, and daggerboards should follow the same rule of only sheathing marine plywood and not attempting to sheathe dimensional wood. If these appendages are to be made from dimensional wood only, I have found that it is best to not attempt to seal with epoxy resin at all, or just do a stabilization sealing with alcohol-thinned epoxy as described above. In any event, paint or varnish provides more than adequate protection for rudders and centerboards in dinghies that don't live in the water for long periods of time.

It is certainly possible to build these structures out of marine plywood, but care must be given to considering the grain strength of the plywood. If the rudder, centerboard, or daggerboard is thin by design, its single layer of plywood might not hold up to the lateral strain, so sheathing the structure would help to reinforce it. In the case of a sailboat, the plan should specify and illustrate the hydrofoil-shaped cross section of the centerboard or daggerboard, which is essential for reducing drag and creating lift. This shape will take considerable planing and sanding to achieve, and you will find that progressively exposing the veneer layers of high-grade plywood will help you to shape the foil evenly and symmetrically. This is another good argument for using plywood.

It is vitally important to sheathe the *interior* of centerboard or daggerboard trunks with the same considerations as the exterior of the hull. And it's important that all plywood surfaces be fully sealed before the sheathing is applied, since these areas will be almost impossible to service later. One alternative for lining the insides of a trunk is ¹⁄₁₆-inch-thick waterproof countertop laminate (formica) glued with epoxy resin to the plywood. Suppliers usually have scrap pieces left over from big kitchen jobs that they are happy to sell very cheaply, and you don't need to care about the color. These laminates provide a slick and durable shield against the board's friction and loading pressures. For the absolute best protection for a structure that—believe me— you don't ever want to have to remove and rebuild, sheathe the interior wall pieces with fiberglass first, then fuse the layer of countertop laminate to it with epoxy. Be sure to take the sum thicknesses of interior laminations into account when you are figuring how wide to make the case. You want at least ⅛-inch clearance on either side of the board so it will not bind, but not much more. Too much clearance will invite rattling or cause increased loading on the centerboard pivot.

When using the countertop laminate, stop a couple of inches from the bottom of the trunk, leaving a sufficient landing for the hull's cloth sheathing to wrap around the edges. This method will facilitate fairing. Glue the laminate in place with epoxy before proceeding with final assembly of the case. Another option would be to carbon-fiber-sheathe the inside of these surfaces. Again, the desired qualities are to have a hard, durable, and chafe-protecting coating that will not need further maintenance or adjustments later in its service life. Assemble the trunk as a standalone structure. Test the fit of the board inside it. In the case of a centerboard, it's a good idea to drill the pivot hole and test it now, as this hole may be very difficult to drill accurately once the case is installed in the boat.

Now, make a template from the bottom of the case and use it as a guide for cutting a slot in the hull. Insert the case in the slot so as to protrude slightly beyond the hull surface. Use heavy fillets and glass cloth laminates to secure it firmly inside the hull. The lateral loads on the case can be very high, so make the junction plenty strong. After these fillets have cured, trim the excess flush with the outside of the hull. When you later sheathe the exterior of the hull (Chapter 18), overlap at least 1½ inches of cloth up into the interior of the case or trunk. The edge of the daggerboard case or centerboard trunk is extremely vulnerable, and it's important to use enough cloth to strengthen this area of the hull. You can also add stainless steel or brass half-oval strips to the leading edges as chafe-resisting areas.

Sailboat ballast keels, such as those on the 21-foot Song Wren and 23-foot Arctic Tern, also need to be built and installed with great attention to structural integrity. My preference is to build the hollow plywood keel separately and bond it to the hull with fillets and multiple layers of heavy biaxial fiberglass. Then when the boat is rolled upright, the keel can be filled with the requisite amount of lead (900 pounds in Song Wren's case) in the form of lead shot. Stainless steel or bronze keelbolts extend through floor timbers deep into the keel, and as the lead shot settles around them, I pour in epoxy to create a solidified resin-lead mass around the bolts. Layer in the ballast, never pouring more than an inch of lead shot without adding epoxy and stirring it all together. Use slow-curing hardener and stage the pour to be certain it penetrates to the full depth of the lead shot. While this process adds the considerable cost of epoxy to the lead, it avoids the very hazardous practice of casting molten lead into a form—not something many backyard builders even want to think about!

Whether it's best to attach all these appendages before or after your exterior hull sheathing will vary with the particular design you are building and with your building schedule. Certain structures such as a centerboard or daggerboard trunk are best built up and installed in the boat before the hull's final exterior sheathing is completed. But I have attached and glassed the exterior stem and keel on vessels after the exterior hull sheathing is completed and then glassed them in place separately with success.

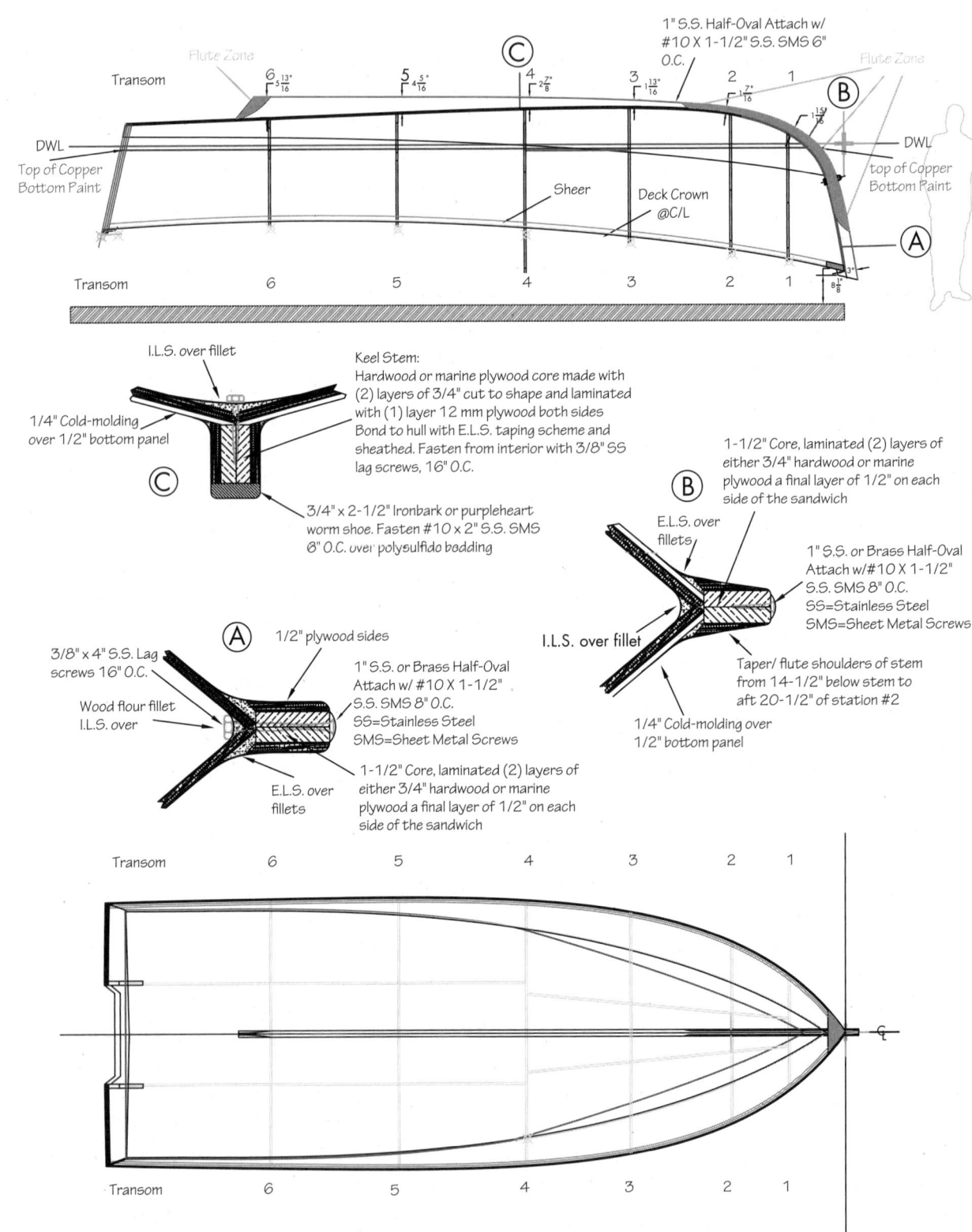

(A)

1" S.S. Half-Oval Attach w/ #10 X 1-1/2" S.S. SMS 6" O.C.

Flute Zone

Transom

Flute Zone

$6\,\frac{13"}{16}$ $5\,\frac{5"}{16}$ $4\,\frac{5"}{16}$ $4\,\frac{7"}{8}$ $3\,\frac{13"}{16}$ $2\,\frac{7"}{16}$

$\frac{15"}{8}$

DWL

DWL

Top of Copper Bottom Paint

top of Copper Bottom Paint

Sheer

Deck Crown @C/L

$3"$

$8\frac{1"}{8}$

Transom

6 5 4 3 2 1

C

I.L.S. over fillet

1/4" Cold-molding over 1/2" bottom panel

Keel Stem:
Hardwood or marine plywood core made with (2) layers of 3/4" cut to shape and laminated with (1) layer 12 mm plywood both sides Bond to hull with E.L.S. taping scheme and sheathed. Fasten from interior with 3/8" SS lag screws, 16" O.C.

3/4" x 2-1/2" Ironbark or purpleheart worm shoe. Fasten #10 x 2" S.S. SMS 8" O.C. over polysulfide bedding

B

1-1/2" Core, laminated (2) layers of either 3/4" hardwood or marine plywood a final layer of 1/2" on each side of the sandwich

E.L.S. over fillets

1" S.S. or Brass Half-Oval Attach w/#10 X 1-1/2" S.S. SMS 8" O.C. SS=Stainless Steel SMS=Sheet Metal Screws

I.L.S. over fillet

Taper/ flute shoulders of stem from 14-1/2" below stem to aft 20-1/2" of station #2

1/4" Cold-molding over 1/2" bottom panel

A

1/2" plywood sides

3/8" x 4" S.S. Lag screws 16" O.C.

Wood flour fillet I.L.S. over

E.L.S. over fillets

1" S.S. or Brass Half-Oval Attach w/ #10 X 1-1/2" S.S. SMS 8" O.C. SS=Stainless Steel SMS=Sheet Metal Screws

1-1/2" Core, laminated (2) layers of either 3/4" hardwood or marine plywood a final layer of 1/2" on each side of the sandwich

Transom

6 5 4 3 2 1

Transom

6 5 4 3 2 1

Figure 17-1A. *A plan drawing of the keel and stem and construction details for them at three points along a typical 30-foot Stitch-and-Glue boat.*

(B)

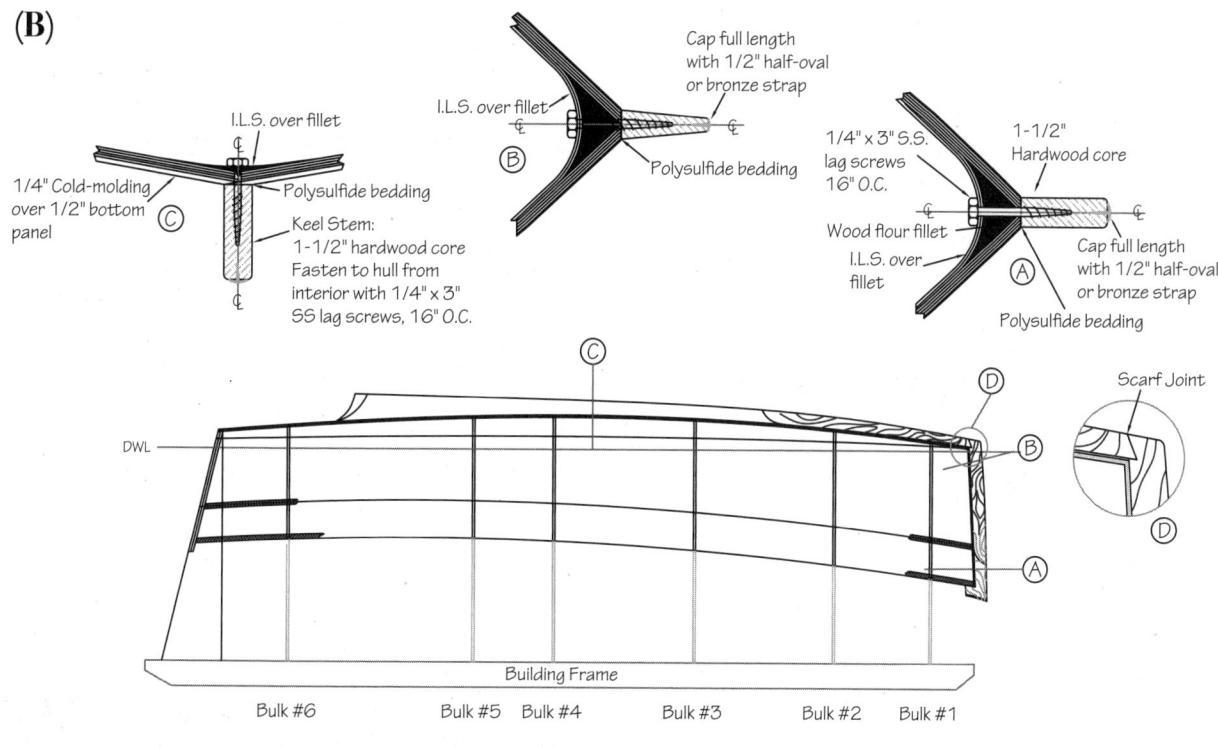

I.L.S. over fillet

1/4" Cold-molding
over 1/2" bottom
panel

©

Polysulfide bedding

Keel Stem:
1-1/2" hardwood core
Fasten to hull from
interior with 1/4" x 3"
SS lag screws, 16" O.C.

I.L.S. over fillet

Cap full length
with 1/2" half-oval
or bronze strap

®

Polysulfide bedding

1/4" x 3" S.S.
lag screws
16" O.C.

1-1/2"
Hardwood core

Wood flour fillet

I.L.S. over
fillet

Ⓐ

Cap full length
with 1/2" half-oval
or bronze strap

Polysulfide bedding

Ⓒ

DWL

Scarf Joint

Ⓓ

Ⓑ

Ⓐ

Ⓓ

Building Frame

Bulk #6   Bulk #5   Bulk #4   Bulk #3   Bulk #2   Bulk #1

NOTES ON ATTACHING STEM & KEEL

STEM and KEEL constructed from 1-1/2" hardwood. I like purpleheart or some very similar hard and durable wood. This structure will be fitted to the boat before glass sheathing of the exterior of the hull, with the idea being that the stem/keel will be simply bedded in place over the sheathed hull on a bed of polysulphide bedding compound. Fasten in place with 3/8" X 3" S.S. lag screws 16" O.C. (as shown in detail "A"). Stem starts rectangular as shown in detail "A" - Stem starts to flute (detail "B") 7" below stem and sheer intersection and is at full taper 13" from top. Taper/flute continues to 15" aft of bulkhead #2 and returns to rectangular section aft. A bronze or stainless steel chafe band from stem head down to and overlapping the entire length of the keel. Use 1/8" thick bronze strap or 1/2" S.S. half-oval for the chafe strip, fasten w/#8 X 1-1/4" S.S. SMS 6" O.C. over bed of 3M-5200 caulk.

Figure 17-1B. *A timber stem and keel detail of the Dunlin 22 design.*

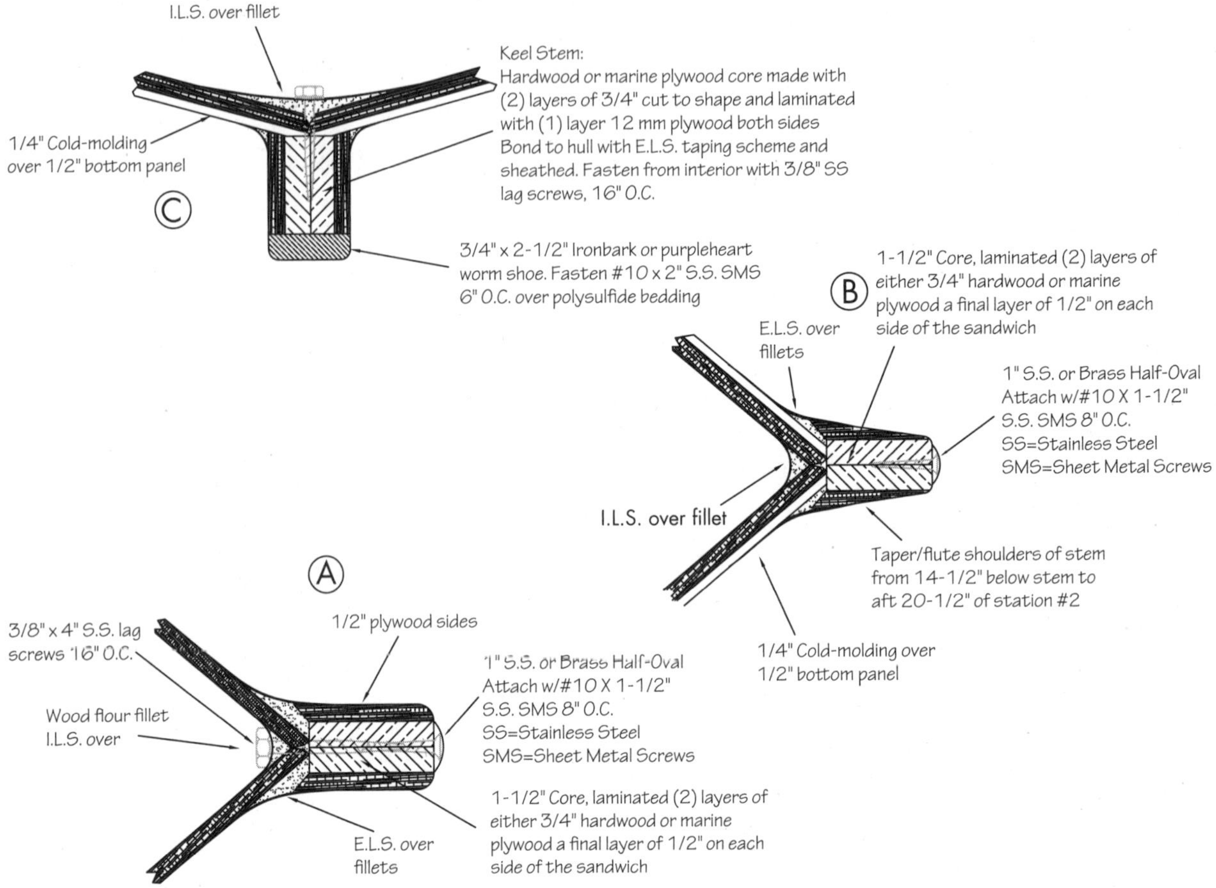

I.L.S. over fillet

Keel Stem:
Hardwood or marine plywood core made with
(2) layers of 3/4" cut to shape and laminated
with (1) layer 12 mm plywood both sides
Bond to hull with E.L.S. taping scheme and
sheathed. Fasten from interior with 3/8" SS
lag screws, 16" O.C.

1/4" Cold-molding
over 1/2" bottom panel

Ⓒ

3/4" x 2-1/2" Ironbark or purpleheart
worm shoe. Fasten #10 x 2" S.S. SMS
6" O.C. over polysulfide bedding

1-1/2" Core, laminated (2) layers of
either 3/4" hardwood or marine
plywood a final layer of 1/2" on each
side of the sandwich

Ⓑ

E.L.S. over
fillets

1" S.S. or Brass Half-Oval
Attach w/#10 X 1-1/2"
S.S. SMS 8" O.C.
SS=Stainless Steel
SMS=Sheet Metal Screws

I.L.S. over fillet

Taper/flute shoulders of stem
from 14-1/2" below stem to
aft 20-1/2" of station #2

1/4" Cold-molding over
1/2" bottom panel

Ⓐ

3/8" x 4" S.S. lag
screws 16" O.C.

1/2" plywood sides

Wood flour fillet
I.L.S. over

'1" S.S. or Brass Half-Oval
Attach w/#10 X 1-1/2"
S.S. SMS 8" O.C.
SS=Stainless Steel
SMS=Sheet Metal Screws

1-1/2" Core, laminated (2) layers of
either 3/4" hardwood or marine
plywood a final layer of 1/2" on each
side of the sandwich

E.L.S. over
fillets

Figure 17-2. *Further detail at three main sections of a stem and keel.*

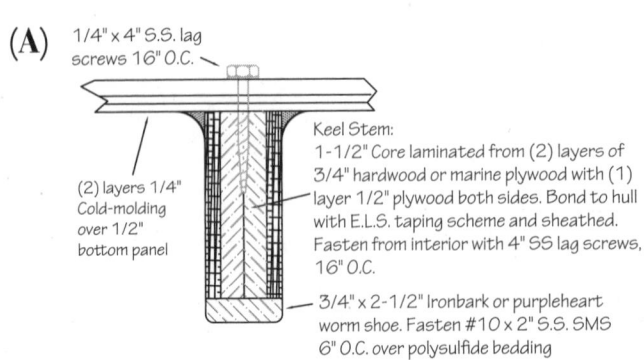

(A)

1/4" x 4" S.S. lag
screws 16" O.C.

Keel Stem:
1-1/2" Core laminated from (2) layers of
3/4" hardwood or marine plywood with (1)
layer 1/2" plywood both sides. Bond to hull
with E.L.S. taping scheme and sheathed.
Fasten from interior with 4" SS lag screws,
16" O.C.

(2) layers 1/4"
Cold-molding
over 1/2"
bottom panel

3/4" x 2-1/2" Ironbark or purpleheart
worm shoe. Fasten #10 x 2" S.S. SMS
6" O.C. over polysulfide bedding

Figure 17-3A. *A hardwood wormshoe on the bottom of the keel can mitigate damage from a grounding.*

Figure 17-3B. *The Godzilli 16 wears a prominent timber stem.*

(A)

Daggerboard cap, 3/4" x 15-1/2" x 6" hardwood, screwed on with (4) #14X2" S.S. SMS

Daggerboard assembled as shown and glassed with (2) layers glass cloth or Dynel cloth set in epoxy. Hardwood cap at the top screwed on with (4) #14X2" S.S. SMS

(B)

Heavy Glassing at edges for Chafe protection at edges

Glassing at edges and a S.S. Half oval set in polysulphite caulk for Chafe protection at edges

Centerboard/Daggerboard
Protection at edges

Figure 17-4B.

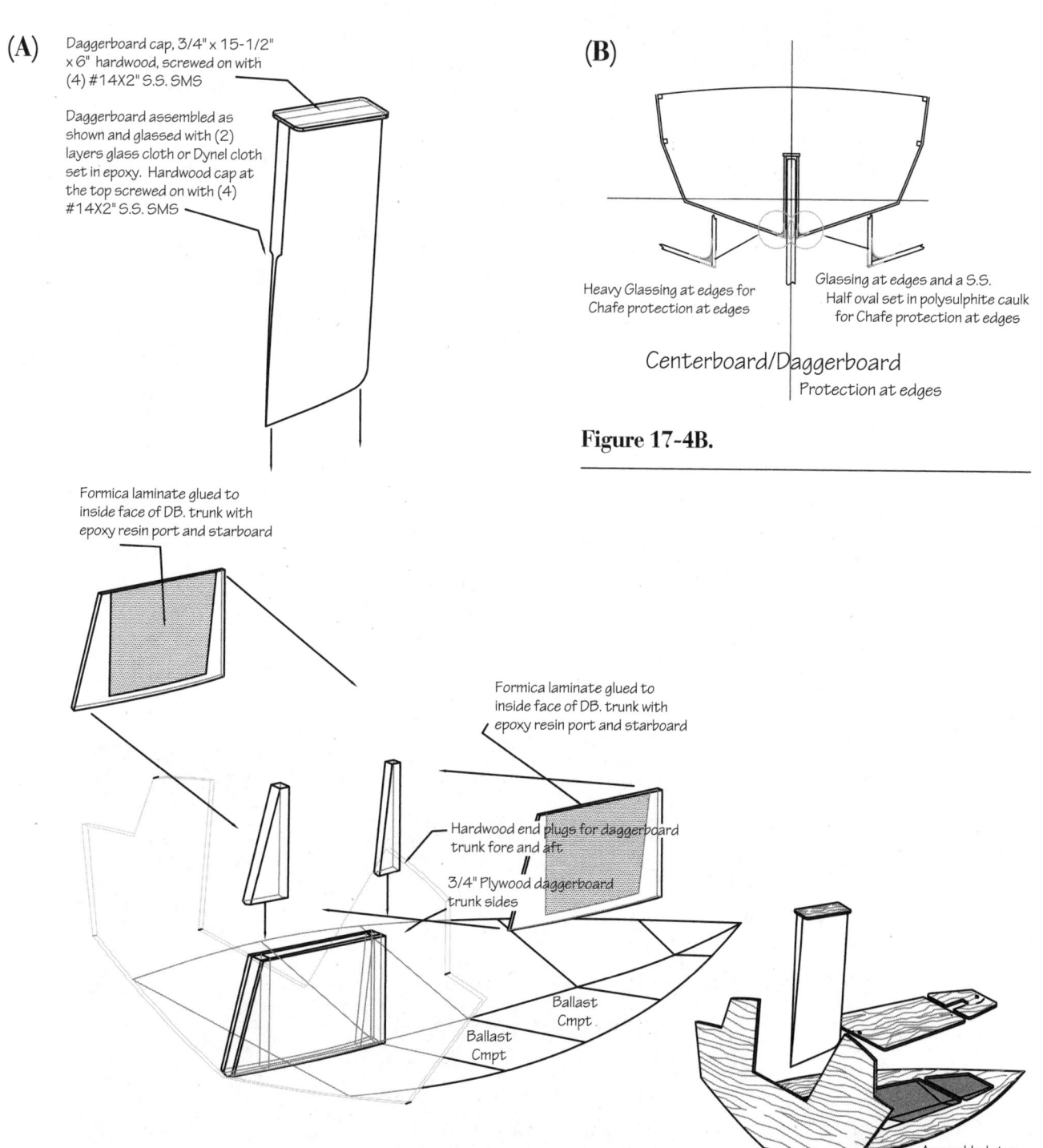

Formica laminate glued to inside face of DB. trunk with epoxy resin port and starboard

Formica laminate glued to inside face of DB. trunk with epoxy resin port and starboard

Hardwood end plugs for daggerboard trunk fore and aft

3/4" Plywood daggerboard trunk sides

Ballast Cmpt

Ballast Cmpt

Assembled view at half scale

Figure 17-4A. *Fitting plan for the daggerboard and its trunk in the tiny Nancy's China sloop.*

Figure 17-5A. *A bow is fitted with the core material for the stem/keel structure.*

Figure 17-5B. *A complete stem and keel are now in place.*

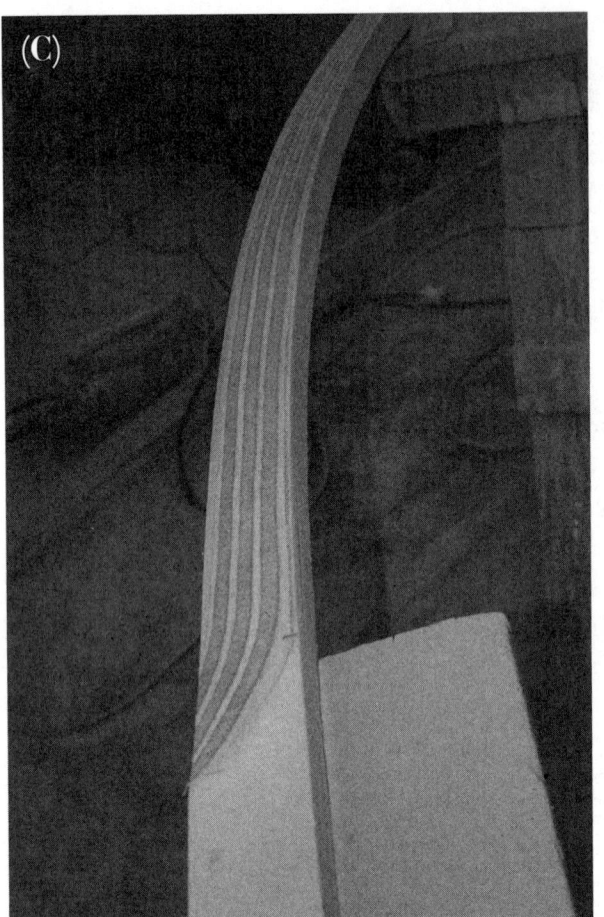

Figure 17-5C. *Planing away some of the stem's material for a fluted cross section will give it a more graceful appearance on the boat.*

Figure 17-5E. *More Spanish windlasses hold the keel in place.*

Figure 17-5D. *A Spanish windlass, which can be tightened with the twist of a dowel, makes an excellent clamp substitute when there's nowhere to fasten a conventional clamp.*

Figure 17-5F. *The stem and keel structure has been glued in place, and the hull is nearly ready for sheathing.*

Figures 17-6A–B. *Glassing the stem and keel effectively welds them to the hull. Begin with full-length fillets that will fill in any gaps between the hull and stem/keel, apply three layers of biaxial cloth set in epoxy, and finish neatly.*

Figures 17-6C. *Filleting and glassing of stem and keel are now complete.*

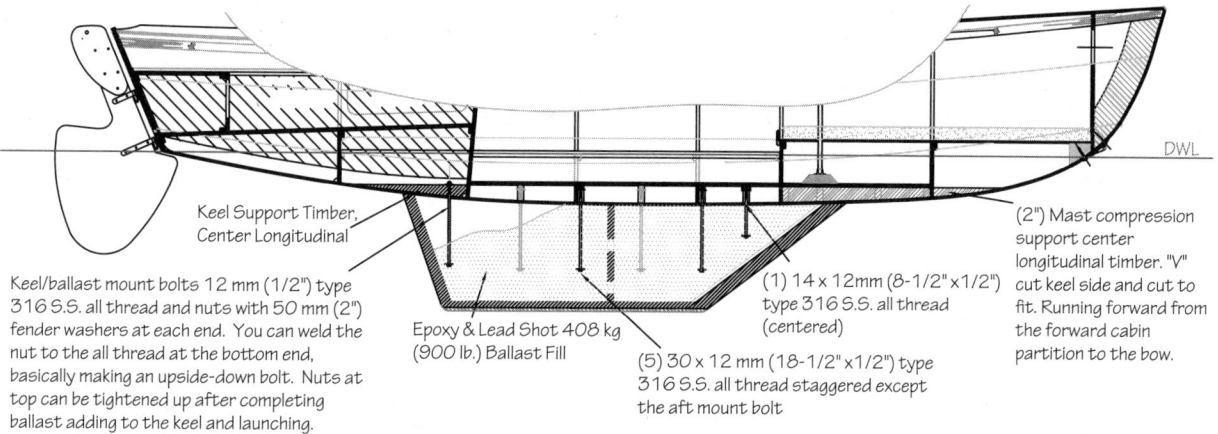

Keel Support Timber,
Center Longitudinal

Keel/ballast mount bolts 12 mm (1/2") type
316 S.S. all thread and nuts with 50 mm (2")
fender washers at each end. You can weld the
nut to the all thread at the bottom end,
basically making an upside-down bolt. Nuts at
top can be tightened up after completing
ballast adding to the keel and launching.

Epoxy & Lead Shot 408 kg
(900 lb.) Ballast Fill

(1) 14 x 12mm (8-1/2" x 1/2")
type 316 S.S. all thread
(centered)

(5) 30 x 12 mm (18-1/2" x 1/2") type
316 S.S. all thread staggered except
the aft mount bolt

(2") Mast compression
support center
longitudinal timber. "V"
cut keel side and cut to
fit. Running forward from
the forward cabin
partition to the bow.

DWL

Figure 17-7. *A hull profile view of the Song Wren 21 illustrates the lead shot/epoxy ballast and keelbolts in a plywood casing.*

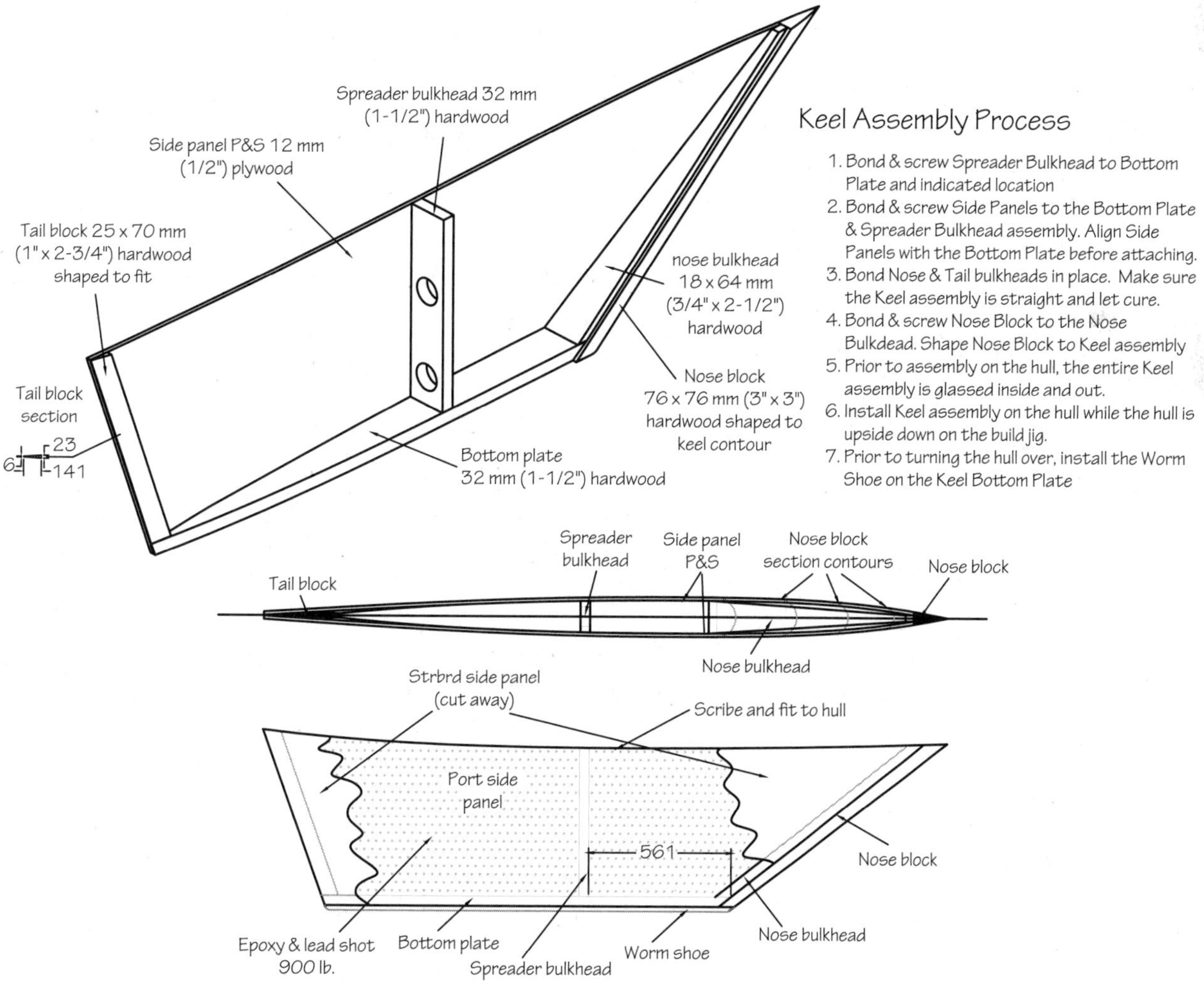

Spreader bulkhead 32 mm
(1-1/2") hardwood

Side panel P&S 12 mm
(1/2") plywood

Tail block 25 x 70 mm
(1" x 2-3/4") hardwood
shaped to fit

Tail block
section

6 — 23
— 141

nose bulkhead
18 x 64 mm
(3/4" x 2-1/2")
hardwood

Nose block
76 x 76 mm (3" x 3")
hardwood shaped to
keel contour

Bottom plate
32 mm (1-1/2") hardwood

Keel Assembly Process

1. Bond & screw Spreader Bulkhead to Bottom
   Plate and indicated location
2. Bond & screw Side Panels to the Bottom Plate
   & Spreader Bulkhead assembly. Align Side
   Panels with the Bottom Plate before attaching.
3. Bond Nose & Tail bulkheads in place. Make sure
   the Keel assembly is straight and let cure.
4. Bond & screw Nose Block to the Nose
   Bulkdead. Shape Nose Block to Keel assembly
5. Prior to assembly on the hull, the entire Keel
   assembly is glassed inside and out.
6. Install Keel assembly on the hull while the hull is
   upside down on the build jig.
7. Prior to turning the hull over, install the Worm
   Shoe on the Keel Bottom Plate

Tail block

Spreader
bulkhead

Side panel
P&S

Nose block
section contours

Nose block

Nose bulkhead

Strbrd side panel
(cut away)

Scribe and fit to hull

Port side
panel

561

Nose block

Epoxy & lead shot
900 lb.

Bottom plate

Spreader bulkhead

Worm shoe

Nose bulkhead

Figure 17-8. *Perspective detail of the Song Wren's ballast keel with its major parts.*

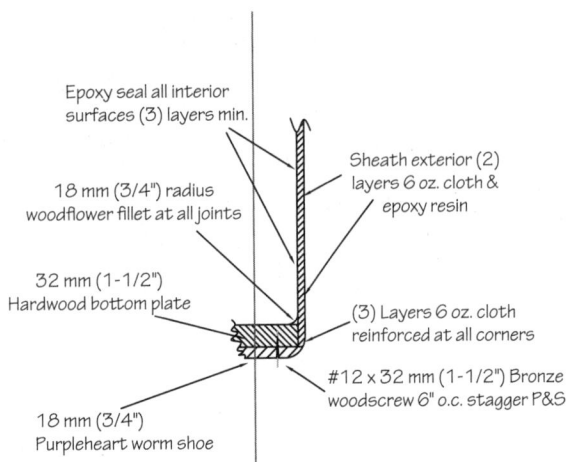

Epoxy seal all interior
surfaces (3) layers min.

Sheath exterior (2)
layers 6 oz. cloth &
epoxy resin

18 mm (3/4") radius
woodflower fillet at all joints

32 mm (1-1/2")
Hardwood bottom plate

(3) Layers 6 oz. cloth
reinforced at all corners

#12 x 32 mm (1-1/2") Bronze
woodscrew 6" o.c. stagger P&S

18 mm (3/4")
Purpleheart worm shoe

Figure 17-9. *Further detail in the ballast keel cross section.*

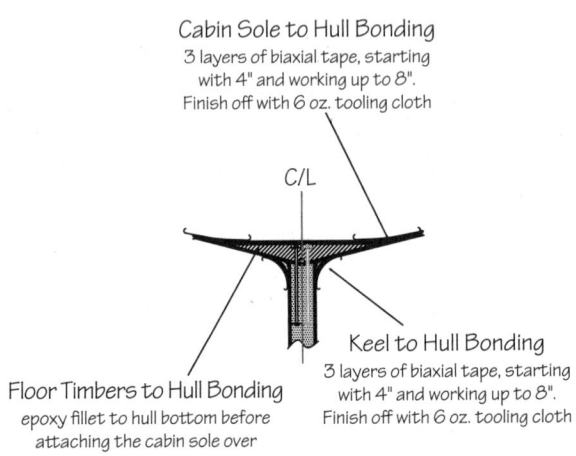

Cabin Sole to Hull Bonding
3 layers of biaxial tape, starting
with 4" and working up to 8".
Finish off with 6 oz. tooling cloth

C/L

Keel to Hull Bonding
3 layers of biaxial tape, starting
with 4" and working up to 8".
Finish off with 6 oz. tooling cloth

Floor Timbers to Hull Bonding
epoxy fillet to hull bottom before
attaching the cabin sole over

Figure 17-10. *This drawing shows the extremely strong glass bonding of the ballast keel to the hull, its holding power augmented by the keel bolts extending through the floor timbers.*

You will almost always want to build the stem and keel and fit them onto the upside-down hull, as attaching them later and trying to glass overhead can make for really short tempers. But also understand that you are working with your own unique building conditions, and you might want to attach a ballast keel way late in the building process as overhead height clearance in your shop might preclude attaching the keel earlier. Work with your own conditions and adjust to suit. If that were the case I would assemble the keel flask and scribe carefully to the upside-down hull, then wait to attach it till after the boat was largely finished and was rolled out of the height-restricted shop.

CHAPTER 18

Sheathing the Exterior

There is no debate about exterior sheathing in our shop. We have seen a lot of boats come back for maintenance or repair, and the best and longest-lived among them invariably have had their hulls and decks sheathed fully on the outside with either fiberglass cloth or Dynel fabric set in epoxy resin or both. A boat lives and works in a relentlessly hostile environment that is always trying to scrape it, batter it, crack it, and breach it. A tough outer skin defends a Stitch-and-Glue plywood hull with a great deal of abrasion and impact resistance. It helps distribute the stress loads at the seams more evenly around the whole structure, and as long as it's not punctured or breached, it provides complete protection against moisture intrusion into the wood. The maintenance of sheathed Stitch-and-Glue boats is greatly reduced, coming much closer to that theoretical ideal of laissez-faire durability, the fiberglass boat.

I use either 4-, 6-, or 8-ounce cloth, depending on the boat's intended service and finish. The lighter cloth will result in a smoother final surface since the thicker yarn bundles of the heavier cloths may tend to print through (show a cloth pattern in the final painted surface) over time. I might use the 4-ounce cloth on smaller boats that are to be painted a dark color because their color aggravates the print-through issue. For workboats and other craft anticipating rougher duty, or for boats to be painted a cooler, lighter color, 6- or 8-ounce cloth works best. As I explained in Chapter 6, a sheathing of Dynel or Xynole polyester cloth can be even more resilient and durable than fiberglass cloth

due to its capability of stretching before fracturing when placed under extreme stress. We also see less print-through in the Dynel-sheathed boats. Application techniques are similar, but the ultimate bulk is different. These synthetics come in light 4-ounce weights that tend to swell to the approximate thickness of 8- to 10-ounce fiberglass when wetted out with epoxy. Since we are trying to protect our wood with as substantial a matrix of epoxy and cloth as possible, this bulk is all to the good.

Before beginning the sheathing, I recommend a couple of preparatory steps. Make sure all hull edges have been faired and rounded over so that the cloth will lie flat over them. A radius of 6 mm or ¼ inch is about the minimum that will allow cloth to lie smoothly on both sides and over the edge; where a larger radius is possible, it will make the sheathing easier. I like to run a single layer of fiberglass tape and epoxy longitudinally along all the exterior hull seams: chines, keel, stem, and transom corners. This adds another measure of reinforcement to these most critical areas and makes for an easier job with the sheathing because any exposed end grain will be thoroughly sealed in this first step. The sheathing can be laid down over these seam tapes while they are still wet, or you can wait until they cure and sand them smooth. Then you can apply your sheathing of the complete hull.

Any of the sheathings—fiberglass, Dynel, or Xynole—can be applied with either wet or dry techniques. The wet method calls for coating the hull with unthickened epoxy, then laying the cloth immediately into and over the top of the wet epoxy.

The epoxy wicks up into the dry cloth, saturating it. Any areas that don't get fully soaked can be painted out with more resin as needed. The wet method works well with small pieces of cloth but becomes progressively messier and more difficult to manage with large sheets.

In the dry method, which I prefer, begin by filling any holes or surface defects on the hull with thickened epoxy. Then seal the entire surface with unthickened epoxy. A couple of coats wet-on-wet make for a good sealing coat. Allow it to cure fully for at least a day. The reason for this pre-coat is to thoroughly saturate the wood with the resin, ensuring the strongest possible chemical bond for the sheathing. After the epoxy sets solid, you can sand the sealed hull lightly with 80–150 grit sandpaper to give the surface mechanical "tooth." When you apply the cloth sheathing, you can focus exclusively on saturating the cloth evenly and smoothly, rather than both cloth and wood substrate at once. This makes the sheathing session much more manageable and much less sticky.

THE DRY METHOD IN DETAIL

The sheathing process will go very smoothly if you prepare and organize for it. Begin by washing the entire epoxy-sealed hull with clean water to remove any amine blush and dry with lint-free, clean cloth wipes. Then sand lightly with 80–150 grit paper. Vacuum or wipe the dust away with clean cloth wipes so that you have a smooth but still uniformly epoxy-sealed hull. Next, cut out your sheathing panels for a complete dry-fitting. The sheathing materials typically come in 38- to 60-inch widths, so cut the panels as long strips parallel to the boat's centerline. (On very large boats where it is just not practical to sheath in longitudinal panels, we sheath in athwartships panels overlapping like fish scales, starting at the stern and working forward. This method requires more sanding on the overlaps for fairing and smoothing but really makes a manageable project out of sheathing.) Each panel should drape over its neighbor with an overlap of 3 or 4 inches. Allow about 10 percent extra length

at the ends of the hull on these longitudinal panels. Wherever possible, make any of the necessary cloth overlaps occur below the waterline to minimize the fairing you will have to do later. If you anticipate rigorous duty for your boat, it's a good idea to also overlap at the chine seams and transom corners. You will then have three layers of cloth reinforcement on the inside and three more outside, for a total of six layers on those joints. These will be *really* beefy joints. Stagger the edges of these overlaps so that they don't coincide with the edges of your reinforcing tape layers that were already done on the exterior seams, which would make fairing more difficult. The goal is to have a single non-overlapped panel on the visible topsides (hull sides) of the boat if possible; this will make life easier later at the final fairing before painting.

Dry-fit the panels and label them so that you won't get confused when you begin the sheathing (Figure 18-1). Some builders have learned the hard way NOT to scribble their locating labels on the glass with a fat black Magic Marker: it can take half a dozen coats of paint before ""STARBOARD TOP" finally disappears. A scrap of masking tape makes a good label; remove it before saturating with epoxy resin.

It's easiest to work with a partner when sheathing. With your partner, grab each end of a cloth panel, unfolding and stretching and pulling it gently to keep it off the floor and suspended just over the hull in its approximate position. Now drop the dry cloth down and in position and smooth out the wrinkles with your hands. (You will of course wear disposable gloves during the whole sheathing process.) By pushing from the middle out to the ends with the heels of your palms, you will be able to smooth out the cloth panel completely. Don't try to *pull* it smooth, as all of these fabrics tend to stretch quite a bit and they will pull and stretch unevenly. Just push with your hands until it is uniformly smooth on the hull (Figure 18-2).

Next, mix and pour small quantities or puddles of unthickened epoxy onto the uppermost parts of the hull, working from the top gradually down until you reach the bottom. With a plastic squeegee,

Figure 18-1. *Here the glass cloth panels have been laid smoothly on the hull and lightly taped into place just before adding the epoxy to wet it out.*

move the epoxy until it uniformly saturates the cloth, using a figure-eight motion to push the resin into the cloth and applying light pressure at first until you get a feel for the job. Begin in the middle of the piece and work toward the bow and stern, from top to bottom. The cloth panel will stretch slightly with the pressure of your application, and this way it has somewhere to go without folding or developing pleats. Work one panel at a time before moving on to the next.

Make sure all the fabric has been completely saturated, but don't leave excess epoxy on the hull because the cloth can then have a tendency to float. Use your squeegee to redistribute or remove any excess epoxy resin. The cloth surface should have

a clear, dull appearance, and the weave of the cloth will still be quite visible and show a bit of a fabric texture. Ideally, wait just a couple of hours, and then apply a second layer of unthickened epoxy over the sheathing, smoothing it out over the entire hull in one complete coat. The bond will be better than if you were to wait overnight and allow the epoxy to cure. If you must wait a day, wash the hull again to remove the amine blush, then sand the cloth sheathing lightly with 80-grit paper. This will ensure that a good mechanical bond augments the epoxy's chemical bond. Now and later when fairing, be careful not to sand too deeply into the cloth, especially at the chines and other corners. Always sand these areas by hand to avoid cutting too deeply. You will know

Figure 18-2. *Another boat showing the glass laid smoothly out and trimmed to fit, ready for epoxy.*

you are sanding too much if the translucency of the cloth panel starts to transition to a silver color. If possible, keep the translucency on everything except for some of the fabric overlaps.

The second coat of epoxy should completely fill the weave of the cloth, but if it doesn't, a third coat should be applied wet-on-wet over the second. As you work these coats, try to leave the surface as smooth and fair as possible. As you will soon learn, sanding and fairing large areas of epoxy is major work. Epoxy resists abrasion, so it resists sandpaper. Many builders have found that a high-density foam roller or a 4-inch foam brush works well to ensure a smooth top layer of epoxy, consistent in thickness. Another good trick is to use a heat gun to warm the epoxy so that it flows better and smooths itself. Some builders lay large panels of peel ply over the entire sheathing surface, working it with a foam roller to create a smooth surface underneath. Whatever fuss and time you expend at this point will be repaid in multiples when you later fair and sand in preparation for painting.

Speaking of multiples: Depending on the size of your boat, you may need to apply two layers of over-all hull sheathing. In our shop, boats over 22 feet usually get an initial sheathing of 6-ounce fiber-glass, followed by a thorough sanding and fairing. Then an outer sheathing of Dynel or Xynole cloth is applied. The principle of this double sheathing is that by going from the most flexible material on the outside (the polyester cloth) progressively into the stiffer fiberglass and finally the stiffest plywood, we are creating the most durable, impact-resistant laminate we possibly can (Figure 18-9).

Figure 18-3. *Here is a typical scheme for overlaying glass panels to provide overlaps on all joints and ensure complete sheathing of all the boat.*

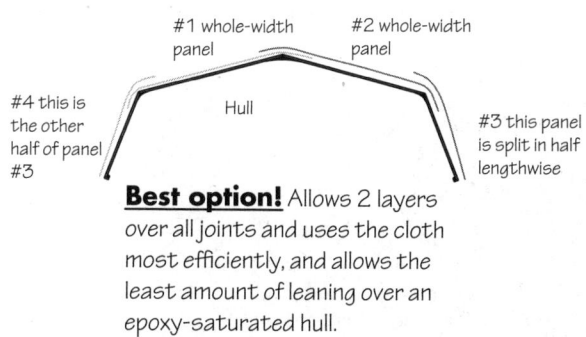

#1 whole-width panel
#2 whole-width panel
#4 this is the other half of panel #3
Hull
#3 this panel is split in half lengthwise

Best option! Allows 2 layers over all joints and uses the cloth most efficiently, and allows the least amount of leaning over an epoxy-saturated hull.

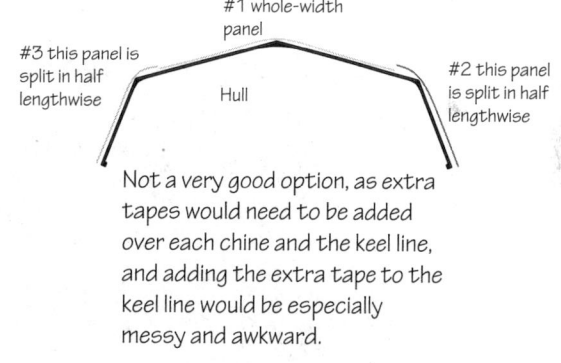

#1 whole-width panel
#3 this panel is split in half lengthwise
#2 this panel is split in half lengthwise
Hull

Not a very good option, as extra tapes would need to be added over each chine and the keel line, and adding the extra tape to the keel line would be especially messy and awkward.

The rule with cloth overlaps is that all chines and keel lines have at least 2 layers of cloth over the axis of the joint. You have many options of how to lay the cloth, but be careful as many of the options will force you to add extra tapes to the joint to get the laminate up to 2 layers.

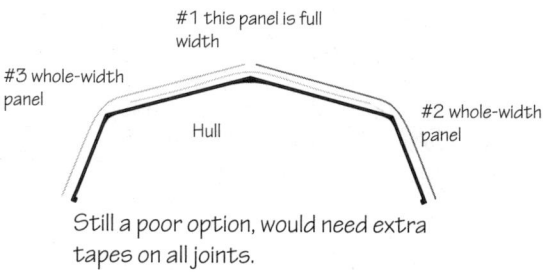

#1 this panel is full width
#3 whole-width panel
#2 whole-width panel
Hull

Still a poor option, would need extra tapes on all joints.

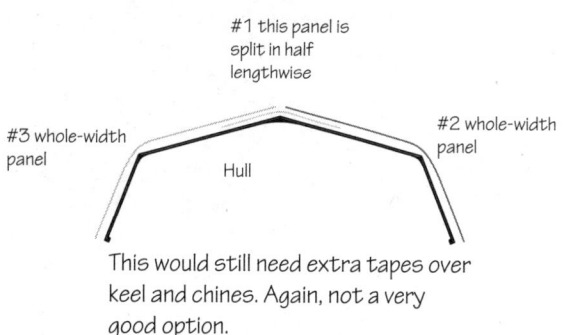

#1 this panel is split in half lengthwise
#3 whole-width panel
#2 whole-width panel
Hull

This would still need extra tapes over keel and chines. Again, not a very good option.

Figure 18-4. *Epoxy has been applied to the aft end of this boat, showing how transparent the cloth becomes with wetting out.*

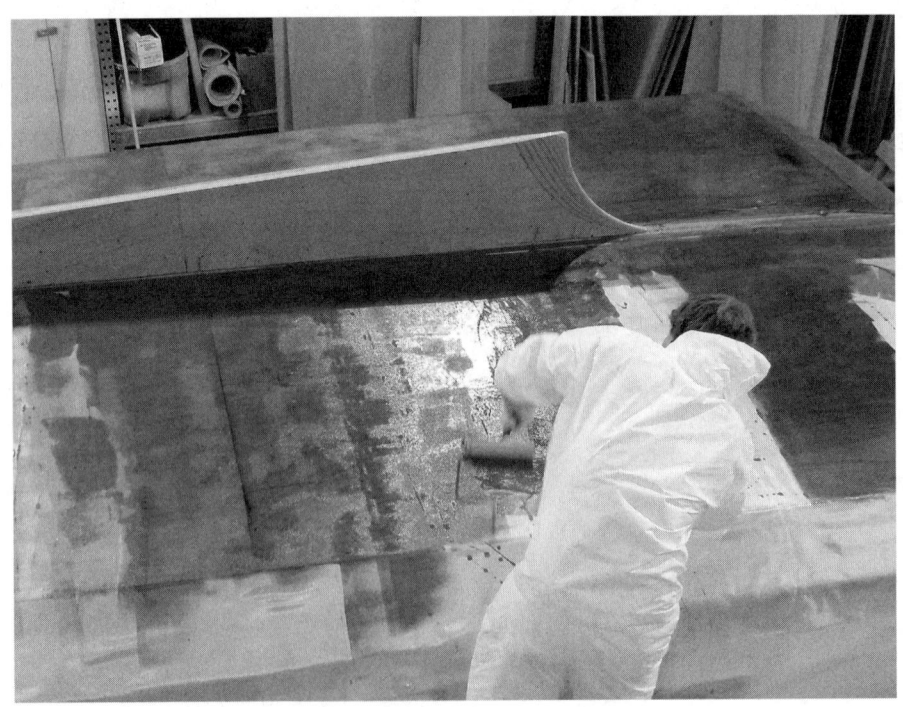

Figure 18-5. *The crew is sheathing this large boat's bottom with vertical drapes of glass cloth to make the glassing easier.*

Figure 18-6. *The earlier horizontal chine reinforcements on this boat are visible through the sheathing.*

Figure 18-7. *Peel ply was applied vertically on this large boat to help the crew reach it all before moving onto the next panel. Note the gymnastics required to stay off the wet epoxied hull!*

Figure 18-8. *Peel ply can easily be pulled off the hull after the epoxy has set. Attempts to re-use it are usually unsuccessful.*

Figure 18-9. *After Peel ply removal, the hull is generally smooth and easy to sand, wasting less epoxy and time in sanding.*

CHAPTER 19

Sanding and Fairing

No way around it: Stitch-and-Glue boatbuilding requires a lot of sanding, and it must not be mindless, zoned-out, rote work. When sanding the hull, from this point on you must do it observantly and carefully to avoid eating into the structural fiberglass fabric. You can use the appearance of the sanded surface as a guide. The epoxy/cloth laminate normally appears to be nearly transparent and when properly sanded creates white dust. But when you begin sanding into the cloth, the surface will become silvery, and you may see the crosshatch pattern of the fabric. Use these as clues to how deeply you are sanding. Your objective is to leave all the cloth and as much of the epoxy as possible on the hull surface. The structural integrity of your boat depends on this, and pumping a lot of expensive epoxy dust into the air is nothing but a waste of time and money.

One thing that will render the work more agreeable is a sander with an efficient dust-vacuum system. Fein and Festool are both fine examples; both use dedicated vacuum units that efficiently extract the sanding dust through holes in the sanding discs. Another factor is attitude. One of my amateur builders wrote a piece in *The New York Times* about the attitude-adjustment technique he used to get through sanding and fairing the hull of his 19-foot boat. Thinking about sanding and fairing the entire hull, he wrote, was overwhelming, so he mentally subdivided it into one-square-foot sections (about 165 of them, by his calculation) and focused on one

at a time. "So instead of contemplating a bleak tundra of tedium, I substituted pleasure in completing square-foot swaths of fair, smooth, ready-to-paint surface," he wrote. Whatever works is good.

Begin the sanding job by washing off any greasy amine blush with clean water and shop towels, and dry the surface before beginning the sanding. This surface cleaning will allow your sandpaper to cut cleanly and clog up much less than if you bypass this step. While we're speaking of clogging, don't fall into the false-economy trap of using every sanding disc until it's worn and dull; change frequently. While past-its-prime sandpaper can still cut, it takes a lot more pressure and is therefore harder on the sander and you than it needs to be.

For the initial and rougher basic sanding, we typically will use a sander/grinder with an 8-inch pad and 80-grit paper. A 5- or 6-inch sander is certainly adequate for smaller boats, though the more expensive variety of sanding tool that will switch between rotary and random-orbital motion is more versatile for this job. Try to work only one area of the disc's surface, lifting one edge of the disc ever so slightly to help keep the paper cool. The greater the friction and heat, the faster the disc will dull. Using very light downward pressure also keeps the sander from walking around and wobbling out of control. Let the tool do the work; if you need to apply a lot of downward pressure, then you most likely need a fresh sanding disc. On the hull exterior, never sand corners, edges, stems, chine or keel seams, or the transom-to-hull sides joint with the disc sander; the epoxied cloth edges can be quickly sanded through, even by the most skilled operator. Confine the disc sander to open, flat surfaces, sanding corners, or joints by hand.

Most basic fairing work will proceed rapidly enough with 80-grit discs. Use epoxy to recoat spots where the cloth weave has been exposed by sanding. Remember that these will show up as more silvery in color. While some epoxy manufacturers claim that recoating without sanding is possible up to 72 hours after the earlier coat has been applied, I feel it's best to sand between *any* epoxy coats that have cured past the tack-dry stage. In our shop we've learned this lesson through disappointing experience.

"Fairing" the surface means filling in gouges and air bubbles and smoothing out the inevitable valleys and ridges of cloth overlaps and other woes. To deal with this catalog of imperfections, you will use epoxy with some kind of filler material, which you will apply with a squeegee. The difficulty of sanding these fillers is proportional to their density. Wood flour is extremely dense and hard and is best reserved for structural fillets or repairs. Microballoon or microsphere fillers are considerably easier for sanding and fairing. Some builders find epoxy fairing putty (e.g., WEST SYSTEM's Microlight or System Three's QuikFair) easier still to work with; it will not drip and sag, and it is quite easy to sand. But do not be tempted to use these less dense fillers for structural work, such as the hull-bulkhead fillets.

Figure 19-1. *A dustless sanding system is not only good for your health but will also make your shop and home much easier to keep clean.*

Figure 19-2. *A Dynel-sheathed hull has had its initial sanding and now can be sealed with several more layers of nonthickened epoxy to build up the surface for further fairing.*

You will have to fair the overlaps of the cloth-sheathing edges. They can be troweled smooth using a plastic squeegeed edging of fairing putty or epoxy mixed with microballoons, sort of like working the edges of plaster in a drywall application in your house (Figures 19-3B and 19-3C). Allow the troweled areas to cure solid, then sand smooth. You could also use a layer of peel ply applied to the cloth overlaps to help smooth out the transition of one panel to another; this helps to save some sanding and fairing.

You will almost certainly encounter many random gouges and divots in the epoxy surface. One trick for repairing these is to use cellophane

Figures 19-3A–C. *After all the epoxy coats and a first thorough sanding, one of the microballoon or fairing compound pastes can be squeegeed onto the hull and sanded to smooth out glass overlap borders and other surface imperfections.*

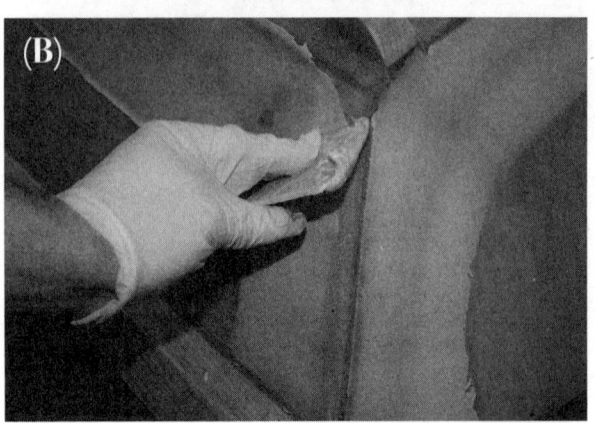

Figure 19-4. *Following fairing and a final sanding, the entire hull is resealed with unthickened epoxy.*

Figure 19-5. *This Bella 12 skiff has been sheathed, faired, sanded, and resealed, and is now ready for its first coat of primer paint.*

Figures 19-6A–C. *These photos show the final sealing coats over a Candlefish 16 hull from different angles.*

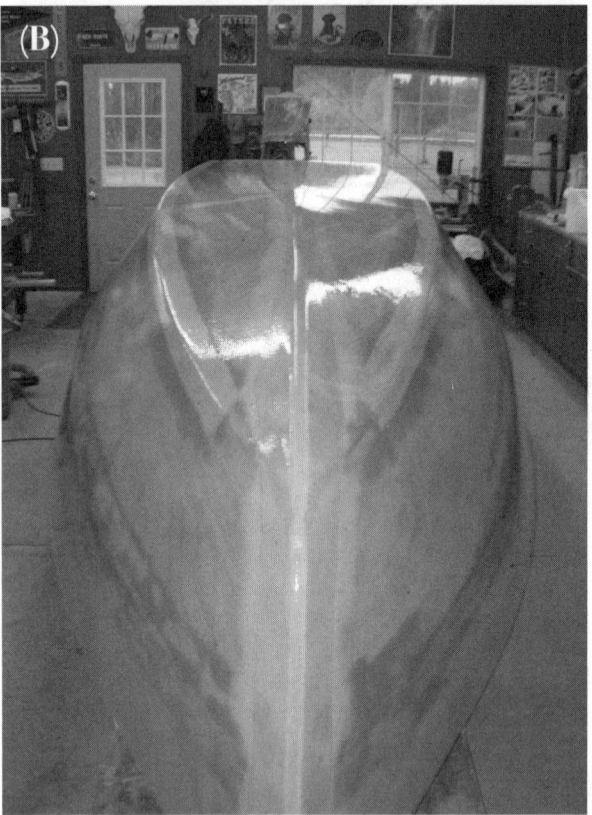

packaging tape, sometimes called gator tape, to hold the filler material in the repair area, preventing the mixture from running and sagging while curing. The tape acts as a sort of poor man's peel ply and yields a smooth surface that requires less final sanding after the patch cures.

After fairing the hull and before painting, recoat the entire hull exterior with a layer or two of unthickened epoxy. This last step thoroughly and uniformly seals the structure, provides additional moisture-proofing, and creates a smooth, stable base for the paint. After thorough curing of the epoxy, lightly sand this last coat with 150-grit sandpaper on a random orbital or small palm sander. Although you'll be tempted to then move to 220 grit—everyone is—a 150-grit finish with a gentle touch on the sander actually will provide a better mechanical "scratch" for excellent paint adherence to the epoxy. As you'll learn in Chapter 25, *between-coats* sanding with 220 grit will (with persistence) get you to a very smooth, lustrous finish.

CHAPTER 20

Marking the Waterline and Painting the Bottom

MARKING THE WATERLINE

The smart move is to mark and paint your boat's bottom before turning the hull right-side up. Before painting, you must locate and mark the waterline, which is much easier to do while the hull is still upside down.

First, let's have a small discussion of what the waterline is. As a designer, I need to assign a waterline to each design. This is the line and orientation where the hull theoretically will meet the water when fully loaded with fuel, water, crew, and gear. In the real world, of course, these loads change and move about quite dramatically, so the designer applies a "best guess" or averaging scenario to determine the floating waterline. Any boat that will live in the water for more than a day or two at a time will need to have that line struck and some sort of protective antifouling paint applied to the hull below that line. In practice, I shift that estimated line skyward by an inch or two, depending on the size of the boat, giving it a painted line that is slightly above the actual float line. I call this the intertidal or true float line. This will allow the boat's antifouling bottom coating to extend slightly above the design waterline for an antifouling splash zone. It guards against the possibility of a grass skirt growing right at the waterline, and it allows for slight variation in the weight of your finished boat (Figures 20-1 and 20-5).

Above this line is an optional painted boot stripe, which parallels the water and accents the topsides color. Above that, finally, is the topsides (hull above the waterline to the sheer) paint of the boat. Most likely your building plans show the position of the estimated true waterline, so by adding an inch or two to arrive at an intertidal line above that, you will be ready to mark the lines on the inverted hull and paint.

You should already have your inverted hull leveled to its design waterline, so you would begin by performing a final check to be sure that nothing has shifted or changed during the construction process to date. The hull must be truly level *both* fore and aft and port to starboard. In your plans, you should see a dimension from the top of the stem or forward sheer to the waterline and a corresponding dimension from the top corner of the transom to the waterline. This dimension is called the "freeboard." Measure from your shop floor (assuming it's reasonably flat and level) up to the stem, and then add the freeboard to determine the waterline where it meets the bow. Since this was done on the hull upside down, you will then subtract one to two inches for the intertidal elevation as I explained above. Do the

Figure 20-1. *The red antifouling paint showing here below the white stripe (boot stripe) is what we call the intertidal zone. It is always best to extend the antifouling paint a couple of inches above the actual float line to keep the boot stripe looking neat and clean.*

same with both corners of the transom. If your shop floor is quite level, you can then mock up a simple marking jig with a sawhorse and some sticks. Move around the inverted hull, marking as many waterline points as you need.

If your shop floor is not level, use a water level or a laser transit. A water level is simply a length of clear hose clamped between the stem and the transom corners. Fill the hose with water until the ends of the water column are at the boat's waterline. Leave one end of the hose clamped in place on the hull and move the other along the hull to mark the waterline everywhere needed. As long as you have some hose extending above the water level at both ends, the water level itself will remain stable. If this method sounds like an anachronism, well, it is—and though a water hose is plenty accurate, a laser level is simpler and easier. Whichever marking

technique you use, the more marks you make, the easier it will be to mask the line for painting.

The next step is to stretch masking tape smoothly along the marks. Fine-line tape will give a sharper paint edge than ordinary painter's tape, but either can be used with success. Pulling long stretches of tape with even pressure, progress from stem to transom. I like to use 2-inch-wide tape along the sides of the boat when possible, but at the extreme ends of the hull fore and aft, you will most likely find that 1-inch tape will conform more easily to all your reference marks. Keep the tape parallel to the markings as you apply it. Lightly press the tape into place; do not rub it firmly until the whole side of the hull has been masked. When you go back to do this, sight along the hull with your eyes at waterline height. If the masking line appears crooked, sight it from as far away as

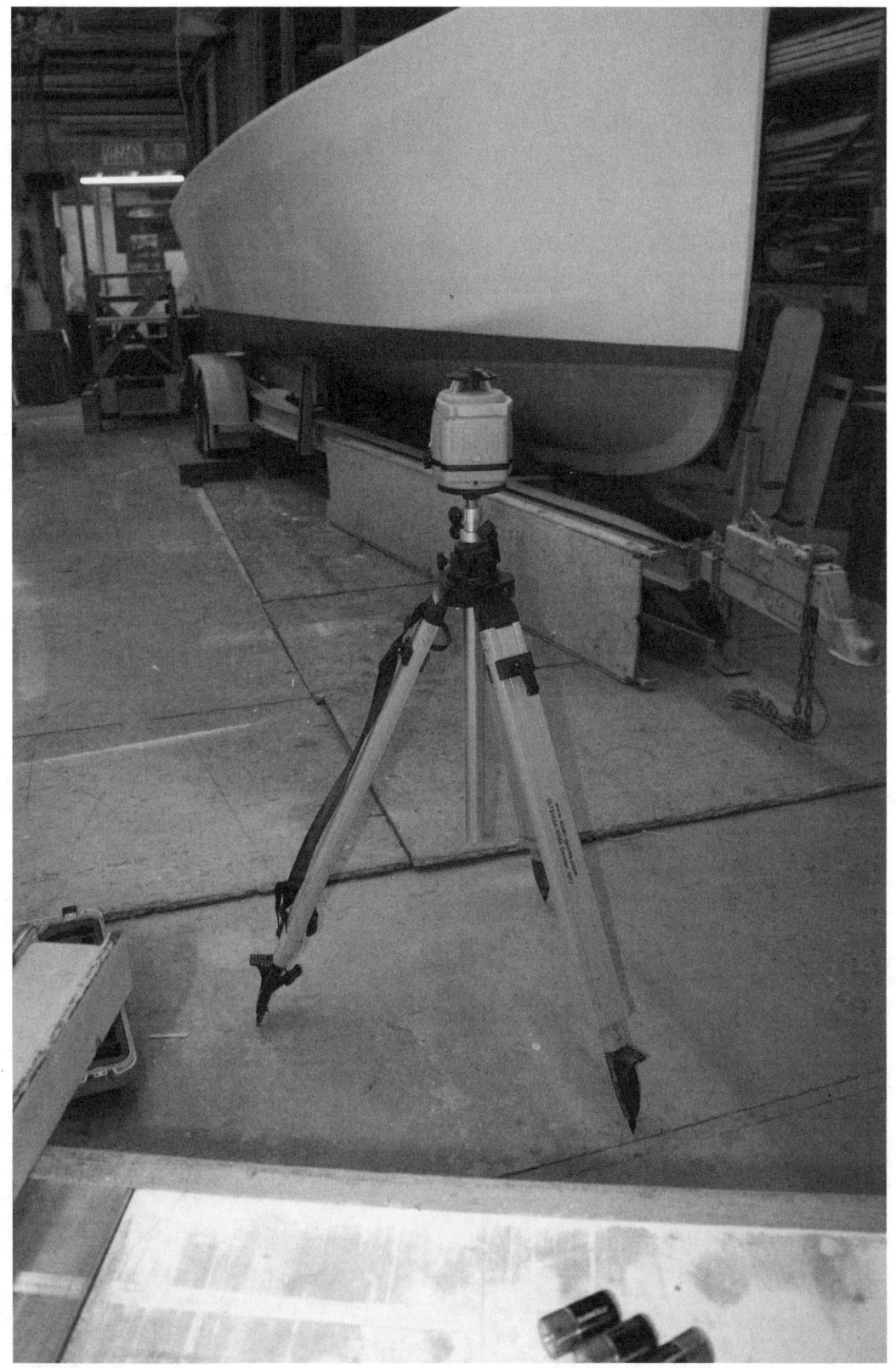

Figure 20-2. *A laser level is easily the best way to mark a waterline.*

possible to double-check. If there is a crooked section, do not rip off the tape, but use it as a reference for a second line of tape applied over the first, with each layer straightening out the line until it is eye-sweet and acceptable. Pulling out a long line of tape and then leaning it into the hull works best. Try to avoid a tendency to dab along with little sections of tape. Instead, allow the tape to help you average out all those waterline marks you previously made on the hull.

BOTTOM PAINTING

Depending on where you will use your boat, the best bottom coating will be an antifouling paint. These are paints that can stand being immersed for long periods of time without blistering or exhibiting other signs of failure. They also will contain some sort of irritant or actual poison to repel the marine animals and algae that want to attach to the bottom of the boat while she floats at her mooring or slip. There are numerous choices. Different states and countries have different regulations, so if you're buying paint by mail, make sure that the formula you're buying is legal where you plan to license your boat. Also, certain paints are recommended for freshwater and others for saltwater. Paints are generally oil-based, although the current trend is toward water-based formulas to conform to new environmental regulations. I have found that you can paint any of these coatings over a sanded epoxy-coated bottom with good results.

Antifouling paints present problems to the boat owner as well as to marine parasites. First, they're expensive, with most currently ranging between $100 and $300 per gallon. Second, they require

Figure 20-3. *The waterline has been established with small pencil marks located by a laser level and is now being taped off in a smooth and level line.*

frequent attention, with hull inspection, cleaning, and sometimes repainting needed at least yearly. And third, just making the choice of which paint to use is difficult. Different marine environments may respond best to different types of paint. For this, local knowledge may be your best resource—other boat owners and the paint department of your friendly neighborhood chandlery. Boaters' online forums can be a useful source of information, and *Practical Sailor* magazine, which does not accept advertising, periodically performs tests and publishes ratings of antifouling paints.

These paints basically divide into "soft" and "hard" categories, with hybrid formulations somewhere between. Soft (or ablative) paint is designed to gradually wear away by friction as the boat moves through the water, continually exposing new particles of copper or other biocides on the surface to repel invaders. These paints are not very effective on either boats that sit idle for long periods at a time or on high-speed boats where the paint is likely to wear away too fast. Hard coatings tend to last longer and perform well at any speed, but they may be problematic on trailered boats because some are not designed to stay out of the water for extended periods. Hybrids attempt to roll the benefits of both types into a single product.

Avoid using conventional marine enamel paint for bottom coatings. Topsides enamels or those paints intended for use above the waterline are not designed to withstand continuous water immersion, and they blister and fail rapidly underwater. Only for the smallest skiff or dinghy would I consider applying a topsides paint to underwater surfaces, and then I would not allow them to stay immersed for more than a couple of days at a time. Keep this in mind if you are building a dinghy to tow behind a cruising boat.

Despite all these provisional warnings, bottom paint is not difficult to apply. Prepare the faired and smooth epoxied hull with a final sanding with 150-grit sandpaper, sanding all the corners by hand.

Figure 20-4A. *After marking the waterline on the Banjo 20, we rolled antifouling paint on the bottom before rolling the hull right-side up.*

Figure 20-4B. *Conversion primer goes onto the bottom before the antifouling paint.*

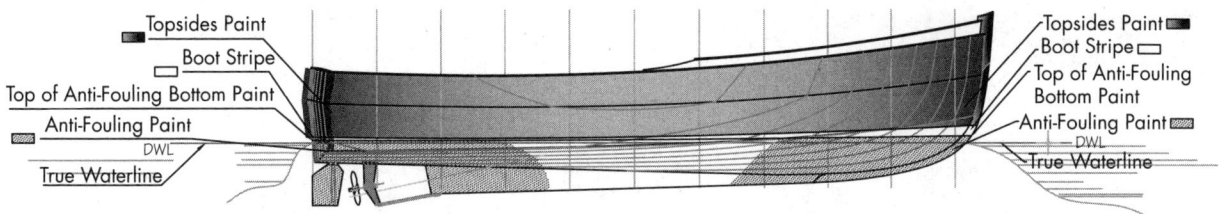

Figure 20-5. *This drawing shows the true waterline, the intertidal zone, boot stripe, and the topsides.*

Dust the freshly sanded surface and clean areas of suspected contamination with water and clean rags, avoiding solvent washes as they just smear around contaminants. Apply an epoxy primer. I use one of two products for this first coat, either Ditzler DP40, which adheres well to the epoxied hull and provides a good base for bottom paints, or Awlgrip 545 Epoxy Primer, which has similar characteristics. (Either of these primers works well as a conversion coat for all paints, either bottom or topsides.) Since the minerals in antifouling paint are heavy, the liquid tends to separate and form a dense sludge in the bottom of the can, which only vigorous shaking and stirring will dislodge; have it done mechanically in the store's paint department shortly before you begin painting. Foam roller application is best, and follow the manufacturer's recommendations for thickness and number of coats. Since this paint is by definition toxic, I would always wear a protective Tyvek suit and respirator.

CHAPTER 21

Rolling the Hull

No matter which way you begin building your boat, upside down or right-side up, you will have to roll the hull over once or twice. But you don't want to do it more times than you absolutely have to—trust me on this—so think through the construction sequence. Complete all the exterior work on the hull while it is upside down, including glassing, fairing, attaching stem and keel structures, and even bottom paint if necessary. The waterline and boot stripe are also easier to mark and paint upside down. If the keel will incorporate ballast, you will pour or install that after the hull is right-side up for the final time. You don't want to have to roll a ballasted hull.

Be sure that the interior structure has been completed to a point where the hull can withstand the strains of rollover without damage. Typically, this means that the vertical interior hull joints and bulkhead joints are filleted and glassed in, and for sure the sheer reinforcements or clamps are in place.

If you're working on a tight budget, a boom crane or other expensive machinery to assist the rollover is probably out of the question. You'll roll the hull the old-fashioned way: with manpower, or as a *WoodenBoat* magazine article calls it, "the buddies and beer method." How many buddies? Here's a simple formula: Take the length of the boat, subtract 8, divide the remainder by 3, and round up to the nearest whole number for the optimal required manpower. For example, if your boat is 22 feet long: $(22 - 8) / 3 = 4.67$. You'll need at least five people. If you have mechanical aids such

as jacks and stands, you might get by with fewer. But there is another rule to keep in mind: Your rolling party will take as much manpower as you have available. If you have five, you'll think you had just enough. If you have eight, you'll think you could never have done the job with fewer. While a lot of small-boat construction is a solitary venture, this is one task where having good and agreeable friends will serve you very well. But also remember that you are the builder and thus the rolling boss; it has to be that you are the final word on any decision, that you are fully invested in keeping all those who are helping safe, and that a successful rollover is the primary objective.

You have two basic methods of rolling the boat over. You can roll the bare hull without jigs or roller frames, or you can fabricate a wooden or metal rolling frame with which to hold the boat and cushion the rolling process. I have rolled boats up to 36 feet long without any jigs with little problem. But it is easier on the nerves to have a rolling frame. It doesn't cost much to build and it can make the job much easier. *WoodenBoat* magazine published a helpful article (July/August 2008, No. 203) illustrating how an ingenious pair of plywood discs can be fashioned to encase a hull and then rotate in place over sets of caster wheels. This method allows rolling a large hull in a small space, an otherwise difficult trick. (You can obtain digital back issues on the *WoodenBoat* magazine website.) Be aware that a rolling frame aids the process but never entirely eliminates the basic job of lifting or jacking up the hull to the balance point, then safely lowering it on the other side. Care and good planning will be necessary no matter what the method.

Obviously, the area for the rollover must be clear of clutter. Have sawhorses, jacks, and blocks ready. You will also need soft things ready at hand to protect the hull at its contact points with the floor. Carpet scraps, movers' pads, cushions, and even old rimless tires will work.

Make sure that one person has been designated as leader, and talk through the rolling process before you start. Who will do what and in what sequence? You want to avoid finding yourself midway through the rollover, with several thousand pounds of weight poised precariously aloft while conflicting opinions about the next step bounce around the walls of the shop.

Figure 21-1A. *Our customer Cyndie Phelps gives us the marching orders to roll her Storm Petrel 33.*

Figure 21-1B. *A shop-made rolling jig takes a lot of the stress out of this roll to right-side up.*

Figure 21-2A. *Lots of friends spread the load and make a rollover into a celebration. This roll used no jigs, just some old cushions on the floor.*

Figure 21-2B. *A block and tackle are rigged to help with the letdown phase.*

Figure 21-2C. *The rolling team pauses at halfway for safety checks.*

Figure 21-2D. *While you can't see the block and tackle on the back side, we are using it to ease the hull down slowly.*

Figure 21-2E. *Now that the hull is safely right-side up, we can work on blocking to level the waterline.*

Figures 21-2F–G. *We block her up level both fore and aft and sideways. We will secure her with jackstands before anyone begins climbing around inside.*

Figures 21-2F–G. (Continued)

As the lift crew raises one side, block the hull into position at intervals, and constantly check to make sure that no person is handling too much weight. When the gunwale is high in the air, and the hull's weight is about to pass the balance point and shift to the other side, move two or more people to the other side to catch and slow the hull as it rolls over. With the hull fully on its side, transfer people around to lower and block the hull in intervals, until the hull is in its new position. If this is a large boat, a good four-part block and tackle with a ¾-inch or larger line will be another valuable friend during this task, allowing lifting and letdown functions to be handled with the same gear. A really large boat might need a couple of these tackles rigged.

A lot of small shops and garages are wood-framed, and their walls and ceiling trusses are not designed to support the strains of tackles to lift or lower a heavy hull. However, it may be possible to anchor a block and tackle off a side wall to act as a safety brake, controlling the speed of rollover and assisting the rollover crew. With a 4:1 tackle and a strong, slightly scared operator hanging onto the rope end, the hull can be slowed down and safety will be assured. Come-along winches also work well for this purpose.

If your crew stabilizes the hull at stem and stern, lifts and lowers in intervals, blocking for safety as you go, and constantly monitors where the weight is shifting, the rollover should go smoothly and safely, and a lot easier than you might suspect. Several YouTube videos are online illustrating boat rollovers, and it can't hurt to study a few as you plan your own.

For hulls larger than 35 feet, you will definitely want to hire a boom crane. There is no point in gambling with the hull in which you've already invested hundreds of hours and with the safety of friends whose helping hands and goodwill you're likely to need again as the project continues. Even with the assistance of a boom crane, things can be very interesting indeed.

Figure 21-3A. *We begin the sequence of rolling a big boat (the MoonRiver 48) by clearing the floor and attaching the rolling jigs.*

Figure 21-3B. *Bolting a support structure to the rolling jig.*

Figure 21-3C. *Attaching crane pick points.*

Figure 21-3D. *A forklift is assisting the crane.*

Figure 21-3E. *The operation seems like conducting an orchestra.*

Figure 21-3F. *Pausing to catch a breath between movements of the rollover symphony.*

Figure 21-3G. *We shift her to one side of the shop so we have room to complete the roll.*

Figure 21-3H. *The MoonRiver requires two cranes and a throng of crew.*

Figure 21-3I. *Now, the crane-powered letdown.*

Figure 21-3J. *Almost there!*

Figure 21-3K. *Even with two cranes, there's still tension in the air when the hull is up on its side.*

I mentioned earlier that there are a few mile markers in the boatbuilding experience. Rollover is one of them. The difference in seeing this hull in a position 180 degrees from where you have been looking at it for weeks or months is one of amazement and wonder. Take time to celebrate with friends, customers, or relatives this amazing process that you are undertaking. Raise a toast to creation and craftsmanship! The memory will stay with you for years.

CHAPTER 22

Bulkheads, Clamps, and Floor Timbers

BULKHEADS

Although a Stitch-and-Glue boat primarily starts with the hull skin and works inward, it must also rely on interior structures for additional hull support (see Chapter 10 for some discussion of these structures). The simplest dinghy gains rigidity from seat thwarts, stern knees, breasthooks, and gunwales. In a large boat, a complicated grid of athwartships and longitudinal bulkheads provides this critical support. These bulkheads need to be fitted into place and bonded with the same type of epoxy fillet/fiberglass tape seams that weld our hull panels together. Great care must be exercised to ensure that the bulkheads are located with precision and in proper relationship to the sheerline and waterline of the boat. (If you're building from a kit, the supplied building jig should already have assured this.) We are also eventually going to add floors, berth flats, galley flats, bookshelves, and so on, to our boat. These structures will also be bonded into the hull so that they contribute to the structure of the vessel as well as provide the architectural definition of the spaces they serve. The unique quality of the Stitch-and-Glue vessel is that every component contributes to the vessel's structural strength, which makes it immensely strong, yet still relatively light. In a properly built Stitch-and-Glue boat, you will hear no groaning or creaking from the structure when running in tough water conditions.

DECK CAMBER

On a decked boat, some of the bulkheads will provide a landing for the deck. The top edges of these bulkheads will take a curve, or crown, so the deck will assume its proper camber when screwed and glued down on them. Your deck installation will follow much later in the construction process, so you do not need to trim the bulkheads precisely to establish the correct crown at this time. You will find the details for this trimming process in Chapter 24. For now, you can leave the curving top edge an inch or two higher than the line shown in the plan. It will not be difficult to trim in place on the boat later.

Boats with a cabin or pilothouse will have one or more bulkheads that extend significantly above the sheer (Figures 22-4A and 22-4B). There are several excellent reasons to build these structures in two halves—one from keel to sheer, and the second from the sheer up. In some designs, a cockpit bulkhead between keel and sheer serves as a landing place for the cockpit sole and side decks. The upper bulkhead forms the aft side of the pilothouse and

(A)

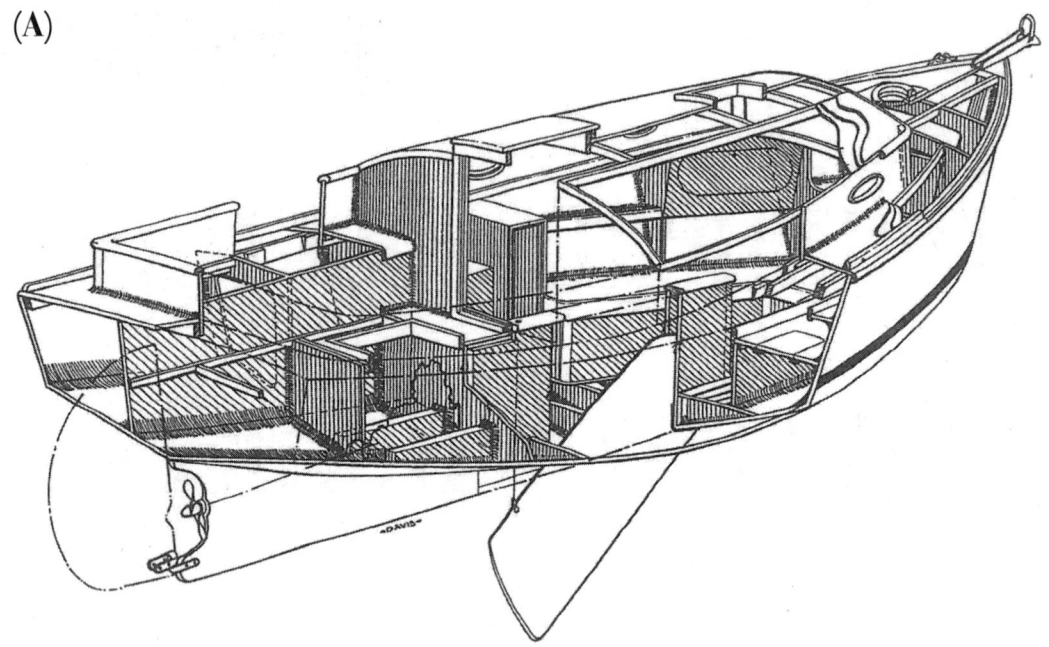

Figure 22-1A. *This cutaway illustration of a 22-foot Stitch-and-Glue cutboat shows how every bulkhead, longitudinal, berth flat, and sole are welded together with epoxy/glass joinery to become parts of the internal structure.*

(B)

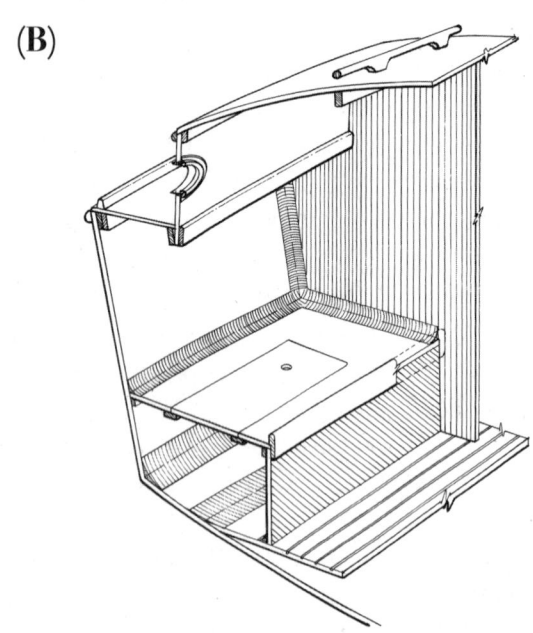

(C)

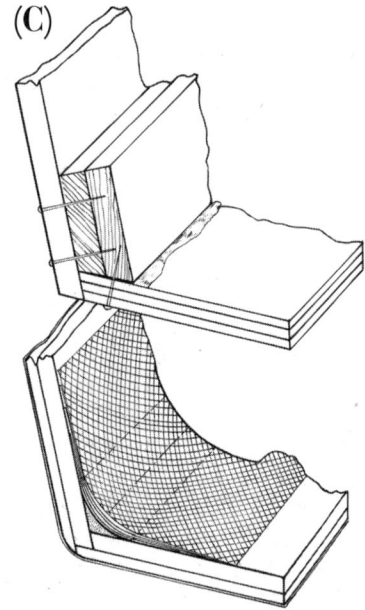

Figure 22-1B. *This detail shows good, clean joinery with ample fillets and cleats for solid attaching of panels.*

Figure 22-1C. *The Stitch-and-Glue chine joint (bottom) is much stronger and easier to keep clean than the traditional chine log joint (top).*

(D)

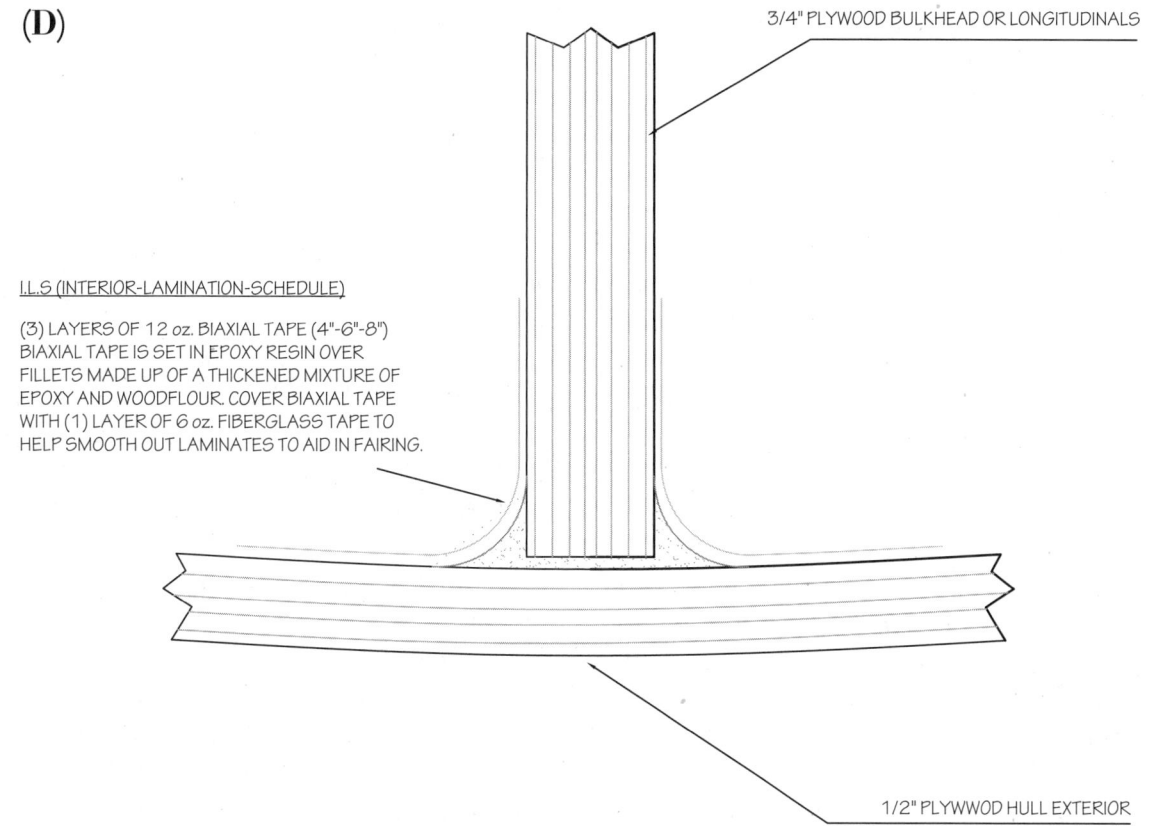

3/4" PLYWOOD BULKHEAD OR LONGITUDINALS

I.L.S (INTERIOR-LAMINATION-SCHEDULE)

(3) LAYERS OF 12 oz. BIAXIAL TAPE (4"-6"-8")
BIAXIAL TAPE IS SET IN EPOXY RESIN OVER
FILLETS MADE UP OF A THICKENED MIXTURE OF
EPOXY AND WOODFLOUR. COVER BIAXIAL TAPE
WITH (1) LAYER OF 6 oz. FIBERGLASS TAPE TO
HELP SMOOTH OUT LAMINATES TO AID IN FAIRING.

1/2" PLYWWOD HULL EXTERIOR

Figure 22-1D. *This bulkhead-to-hull plan detail calls for a beefy lamination of three biaxial tape layers.*

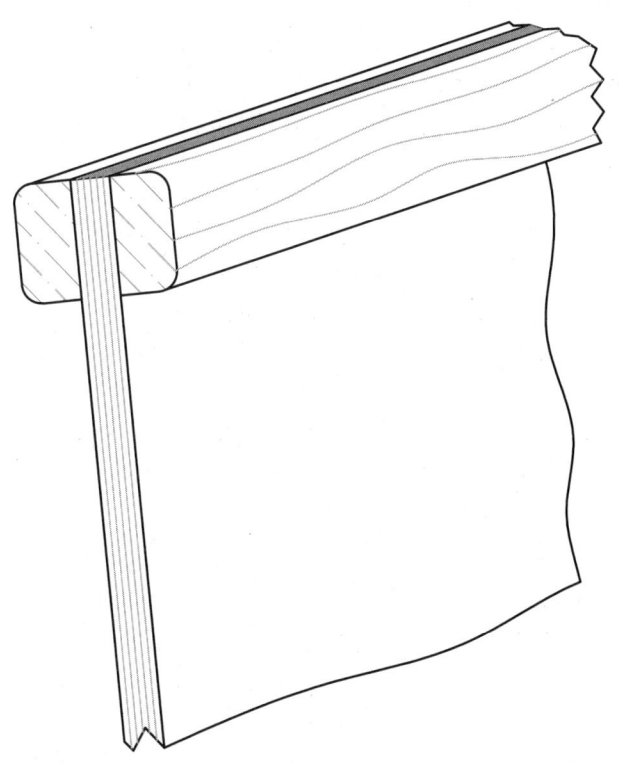

Figure 22-2. *On smaller open boats, a solid gun-wale/clamp system works well.*

is sandwiched forward of the cockpit mating bulkhead. I fasten the two bulkheads together with screws and build an epoxy-tape seam on the after side of the cockpit bulkhead and the forward side of the aft pilothouse bulkhead. This is easy and strong and provides landings for structures both in front of and behind the bulkheads. Another method for extending these bulkheads to their desired full height is aided by puzzle scarf joints CNC-cut ahead of time. The shorter puzzle joint works to keep the upside-down hull from being too tall for working on. After rolling right-side up, the extended portion can be added for the full height of the aft pilothouse bulkhead.

SHEER CLAMPS

While the chines, keel line, and bulkheads can be easily welded or fused together with epoxy/taped seams, there isn't a convenient way to fasten the hull-to-deck joint with glass and epoxy fillets. Instead, laminated wooden sheer clamps provide stiff, fair landings for the decks. In open boats with unprotected sheers that are not braced to any other structure, it is important to reinforce this area with gunwale sheer clamps. There are several types of sheer clamps; we'll discuss the gunwale/clamp first.

Glued and fastened securely on the sheer of an open skiff, dinghy, or open rowing boat, a gunwale/clamp looks good and finishes off the boat nicely (Figure 22-3). If the open boat is a sailing design or will be subject to a great amount of strain from oarlock fittings, I sometimes will glue small spacer blocks to the sheer before fastening an inwale into place (Figure 22-4). This makes the inwale act as a girder and helps to further stiffen the clamp.

The second type of sheer clamp is properly termed an exterior clamp, and it will stiffen and fair the sheer of a hull that has decks and bulkheads. While the chine is held fair in relation to the bottom of the boat, the sheer is not. If there are bulkheads for interior structures, the unsupported sheer edge of the side panels may actually stretch around the bulkheads, creating a somewhat scalloped look when viewed from above (Figure 22-8). I don't often use an exterior sheer clamp on boats longer than 28 feet, however, because it's hard to make the clamp structure with adequate support for the strains of a large boat. Making it strong enough would also make it look out of scale and too heavy. The decks must fasten over the topside planking and the clamp, and when decks are sheathed, the cloth would wrap over and down onto the face of the exterior clamp, which is typically dimensional wood. We have talked previously about the issues of trying to glass over large pieces of dimensional wood, and this would be an example. Unfortunately, because of the amount of glass that needs to be applied to reinforce this edge, it is difficult to bright-finish the sheer of the boat, as can easily be done when a gunwale-type clamp is used.

Figure 22-3. *A stronger gunwale/clamp system for an open boat would use spacer blocks to create kind of a girder beam at the sheer—more work to make and finish but very nice-looking, functional, and well worth the effort.*

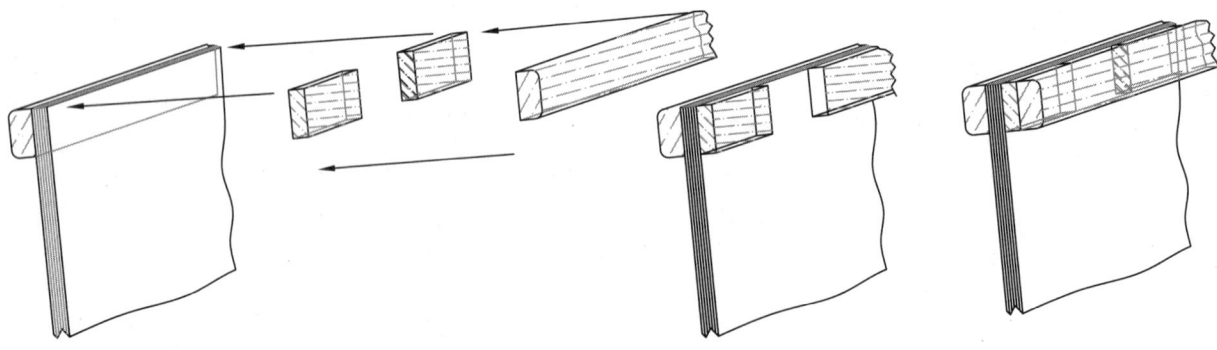

Figures 22-4A–B. *The tall pilothouse bulkhead on the Surf Scoter 23 is built in two pieces with the upper section installed after rolling right-side up.*

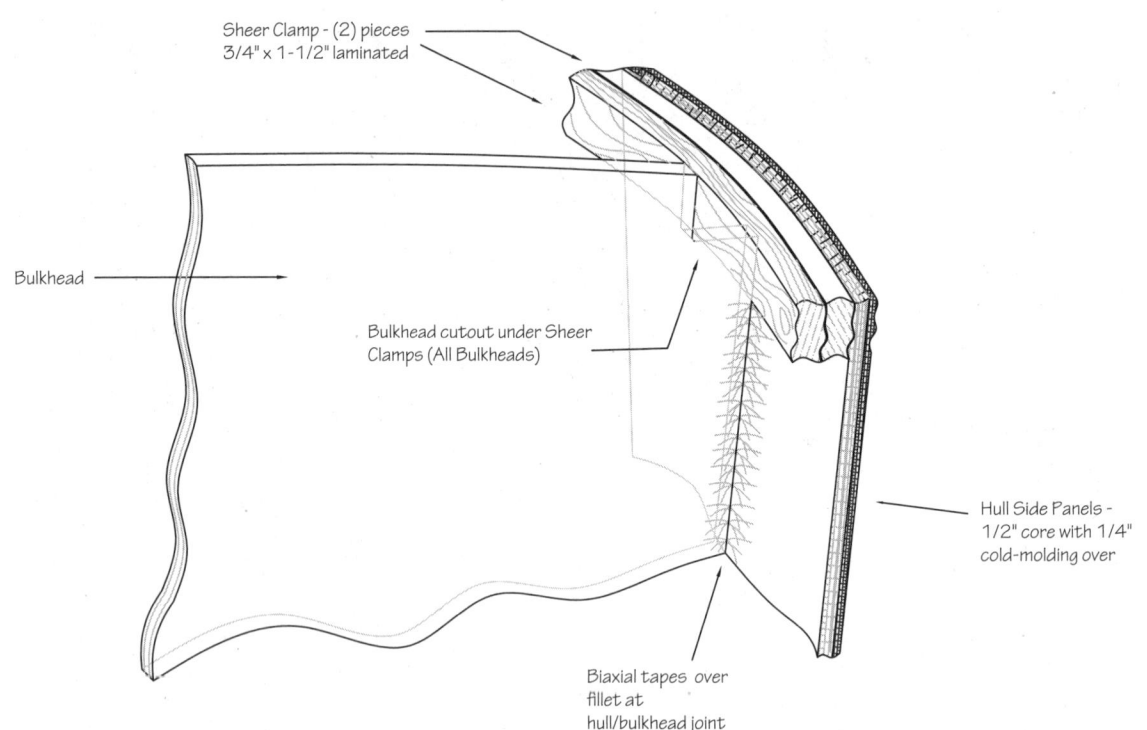

Sheer Clamp - (2) pieces
3/4" x 1-1/2" laminated

Bulkhead

Bulkhead cutout under Sheer
Clamps (All Bulkheads)

Hull Side Panels -
1/2" core with 1/4"
cold-molding over

Biaxial tapes over
fillet at
hull/bulkhead joint

Figure 22-5. *A strong sheer clamp keeps the sheer smooth and fair and provides a solid landing for the decks.*

(A)

Figures 22-5A–E. *Several overall views and a detail of the MoonRiver 48 show the structural interplay of the sheer clamps, stem, bulkheads, and longitudinals.*

Figures 22-5A–E. (Continued)

Figures 22-5A–E. (Continued)

Figures 22-5A–E. (Continued)

Figure 22-6A. *Intermediate chine and sheer clamps are in place on a Surf Scoter 23 before attaching the side panels.*

Figure 22-6B. *Scarfing sheer clamps to length before installation.*

Figures 22-6C–E. *Clamps are laminated in place on the building jig setup so they can be easily bent into proper form.*

Figures 22-6C–E. (Continued)

I use a variation of the exterior clamp on my Lichen design that incorporates a wider sheer clamp section: 4 inches wide, ¾ inch thick, and made of marine plywood. To this I attach a smaller, ⅝-inch × 1½-inch teak rubrail at the top of the sheer to finish off the joint and protect it from chafing wear (Figure 22-10).

A third type of clamp, used more often for boats longer than 16 feet, is an interior clamp laminated from two or more layers of ¾-inch dimensional wood fastened through the topsides of the hull. The deck fastens securely to the interior clamp, and the sheathing runs over the deck edge onto the topsides marine plywood. A small hardwood rubrail can then be fastened at the sheer, just below the roundover at the deck edge, protecting the sheer from chafe. If you want toerails, they can be fastened over the deck into the clamp with long, countersunk screws (Figure 22-10). On smaller (up to about 20 foot long) decked boats, the simple exterior sheer clamp and decking extending over are used (see Figure 22-7). This is a simple and elegant way to build these types of designs.

Always install both port and starboard sheer clamps during the same work session. These clamps are stiffeners and could pull a boat out of shape if the work isn't balanced side to side. Also try to run these clamps full length along the sheer, through notches in the bulkheads, so that they will pull the plywood sheer into a fair curve. Often I will laminate one layer at a time, both sides, to make the clamps easier to handle, using C-clamps and fasteners to hold them in place until the epoxy sets up.

Figure 22-7. *An exterior gunwale clamp is very useful on smaller decked boats.*

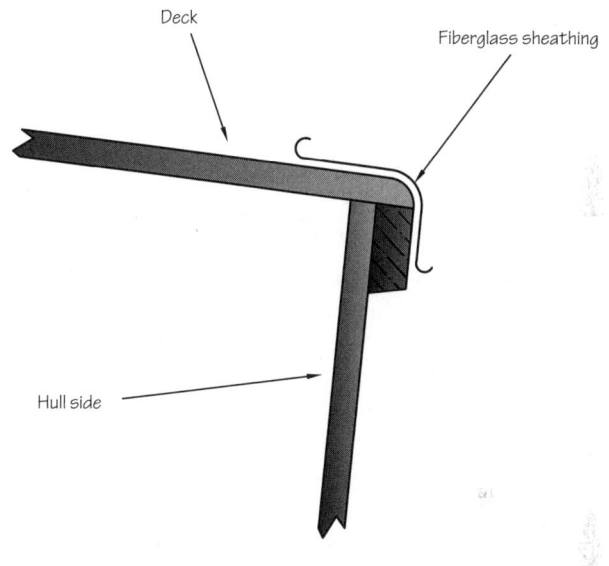

Deck

Fiberglass sheathing

Hull side

Figure 22-8. *A strong sheer clamp assures a sweet and fair curve. Without it, we could expect ugly scalloped sections between bulkheads.*

Top View

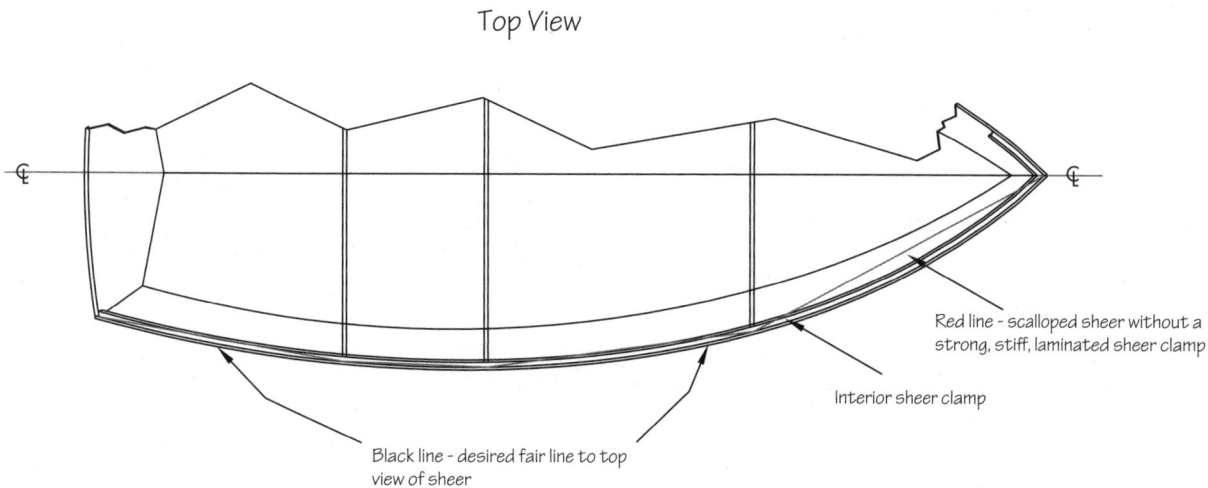

Red line - scalloped sheer without a strong, stiff, laminated sheer clamp

Interior sheer clamp

Black line - desired fair line to top view of sheer

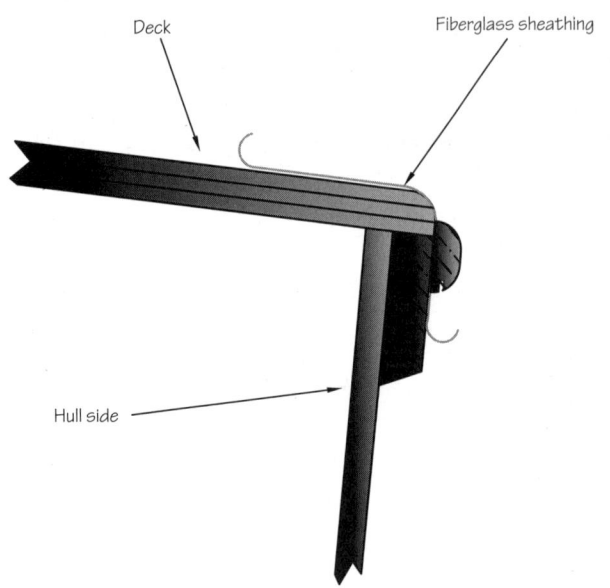

Figure 22-9. *This version of the exterior gunwale clamp can be used for decked boats up to about 26 feet. Note the guard that will be mounted on it to keep the hull/deck joint protected.*

I glue in a single layer to port and one to starboard, clean up the excess glue, and the next day glue in the second layers, port then starboard. If you're building upside down, you could prelaminate these clamps in one piece port and starboard, using the upside-down bulkheads as forms. Some packaging tape will keep the laminations from sticking to the bulkheads. After the glue sets up hard, you can remove them from the bulkhead setup, mill the edges smooth, and finally reinstall on the setup before stitching on the side panels. That way, when you stitch the side panels, you can coat the face of the sheer clamp with thickened epoxy and then screw through the side panels and into the sheer clamp structure. Cleaning up excess glue completes this process, keeps things neat and clean, and eliminates excess sanding.

The laminations you use for making sheer clamps should be full-length stock if possible, or else scarfed to length. I like to use a supple and air/dried softwood. Here in the Northwest, Port Orford or yellow cedar or Douglas fir are good species to use.

Figure 22-11. *When attaching a hull panel to an intermediate sheer clamp, you will need to plane a bevel as shown to provide a good landing.*

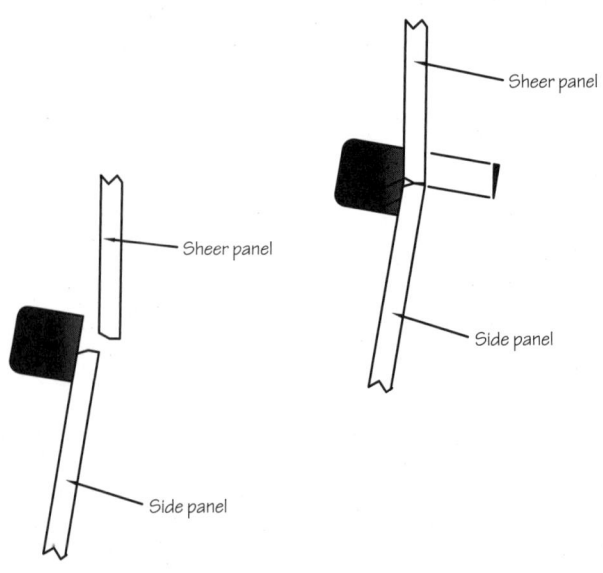

Figure 22-10. *A laminated sheer clamp with guard on the outside and toe rail above.*

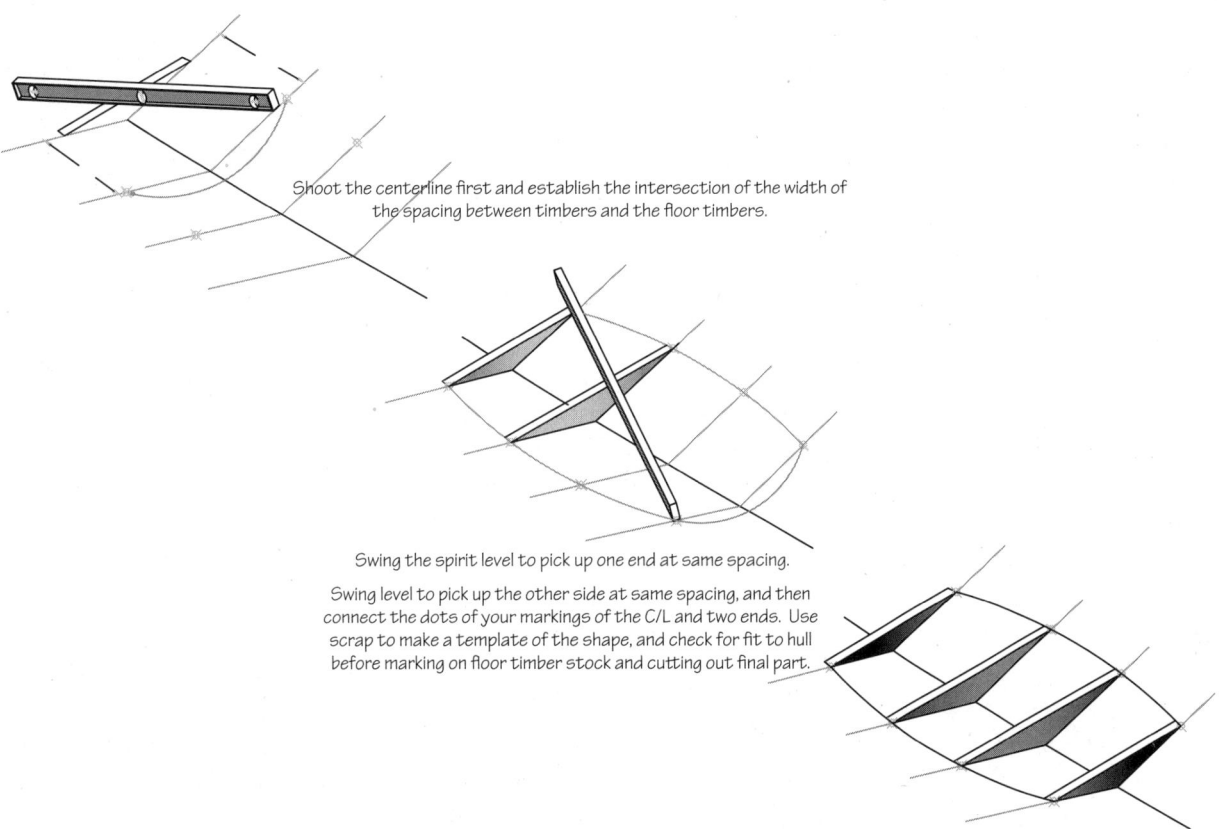

Shoot the centerline first and establish the intersection of the width of the spacing between timbers and the floor timbers.

Swing the spirit level to pick up one end at same spacing.

Swing level to pick up the other side at same spacing, and then connect the dots of your markings of the C/L and two ends. Use scrap to make a template of the shape, and check for fit to hull before marking on floor timber stock and cutting out final part.

Figure 22-12. *Laying out floor timbers always starts at one end. Use a spirit or digital level and lay out one at a time.*

Multi-chined vessels like the Surf Scoter will sometimes use an intermediate sheer clamp system to pull the edges of the hull panels into fairness and to help reinforce the rest of the structure (see Figures 22-6C–E). Just like the sheer clamp itself, notches in the bulkheads can be cut, and these can be laminated in place using the bulkheads as forms. Packaging tape in the notches will keep the epoxy used in lamination from sticking to the bulkheads, allowing you to remove, clean up, and reinstall the clamps with glue before finishing up the stitching of the side panels.

FLOOR TIMBERS

Many designs feature a cabin sole (floor) set very near the actual bottom of the boat. If the bottom is a V shape, I often install this cabin sole on a grid of timbers set at some distance apart from each other. These timbers will be fastened and glued into the bottom of the boat. Don't fillet and tape them because composite taping structures tend to pull away or crack when used to secure large pieces of dimensional wood. If you are building a sailboat there may also be a mast step and/or keel bolts built into this structure, transferring the strains (compression) of the mast and ballast keel into the grid. The cabin sole will be bonded onto this floor timber grid with glue and fastenings, and its faying edges will be bonded at the outer limits with a composite fillet and tape seam joint. I prefer good, stout floor timbers built of a durable and stable wood. The mahoganies work very well for these structures, but fir and yellow cedar and other species can be used successfully. Choose a wood that seals well with epoxy and is considered durable in marine use. (The bilge is going to have water in it some time or another, and your only recourse is to use first-class materials and to epoxy-seal the holy heck out of them!)

Because of the odd shapes and the expense of dimensional wood, I use templates made from thin scrap plywood to pattern the floor timber stock. If you have a laser level, this is an easy job and needs no description, but if not, use the method below to lay out the pieces.

Pick a starting place, usually next to a bulkhead. Your plans should state the depth of the floor timber at that point (most often as an elevation related to the designed waterline of the boat). After marking the appropriate height on the bulkhead, use a small spirit level to mark a horizontal line on the bulkhead parallel with the athwartships waterline. Your floor timbers will probably have some uniform centerline-to-centerline spacing, which your plans will specify. Mark the centerline for each floor timber forward of the aft bulkhead, and mark another set of layout points both port and starboard where each floor timber will lie. With a pencil, connect the layout marks (Figure 22-13).

Go back to your original horizontal marking along the aft cabin bulkhead and tack a small horizontal cleat on that line. With a batten extension, hold one end of the level on the cleat and mark an intersection with the edges for each of the floor timber layouts. Keep an eye on the level's bubble to pick off the ends of the floor timbers where they would terminate against the bottom of the boat. Cut several small battens the same width as the newly marked floor timbers and place them across the hull at the marks. Measure the depth of the "V" at the keel centerline, noting this dimension on a scrap of paper. Measure each of the beams (widths) and note them on your list. Draw in the floor timber widths and the centerline depths onto your pattern stock, cut out the patterns, and check for an accurate fit in the hull.

The floor timber templates may not fit precisely if you have already taped the seam in the keel of the boat. To compensate, set your pattern in place, level athwartships. With your pencil compass held exactly vertical and set at about ¼ inch or at the major height of the gap between hull and pattern, scribe the outline of the true bilge of the boat onto the pattern. Repeat this procedure for each of your patterns, and cut out along the scribed line. Your patterns should then fit in place but will be ¼ inch short of the true height. When you transfer the pattern to your floor timber stock, add ¼ inch to compensate for the extra height. I usually give these patterns ¼ inch or so extra height anyway to allow me the luxury of scribing them down into perfect position and height.

After cutting out the floor timbers, cut a limber hole in each one so that bilge water will be able to flow through and settle at the lowest point in the bilge. Locate the limber holes off center if there will be keel bolts going through the centerline of the floor timbers. Cut the limber holes large enough to accommodate any anticipated bilge pump hoses and other plumbing that might need to run through the bilge.

Preseal the floor timbers with at least two heavy coats of unthickened epoxy before installation. When you install the timbers, bond them in place with ½-inch-deep wood flour fillets backed up with #14 × 1½-inch screws from outside the hull if at all possible. Once the grid is installed, I roll another full coat of epoxy over the whole bilge area—floor timbers, panels, and all—after installation to ensure adequate protection from moisture. If you're still worried about the certainty of the sealing, roll on another coat for good measure. There is no such thing as too good a job of sealing the bilge.

For large boats with a lot of potential keel strain, I often sandwich dimensional floor timber stock between ½-inch marine plywood. Then I bond in these floor timbers with epoxy and glass composite seams to integrate them into the hull structure. These bonded-in floor timbers will greatly contribute to the structural strength of your boat and will make for a very long-lasting vessel.

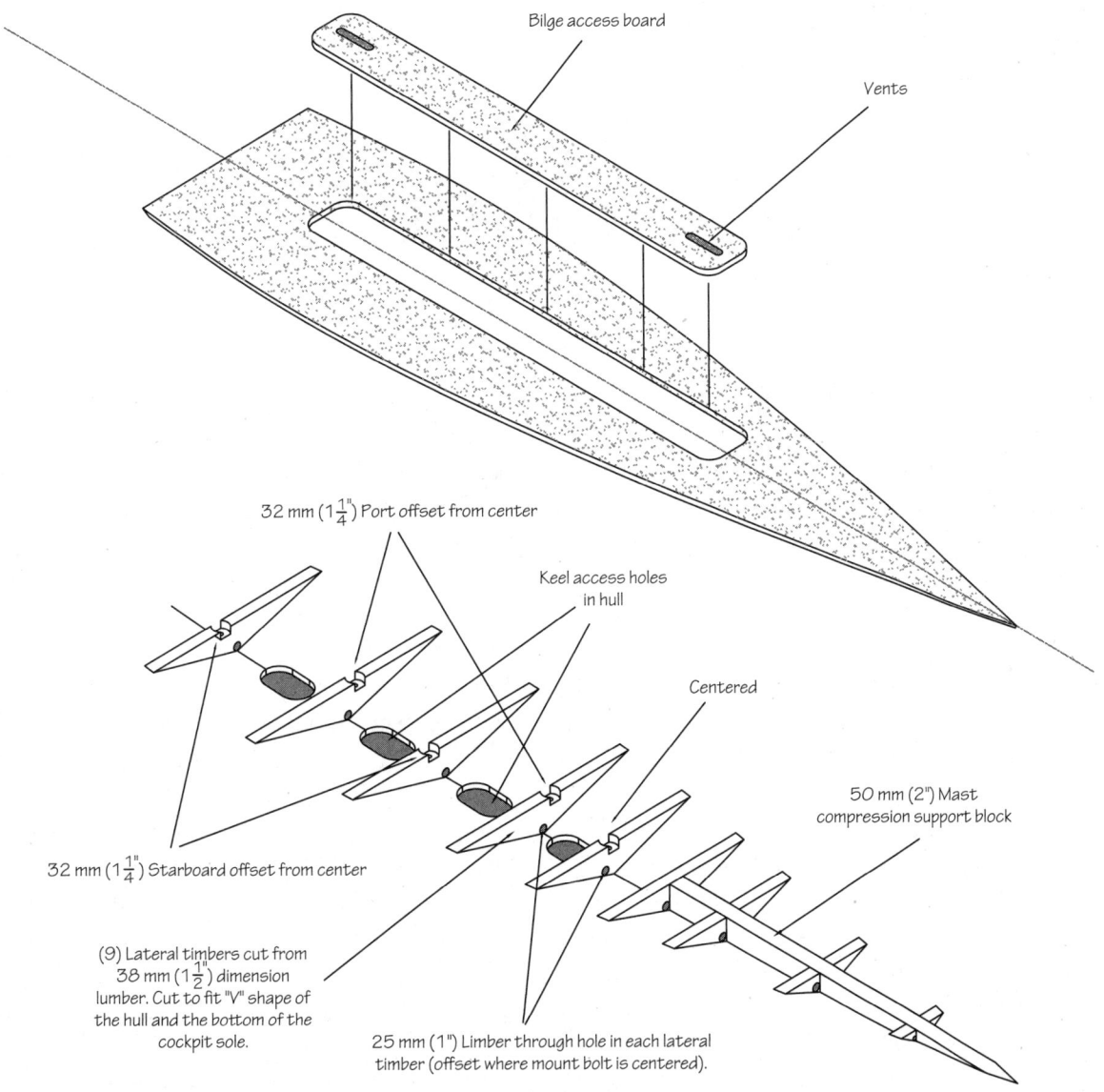

Figure 22-13. *Exploded drawing of the floor timber grid with cabin sole and inspection hatch.*

229

CHAPTER 23

Interiors, Hatches, and Doors

Once you have savored the exuberance of completing the hull itself, it's time for a dose of reality. Climb into the boat with an armload of plans, framing square, tape measure, felt-tipped pen, masking tape, and a couple of stiff battens. You are about to design and build the interior.

If your boat's cabin is large enough to be anything more than a bare-bones cuddy for storage and retreat from nasty weather, it's worth devoting thought and care to making it an inviting and comfortable environment. Here are some principles to consider:

- Storage. Clutter and dishevelment are miserable companions in a boat, so contemplate all the things you will need to organize and store, and consider what opportunities the cabin space might offer for lockers, drawers, bins, shelves, trays, and pigeonholes. No boat has ever had too much storage space. Think generously and creatively. For example, a couple of wooden pockets screwed to the inside face of the aft cabin bulkhead beside the companionway can corral the crew members' cell phones, which will be easily reachable either from the cockpit or inside the cabin.

- Flexibility. Many interior components of a Stitch-and-Glue boat are intended to contribute to the hull structure and must be glued and filleted permanently in place. But some bits such as bookshelves and small storage bins are better screwed to concealed wooden cleats to make future rearrangements possible.

- Safety. Think ahead to a day on gnarly water and imagine every place where a lurching body could meet a boat part. Then think rounded corners instead of square, and avoid projections that are likely to poke or bash.

- Comfort. Build mockups of furnishings in place to test their comfort and utility—for example, seatback angles and light placements. A reading light installed in the wrong place can be a glaring annoyance instead of a welcome luxury.

- Design integrity. Since your cabin interior will stay dry—let's hope—and protected from the weather, you have more latitude in choosing materials and decoration inside than on your boat's exterior. Interesting wood contrasts, like yellow maple played against red-brown mahogany, can be very effective. But be wary of excess: these are small rooms, and too many ideas vying for attention will read as clutter.

Before beginning interior work, your right-side-up hull should be level to the waterline both fore-and-aft and athwartships. It's a good idea to recheck these levels one more time. This is also the time for a mental tune-up. Don't be overwhelmed by the number of projects ahead, and resist making a list of them. Just pick them off one by one. Guard against post-rollover stall-out. Each step of the remaining construction needs to happen in its proper order and be given sufficient time and effort. When I was younger, I enjoyed long-distance running, which any runner will tell you involves mind over matter. Strength and stamina are secondary to mental resolve, and the key is to stay focused and not be distracted by the distance that remains. Find your rhythm, keep a steady pace, and take one step at a time.

MAST STEP

If you are building a sailboat, you must construct a structure that bears the weight of the mast and withstands the compression loading and strain of the mast heel. If your design calls for a deck-stepped mast, you'll need to install a structure called a compression post from the deck to the keel to distribute the compression loads. This vertical post will go from your floor timber grid to the deck beam that the mast will ride atop, or all the way up to the cabin roof. My plans often call for a stainless steel tube with a flat plate welded to each end and bolted in place, and if you don't like the industrial look, you can encase it in a handsome wooden post, though the increased size might get in the way of moving about thru the cabin. If it is a keel-stepped mast, the step will be a kind of mortise for the mast base to tenon into. In either arrangement, the compression (downward) forces from a mast are considerable, so build sufficient strength into the step and distribute the load over an appropriately large area of the hull structure. If your mast is keel-stepped, provide a drain hole from the bottom of the mortise into the bilge. No

matter what kind of mast boot you put at the partner, some moisture will always try to run down the mast and into the mortise. It is imperative that you allow that water to drain away.

CABIN SOLES

I like to have continuous access into the bilge, 6 to 8 inches wide, for the entire length of the sole if at all possible in the build. This panel or panels can simply lie flat on the timbers without fasteners. Elsewhere, use a polyurethane adhesive to fasten epoxy-sealed cabin soles to the floor timbers; adhesive caulking is less messy than epoxy and has good gap-filling capability. This is the perfect application for it as an adhesive. Place two beads on top of each floor timber and gently lay the cabin sole over the floor timber grid. Then add at least four fasteners per floor timber, evenly spaced. Screwing the sole piece onto the floor timbers and epoxy glassing the edges onto the hull create an extremely strong floor timber/cabin sole gridwork. If you're building a larger powerboat, the cabin sole is most likely hung on longitudinal bulkheads, and consideration for the cabin sole is simpler than for the sole that is lower and bonded into the bottom of the boat.

You don't have to use varnished or painted plywood for your cabin sole. Some of my customers who've wanted classier flooring have ripped and resawn thin veneer-like strips of mahogany or Douglas fir—sometimes both for contrasting color—and glued them onto a plywood substrate for a look similar to the nautical tradition of a teak-and-holly sole. As long as your cabin stays dry, this will look beautiful and take little maintenance.

Besides providing for occasional hands-on access to the bilge, you must create a path for ventilation even when the access plate is closed. You can drill several holes in the plate or install a small louvered vent, which is inexpensive and commonly available at chandleries. Whatever cuts or holes you make in the sole, be sure to seal any plywood edge grain you expose with epoxy, and by all means

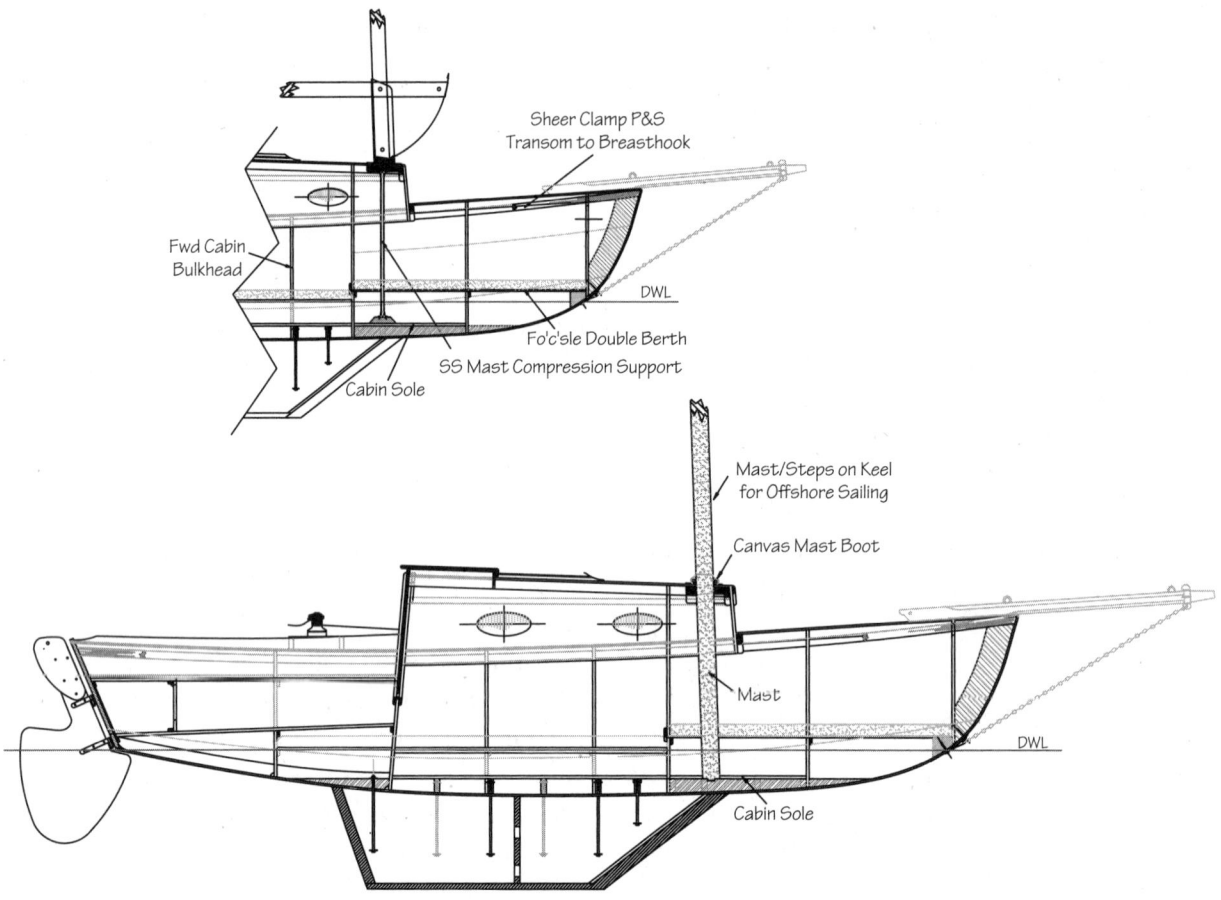

Figure 23-1. *A mast stepped on the deck or cabin top needs a compression post to take the vertical load to the floor timbers. A keel-stepped mast below transfers the load directly.*

epoxy seal the bottoms of the floorboards before installation.

Cockpit soles differ from cabin soles in that the cockpit sole almost always attaches to or rests on cleats fastened to the bulkheads and hull sides (more like the large boat cabin sole). The simple reason is that the cockpit sole rides higher in the boat. If your cockpit is to be self-bailing, the sole will be well above the waterline, and water must be able to run off and out of the boat without puddling. Unless you use a watertight hatch, you will need to leave it fully intact without openings. This also means that the sole must be firmly bonded to its cleats with a watertight seal. What about the space left between the sole and hull bottom? In my early years of boatbuilding, I have at

times filled it with two-part polyurethane foam that would expand to completely fill the space. Long-term experience has taught me that this isn't a good practice. The foam traps moisture, which eventually becomes a problem. Also, providing flotation below decks is not a good idea. Adding flotation anywhere but up as high as possible under the deck simply makes the boat unstable if the boat fills with water by raising the center of gravity. If you want to leave this space as an open compartment, provide access in the form of a watertight screw-in plastic deck plate, accessed through holes in the vertical bulkheads at the ends of the void. Be sure to vent any of these stowage or just dead space areas in as many ways as you can. Drill a small hole in the vents to allow expanding air on a hot day to escape.

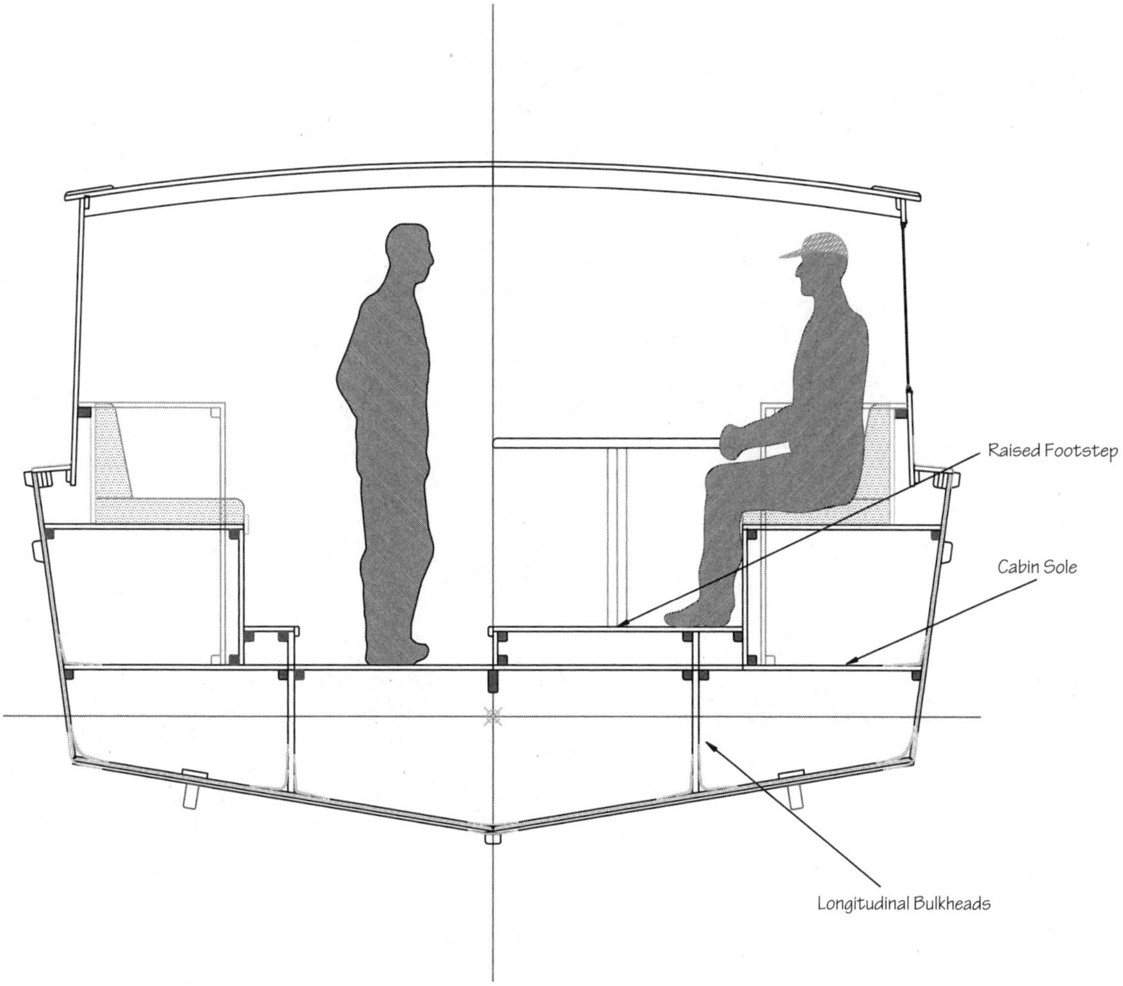

Raised Footstep

Cabin Sole

Longitudinal Bulkheads

Figure 23-2. *A larger boat's cabin sole is typically fastened to the longitudinal bulkheads, not to the floor timbers like on a sailboat or a smaller boat.*

Figure 23-3. *These watertight hatches in the cockpit sole of the Surf Scoter 23 allow bilge access.*

When the boat is left for a few days between uses, the deck plate can be removed or taken out for winter storage, and the compartment can be left open for ventilation until the spring launch. Use this deck plate access system for any watertight compartments you build into your boat. You don't want any permanently sealed structures inside the hull that you can't get into and ventilate.

In smaller boats, it may be impractical to build a cockpit sole high enough above the waterline to provide drainage to the outside of the boat. You have two choices in this case: (1) You can drain it to the bilge under the sole so that the crew will not find their feet sitting in pools of water and pump

Figure 23-4. *This is a cabin sole shown upside down before installation. Note the bevels at its edges and thorough epoxy sealing.*

it out manually whenever enough collects. (2) Or you can limber it into the boat's central bilge to be evacuated by a manual or electric bilge pump.

You can build a simple floorboard-type sole of ¾-inch plywood with drainage holes drilled at intervals. A complicated, though very elegant-looking and traditional, sole can be made as an interlocking lattice of teak or mahogany dimensional lumber. Be forewarned that loose change, jewelry, and Dorito crumbs will inevitably fall through a lattice. With either type of sole, you will make it so that it can be lifted out for access for cleaning, and be sure to provide an opening large enough for a manual bilge pump hose to fit through it.

While we are considering cockpits and drainage, we need to discuss a bit of fussy construction under the cockpit seating. Since the cockpit is going to get rained on and if you want to have access lockers under the seats, you will need to build a drip gutter system to channel water into the bilge (Figure 23-5). This may or may not be shown on your plans; not all of them go to this level of detail. The gutters must be wide enough that they won't easily clog with debris and sloped downward enough to assure drainage and keep your cockpit lockers dry. These gutters are not difficult to build, but they may take some time. It wouldn't be a bad idea to temporarily screw in the system and test it with some "rain" from a garden hose before permanent installation.

Remember: All parts of all these compartments and structures must be scrupulously epoxy-sealed before installation.

LONGITUDINAL BULKHEADS

The longitudinal bulkheads form part of the framing structure of larger Stitch-and-Glue designs, particularly those with decks. These are nothing more than large bulkheads turned lengthwise in the boat.

The Surf Scoter cockpit has two longitudinal bulkheads running from the transom to the rear of the pilothouse bulkhead. These form the sides of the outboard motor well, the sides of the fuel tank compartment, and the adjoining side supports for the stern seats, and they provide bearing and fastening for the cockpit sole. The thrust of the outboard is dispersed throughout the boat's structure by these two cockpit longitudinals. All longitudinals are bonded into place with taped seam fillet joints with the same care as in the major athwartships bulkheads (Figure 23-6).

Your plan will show exact locations for floor timbers and longitudinal bulkheads, and in most cases the longitudinals were included in the original bulkhead/longitudinal setup before you started

Figure 23-5. *Cockpit seats that open will need drainage gutters as the seams cannot be made watertight.*

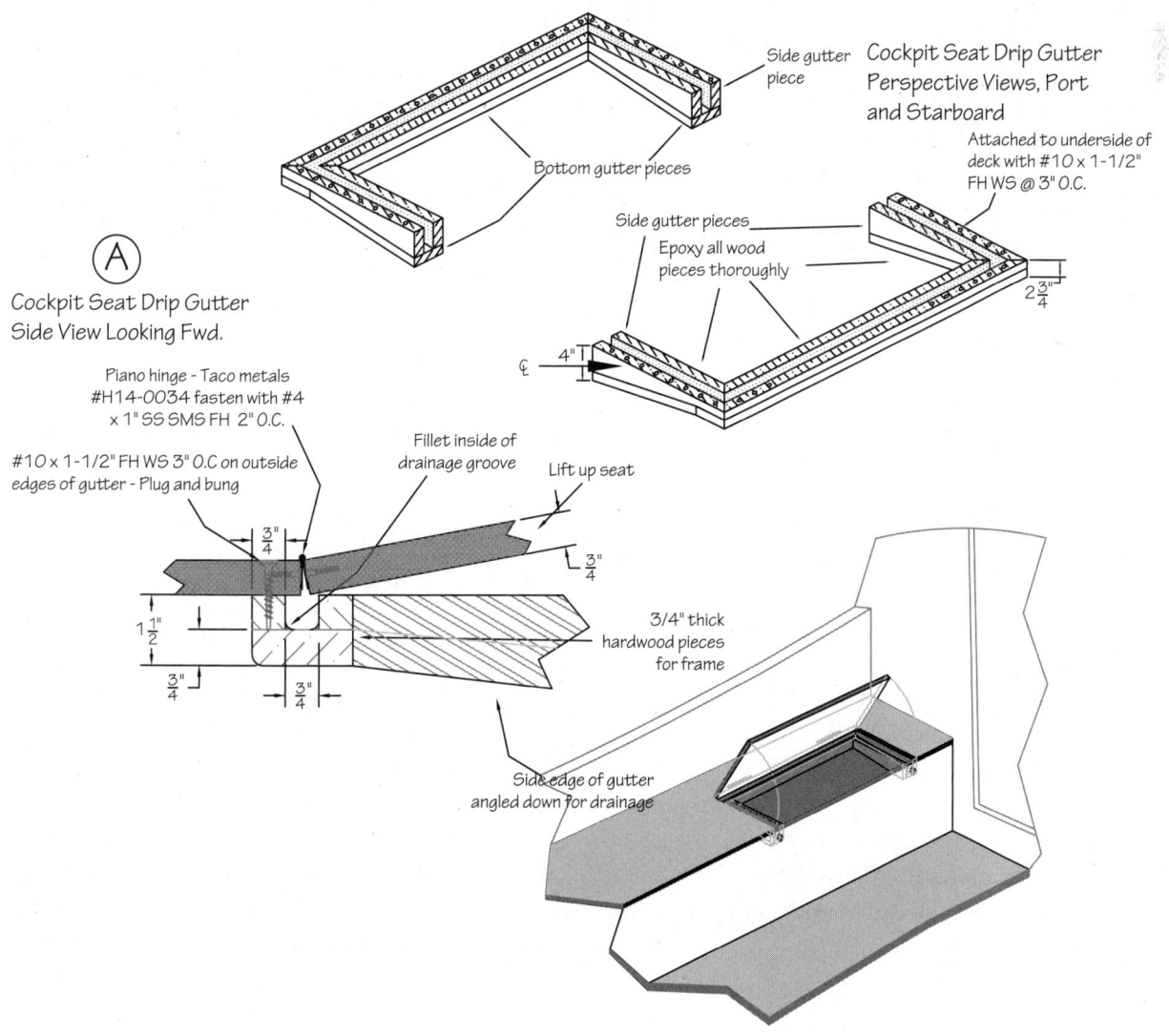

Figure 23-6. *A Surf Scoter's cockpit longitudinals form both the support for the sole and reinforcements for the outboard mount.*

planking your boat. Some advance planning is in order, however, before you begin measuring and cutting. Depending on what systems you plan to install in your boat, you may need to adapt access for their components. Numerous items may have to fit into the spaces within this structural grid: fuel tanks, water tanks, electric bilge pumps, batteries, ballast, and even potentially the engine. It's a very good idea to buy in advance or at least spec out all these so that you know their exact dimensions and where each will take up its residence. Some items may be easier to install before completing and closing up the compartments.

BERTH FLATS

Berth flats are like large, flat, horizontal bulkheads or longitudinals (Figure 23-7B). In most Stitch-and-Glue boats, they are positioned between major athwartships bulkheads. Their outboard edges almost always contact the hull sides, while their inboard edges most often terminate atop longitudinal berth bulkheads. The berth flat, if it is fastened to the hull and to athwartships and longitudinal bulkheads, can add a great deal of strength to the boat. It helps to transfer racking strains in the boat's structure to other strong points in the hull.

Figure 23-7A. *The bottom of a cockpit bilge before epoxy sealing and installation of sole and seating.*

Figure 23-7B. *Berth flats with cutouts before installation.*

Figure 23-7C. *The entire bilge has been sealed and is ready for soles and berth flats.*

Figure 23-7D. *Cleats are bonded to the bulkheads and longitudinals to provide solid landings for the soles and berth flats.*

Figure 23-7E. *Installing cockpit sole panels.*

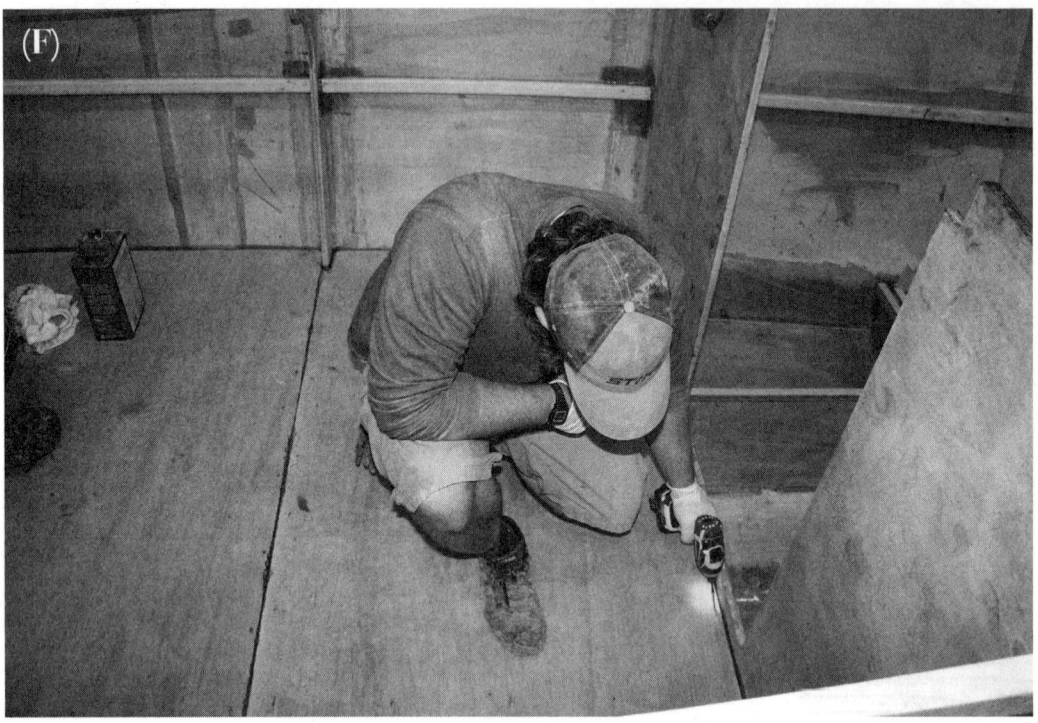

Figure 23-7F. *Bonding cockpit sole panels.*

Berth flats can be built of ½- or ¾-inch plywood. If the flat is also functioning as a settee in the main part of the cabin, opt for the heavier stock. If the flat is in a forward V-berth area in a small boat that will be used only for sleeping, ½-inch plywood will be more than adequate.

The berth flat should be bonded to any surrounding structures with epoxy fillets and taped joints to make it an integral part of the boat's structure. You can gain access for stowage and ventilation through the longitudinal bulkheads with pigeon holes (holes in the vertical bulkheads or longitudinals), or you can close them off with simple doors or through rectangular cutouts in the horizontal flats. Again, always be sure to provide adequate ventilation. Never build a compartment that has no access point or ventilation into it. These access points do not need to be watertight and can have looser fitting duckboards or hatches for access.

CEILINGS

In boat language, a ceiling is not the panel over your head but an inner liner for the hull sides. Open boats almost never bother with ceilings, but cruising boats with cabins often have them. If your boat is to have a simple workboat appearance, just painting the inside of the hull will be fine. If you aspire to more of a yacht-y mood, a ceiling will add a considerable touch of class to even a small boat, and it has some insulation value (Figure 23-9). In a Stitch-and-Glue boat, it is not very difficult to build, and it doesn't have to add much weight.

The ceiling should be spaced away from the hull with vertical battens on centers 16 to 24 inches apart. The ceiling itself will be longitudinal strips of dimensional lumber resawn thin enough to bend easily with the contours of the hull, perhaps ³⁄₁₆ to ⅜ inch in thickness and usually 2½" to 3" widths.

Figure 23-8. *The cockpit sole is now installed and ready to bond to the hull sides and bulkheads with epoxy fillets and glass tape.*

Figure 23-9. *The interior of the Pyladian 31 has a yellow cedar ceiling and fiddles to hold the berth cushions in place.*

ceiling would otherwise meet the bottom hull panels or a narrow space between each strip. The ceiling should not be permanently glued to the battens. Use small stainless steel or bronze screws so that the strips can easily be removed for hull inspection or repair.

FIDDLES

Fiddles retain things that are stowed on shelves or flats, be they cushions, books, or pots and pans. They also trim, finish, and accent a boat's interior.

I use dark hardwood for fiddles, and I don't usually epoxy-seal or varnish them because over time they get scratched and gouged as the boat

Vertical-grain fir is attractive and light in weight and color, though fairly expensive. I personally favor Port Orford or Alaska yellow cedar. Even some of the mahoganies might be used, albeit they would not add much brightness to the boat's cabin. Be sure to allow a ventilation path for air to pass behind the ceiling, such as an open slot where the

is used. If you use a blond-colored wood, deep scratches will generally stain dark and become very noticeable, which is the most compelling argument for using darker hardwoods. Finish and protect the wood with several coats of Deks Olje®, SeaFin®, or a similar oil. A few days after applying the last coat of oil, I apply a coat of Trewax furniture wax and buff the pieces. This finish looks a lot like hand-rubbed varnish but can be easily touched up with more oil and wax. An annual waxing keeps the fiddles looking soft, smooth, and shipshape. And by using darker mahogany-type hardwoods, the inevitable gouges and scratches that will come with use and age won't be as noticeable as they might in a light-colored wood.

I attach fiddles with countersunk holes and stainless steel sheet metal screws 8 to 12 inches apart. Fill the countersunk holes with wood bungs secured with quick-setting epoxy or a cyanoacrylate adhesive (instant glue) of the gap-filler type, then sand them smooth. You can buy ready-made bungs of various woods such as teak and mahogany, but they are easy to make yourself with a drill press or cordless drill and a tool called a plug cutter, using scrap wood from the fiddle stock for matching color and grain. When installing, make sure the grain in the bungs parallels the grain in the fiddle. The parallels are less distracting to the eye, and more importantly, they form a permanent testimony to the care you've taken in building your boat.

Figure 23-10. *Bungs are wooden plugs glued into the countersinks over mechanical fasteners. When used on wood that is to be finished bright, such as this caprail of the Kingfisher 33, take care to align each bung's grain with the wood around it.*

Figures 23-11A–B. *After the bung is glued into place you can trim it with a pull saw or chisel and sand it smooth with the surrounding surface.*

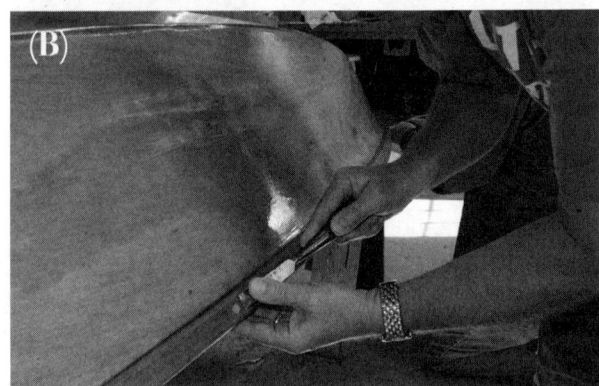

MORE ON ACCESS AND VENTILATION

The ability to access every section and compartment of the boat is important for inspection, maintenance, and repair in the event of hull damage. Complete access also helps in another critical area: ventilation. Basic rule: You can never have too much ventilation in a boat. The marine environment can range from chilly fog to steamy humidity. At either extreme, or anywhere in between, your boat has to deal with a lot of moisture. Provide multiple openings in every compartment for ventilation. For truly efficient ventilation, the vents should be in the ends of the compartment to establish an air-flow pattern, providing fresh air supply at one end and flushing stale air at the other, a breathing effect of sorts.

I use several methods to provide adequate ventilation and access. In vertical bulkheads, I cut large access openings called pigeon holes. The doorless pigeon holes not only provide good ventilation and access, but organize gear efficiently. Doors require ventilation slots with small bronze, stainless steel, or wood covers. If you're inclined to make your own vents, you can cut them from scrap ¼-inch marine plywood. Vent holes for wood covers can be gang-produced on a drill press with a sharp, bradpoint drill and screwed over a hole cut into the door or bulkhead. Occasionally, I cut out my bird-shaped logo, particularly if the boat has a painted interior. Or a fish or whale shape can look nice. You can make all sorts of decorative ventilation patterns; use your imagination.

For storage access to the horizontal berth and sole flats, use a rectangular cutout in the plywood with a couple of 1½-inch-diameter finger holes. Bolt a couple of wood cleats on at least two sides of the flat to support the cutout lid. Latches are unnecessary in most cases, since gravity and the weight of the cushions keep the cutouts in place. (On an offshore boat, however, I'd latch down the cutouts as you would not want them to open in rough seas while the boat is being tossed about.)

Every cabin should have at least two sources of fresh air. In a boat as small as our 15-foot Nancy's

Figure 23-12. *Bronze cowl vents on the Godzilli 16 allow air to circulate inside.*

China, they might include one cowl-type ventilator at the bow (which can also be used as a chainfall for the anchor rode) and a series of vent holes drilled in the companionway drop slides. A larger boat like the 22-foot Surf Scoter might want two cowl vents on the pilothouse roof and one cowl vent in the forward hatch. Two solar-powered ventilators would be even better; use one as an intake on the forward hatch and the second as an exhaust vent in the pilothouse roof.

Don't skimp on the ventilation system for your boat; provide for worst-case scenarios under different weather and wind conditions. And remember, a boat is very much like a house: A lived-in and used house keeps a lot better than one that is closed up and stale for long periods. So stir your boat's air frequently by using her.

HATCHES

Hatches are like large opening doors or windows into the decks, cabintops, or even the cockpits of a boat. In the world of boatbuilding, like the world at large, there are two kinds of people: those who do things the easy way and those who instinctively gravitate to the hard side of the spectrum. The easy solution for hatches is to buy commercial aluminum-frame units. They are easily installed, seal well, and usually have good, sturdy hinges.

Many have optional bug screens and even privacy shutters. Their main drawback is cost. Some purists also object to the presence of aluminum trim on a wooden boat, but I feel that the observer's eye easily looks past such manufactured hardware, as it does with anchors and outboard motors.

However, you can build your own wooden hatches. The advantages are lower cost and a more traditional look. The drawback is that you will spend more time both building and maintaining them.

One type is a simple hardwood box frame with finger-jointed or dovetailed corners, fitted with a thick and strong translucent Lexan (polycarbonate) or solid wood top. The box will be sized to fit snugly over a coaming that will keep rain and spray out of the opening.

If you choose a Lexan top, overbore pilot holes in the Lexan at least one drill size larger than the fastener diameter. Polycarbonate expands and contracts with temperature changes, and overboring the pilot holes allows the Lexan to move without cracking around the fasteners. Bed the Lexan in a polyurethane bedding compound. Applying a clear primer to the Lexan will help the caulking compound adhere to it. Set the screws with hand pressure evenly around the hatch frame. Fasteners every 3 inches is the minimum for bedding ¼-inch Lexan; ⅜-inch Lexan requires a fastener every 4 inches. If you don't like the look of the Lexan edge, trim it with a half-round of hardwood. If the clear hatch does not ensure sufficient privacy, light sanding with 220-grit sandpaper followed by a scuffing with Scotch-Brite pads will frost it, while still allowing plenty of light below.

For a solid wood top, build the frame first, routing its inside perimeter pieces with a rabbet bit. Set the hatch frame upside down over a piece of ½- or ¾–inch plywood stock and mark the outline of the frame. Cut along the outline, carefully dry-fit the piece, and epoxy the top into place.

I like to make a solid topped hatch cover more attractive by cutting a series of small grooves in the plywood top piece with a table saw and a kerf or dado blade. Gluing a series of pieces of contrasting wood into the grooves creates an inlaid hatch with

the watertight integrity of solid plywood. Carefully seal the hatch with epoxy and apply a minimum of six coats of varnish.

With minor variations, the same scheme can be used to build curved hatch tops. The saw kerfs allow the insert plywood to bend in an arc, and the contrasting wood splines lock the shape into a permanent curve. For hatch slides, dadoing pieces of dimensional hardwood and attaching some

Figures 23-13A–B. *Companionway drop slides and hatch on a Winter Wren sailboat.*

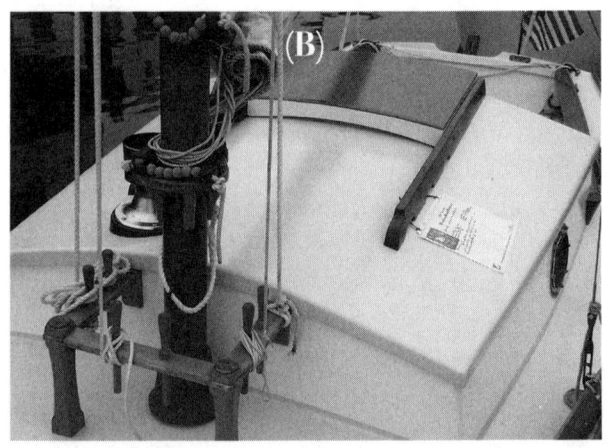

(A)

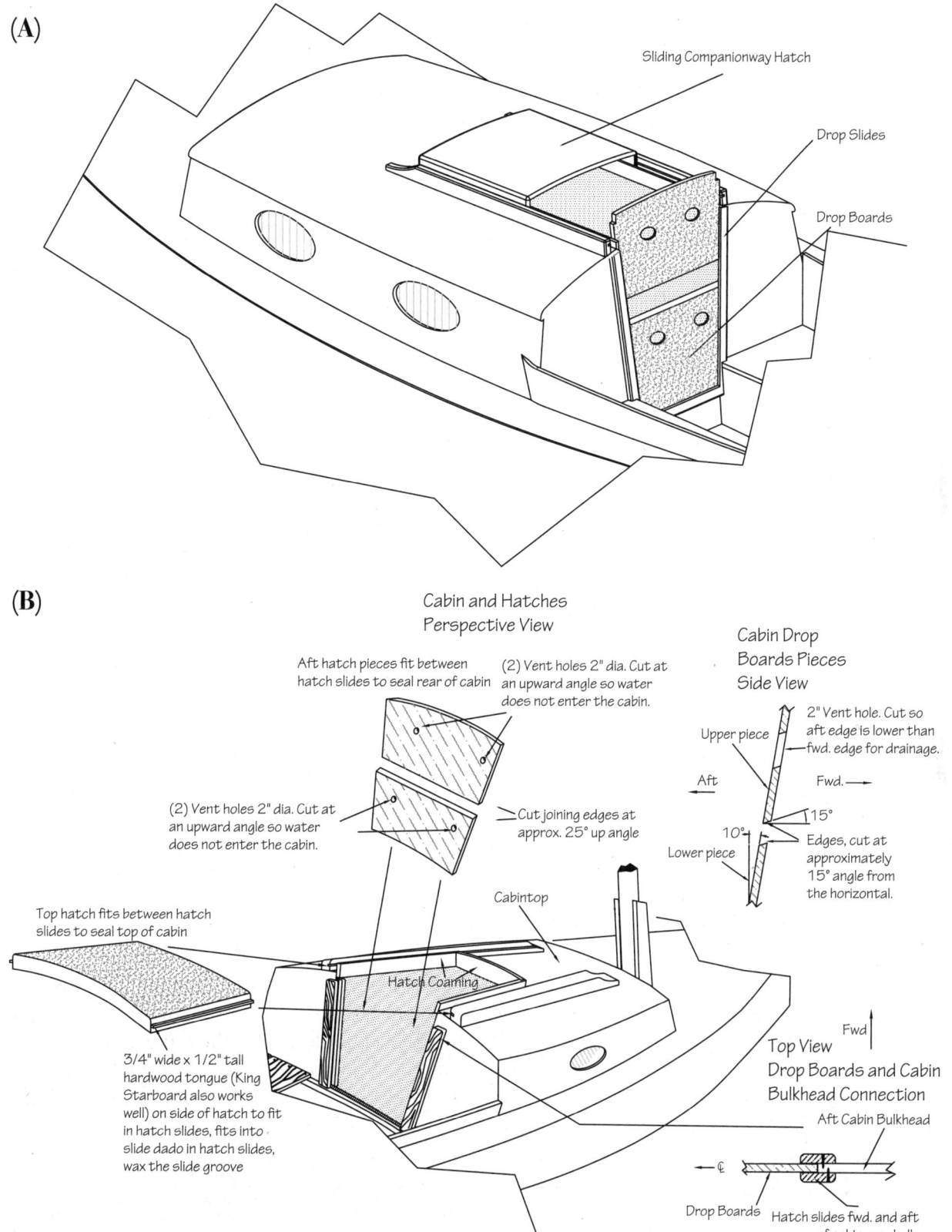

Sliding Companionway Hatch

Drop Slides

Drop Boards

(B)

Cabin and Hatches
Perspective View

Aft hatch pieces fit between
hatch slides to seal rear of cabin

(2) Vent holes 2" dia. Cut at
an upward angle so water
does not enter the cabin.

Cabin Drop
Boards Pieces
Side View

2" Vent hole. Cut so
aft edge is lower than
fwd. edge for drainage.

Upper piece

Aft   Fwd. →

15°

10°

Lower piece

Edges, cut at
approximately
15° angle from
the horizontal.

(2) Vent holes 2" dia. Cut at
an upward angle so water
does not enter the cabin.

Cut joining edges at
approx. 25° up angle

Cabintop

Top hatch fits between hatch
slides to seal top of cabin

Hatch Coaming

3/4" wide x 1/2" tall
hardwood tongue (King
Starboard also works
well) on side of hatch to fit
in hatch slides, fits into
slide dado in hatch slides,
wax the slide groove

Fwd

Top View
Drop Boards and Cabin
Bulkhead Connection

Aft Cabin Bulkhead

℄

Drop Boards   Hatch slides fwd. and aft
of cabin rear bulk.

Figures 23-14A–C. *These sailboat plan details illustrate construction of drop slides and hatch.*

(C)

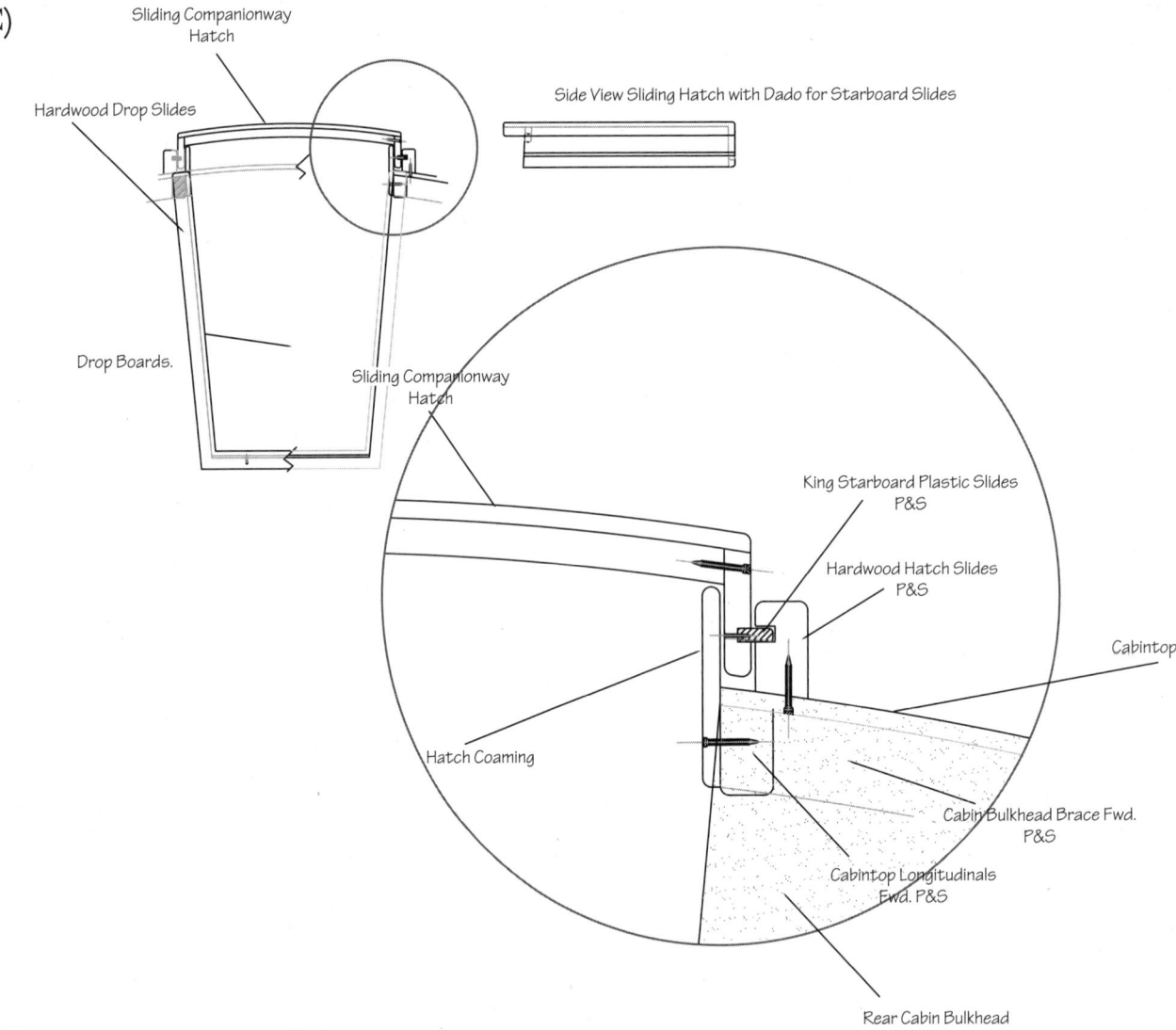

Sliding Companionway Hatch

Hardwood Drop Slides

Side View Sliding Hatch with Dado for Starboard Slides

Drop Boards.

Sliding Companionway Hatch

King Starboard Plastic Slides P&S

Hardwood Hatch Slides P&S

Cabintop

Hatch Coaming

Cabin Bulkhead Brace Fwd. P&S

Cabintop Longitudinals Fwd. P&S

Rear Cabin Bulkhead

Figures 23-14A–C. (Continued)

small slugs of UHMW (ultra-high-molecular-weight) or Starboard plastic will give you a smooth slider.

For hinged hatches on the foredeck and cabintop, use two sets of hatch hinges, one placed forward and one set aft. Hatch hinges have easily removable pins so that the hatch can be raised in either direction, allowing for the best ventilation by acting as a wind scoop.

To reinforce the hatch opening, I use a two-part carlin and coaming system. Before the decks are laminated into place, beams are installed to define the perimeter of the opening. The deck is then glued and fastened over the framing. After the decks have been sheathed, the hatch opening is cut out, a task easily accomplished if you first drill small pilot holes from below, up through the deck at each of the four corners. Frame the opening on the top side with a hatch coaming fastened directly into the sides of the framework. The hatch cover overlaps the coaming, and with a bit of foam-tape weather stripping on its bottom edges, it works very well. For offshore sailing, you might fit another wooden framework around the outside of

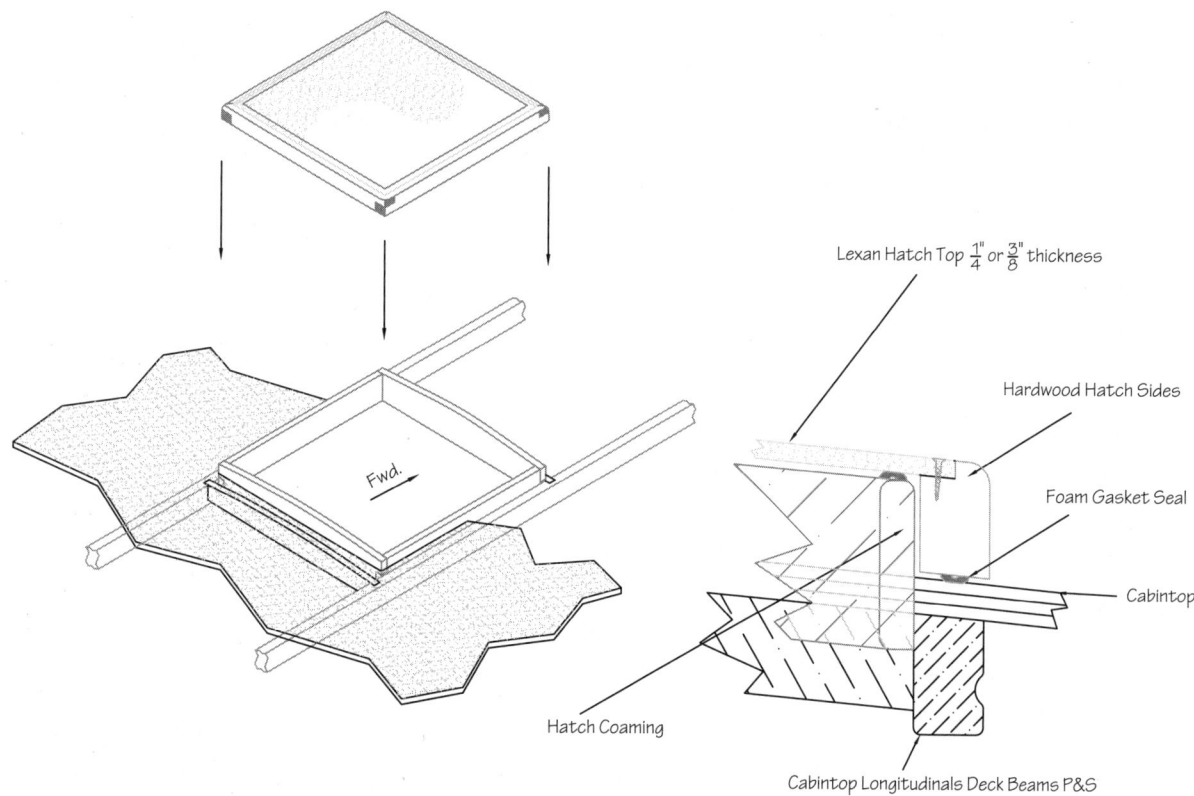

Lexan Hatch Top $\frac{1}{4}$" or $\frac{3}{8}$" thickness

Hardwood Hatch Sides

Foam Gasket Seal

Cabintop

Hatch Coaming

Cabintop Longitudinals Deck Beams P&S

Fwd.

Figure 23-15. *Details of a hinged foredeck hatch.*

the hatch; generously scuppered, this functions as a wave break and protects the hatch.

DOORS

For stowage areas too large or too visible for pigeon hole access, you may have to build cabinet doors. Most of the cabinets in our Stitch-and-Glue boats are built from sheets of plywood that run into corner blocks at their edges. This way, the plywood removed from the cutout in the cabinet front can be returned to its place of origin as a cabinet door. If you keep track of the cutouts, your doors will even match the grain of the surrounding bulkheads. Flat hinges seem to work best if they are screwed or bolted through the door casing.

For companionway doors, I stick to three choices: drop slides for sailboats, sliding doors for pilothouse boats, and hinged doors for boats

without room for a sliding door. Each has advantages and disadvantages. Drop slides for a sailboat companionway are probably the easiest, strongest, and safest alternative. Cut the door shape out of the bulkhead. Frame the opening with hardwood slides, tapering the sides of the opening slightly (at least ½ inch narrower at the bottom than at the top). Be sure to taper the interior edge of the hardwood strips to allow the plywood cutouts to slide easily; if you don't, the drop slides jam and stick.

Cut the drop boards into easily handled sizes, with a 25-45-degree bevel in the joints between boards to keep rainwater out of the boat. Be sure to scupper (drain) the bottom gutter so that the bottom slides don't sit in a pool of water.

On pilothouse boats, I make sliding doors whenever possible. They work well if the sliding apparatus is at the top of the door. When a door slides on a bottom track, it always seems to stick and jam.

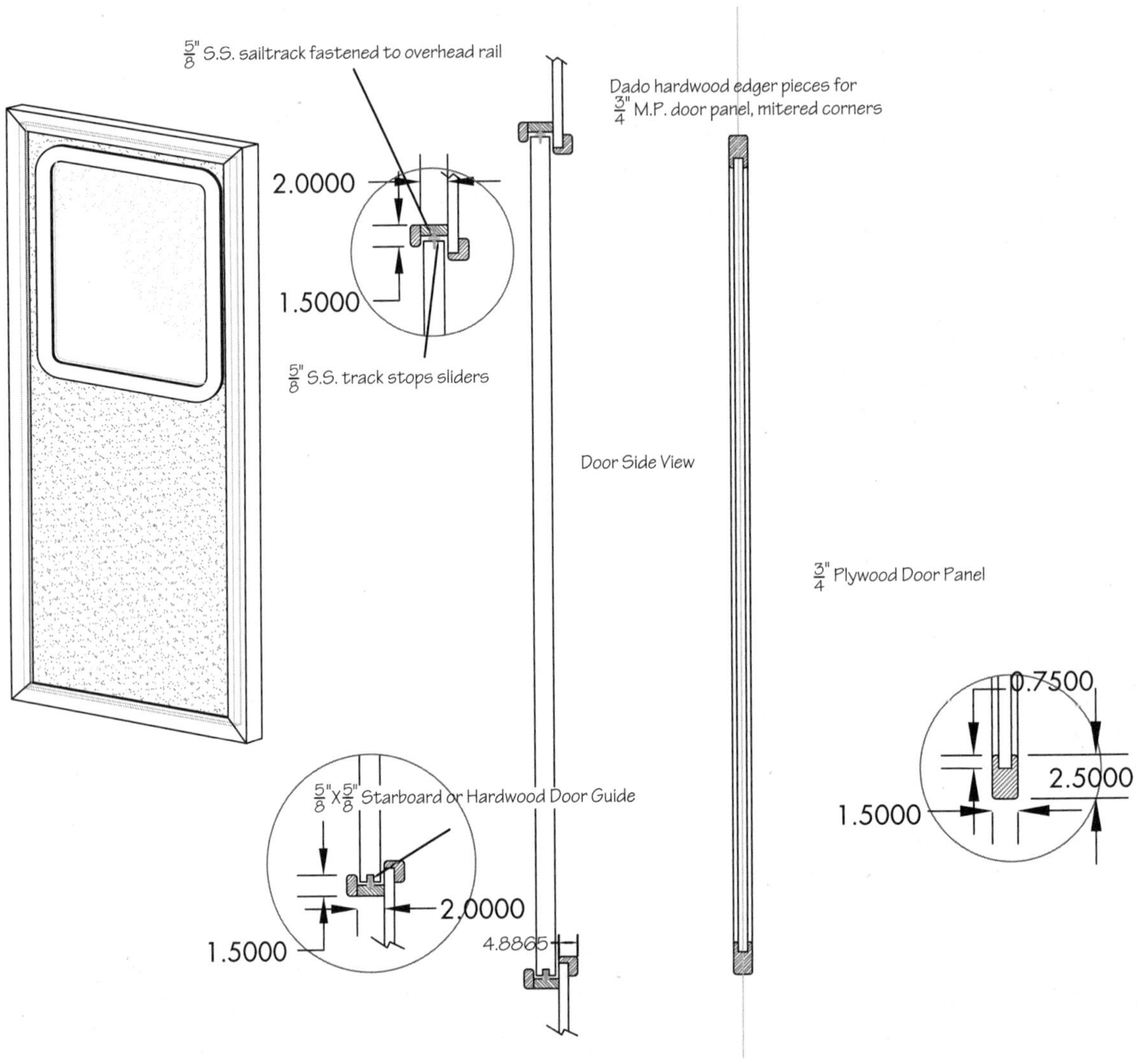

Figure 23-16 (diagram)

$\frac{5}{8}$" S.S. sailtrack fastened to overhead rail

Dado hardwood edger pieces for $\frac{3}{4}$" M.P. door panel, mitered corners

2.0000

1.5000

$\frac{5}{8}$" S.S. track stops sliders

Door Side View

$\frac{3}{4}$" Plywood Door Panel

$\frac{5}{8}$" X $\frac{5}{8}$" Starboard or Hardwood Door Guide

2.0000

4.8865

1.5000

0.7500

2.5000

1.5000

Figure 23-16. *Details of a sliding pilothouse door.*

With the sliding mechanism on top, the bottom track simply keeps the bottom of the door from swinging out. Usually, there's not enough beam in the boat to remove the door by sliding it out at one end, so make the door track removable by removing the screws in the upper sliding track. Again, use StarBoard® or UHMW plastic for the slider slugs, as they exhibit little friction in use and are easy to cut out and mill with standard woodworking tools. Use the bulkhead cutout for the door blank, and frame the door with 1½-inch-square stock dadoed out for the bulkhead thickness.

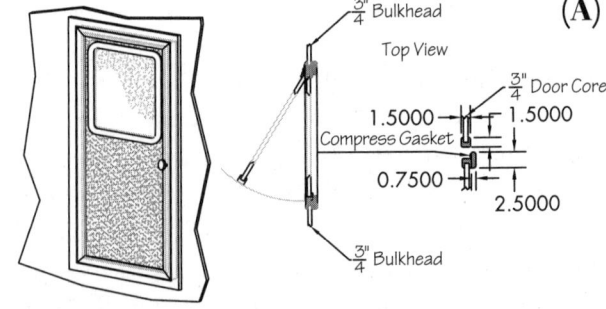

(A)

$\frac{3}{4}$" Bulkhead

Top View

$\frac{3}{4}$" Door Core

1.5000

1.5000

Compress Gasket

0.7500

2.5000

$\frac{3}{4}$" Bulkhead

Figures 23-17A–D. *This drawing and the following photos illustrate construction and installation of a hinged pilothouse door.*

Figures 23-17A–D. (Continued)

(D)

Figures 23-17A–D. (Continued)

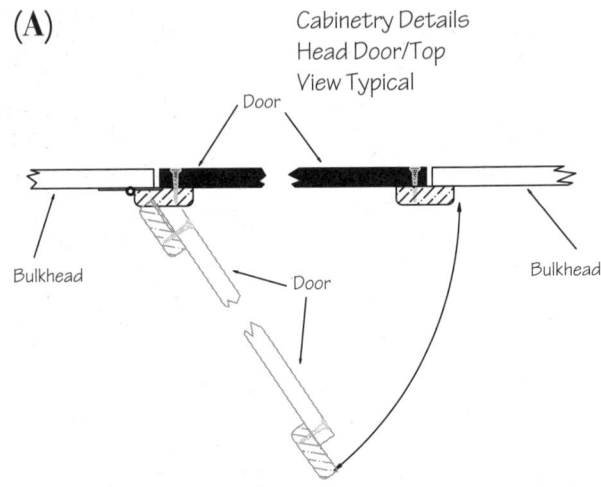

(A)

Cabinetry Details
Head Door/Top
View Typical

Door

Bulkhead   Door   Bulkhead

(B)

Figures 23-18A–B. *Construction and installation of an interior cabinet door.*

I make hinged pilothouse doors similarly to interior doors by building a hardwood jamb around the bulkhead door cutout. You want to be sure the finished door is larger than the cutout, so use a mitered trim around the perimeter of the cutout just as you might have for the interior cabinet doors. Hang the door with removable hinge pins to allow unshipping in nice weather (see Figures 23-17A–B).

Finally, even though Stitch-and-Glue boats are intrinsically simple, your detailed craftsmanship with these components, keeping a weather eye on the grain and color of the wood, can really show off your hard and meticulous work.

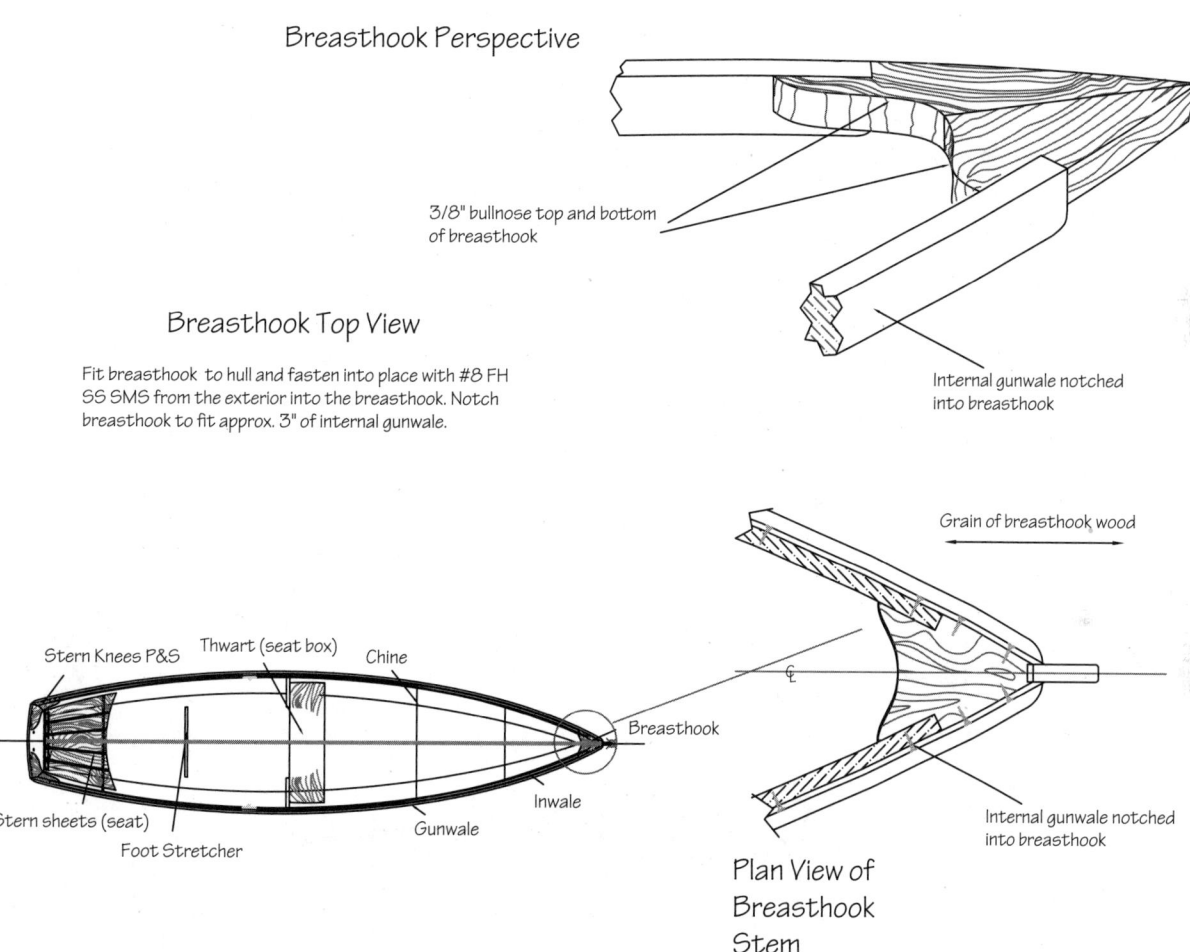

Breasthook Perspective

3/8" bullnose top and bottom of breasthook

Internal gunwale notched into breasthook

Breasthook Top View

Fit breasthook to hull and fasten into place with #8 FH SS SMS from the exterior into the breasthook. Notch breasthook to fit approx. 3" of internal gunwale.

Grain of breasthook wood

Stern Knees P&S
Thwart (seat box)
Chine
Breasthook
Inwale
Stern sheets (seat)
Foot Stretcher
Gunwale

Internal gunwale notched into breasthook

Plan View of Breasthook Stem

Figure 23-19. *A breasthook at the bow of an open boat can be a nice place to show off your woodworking talent.*

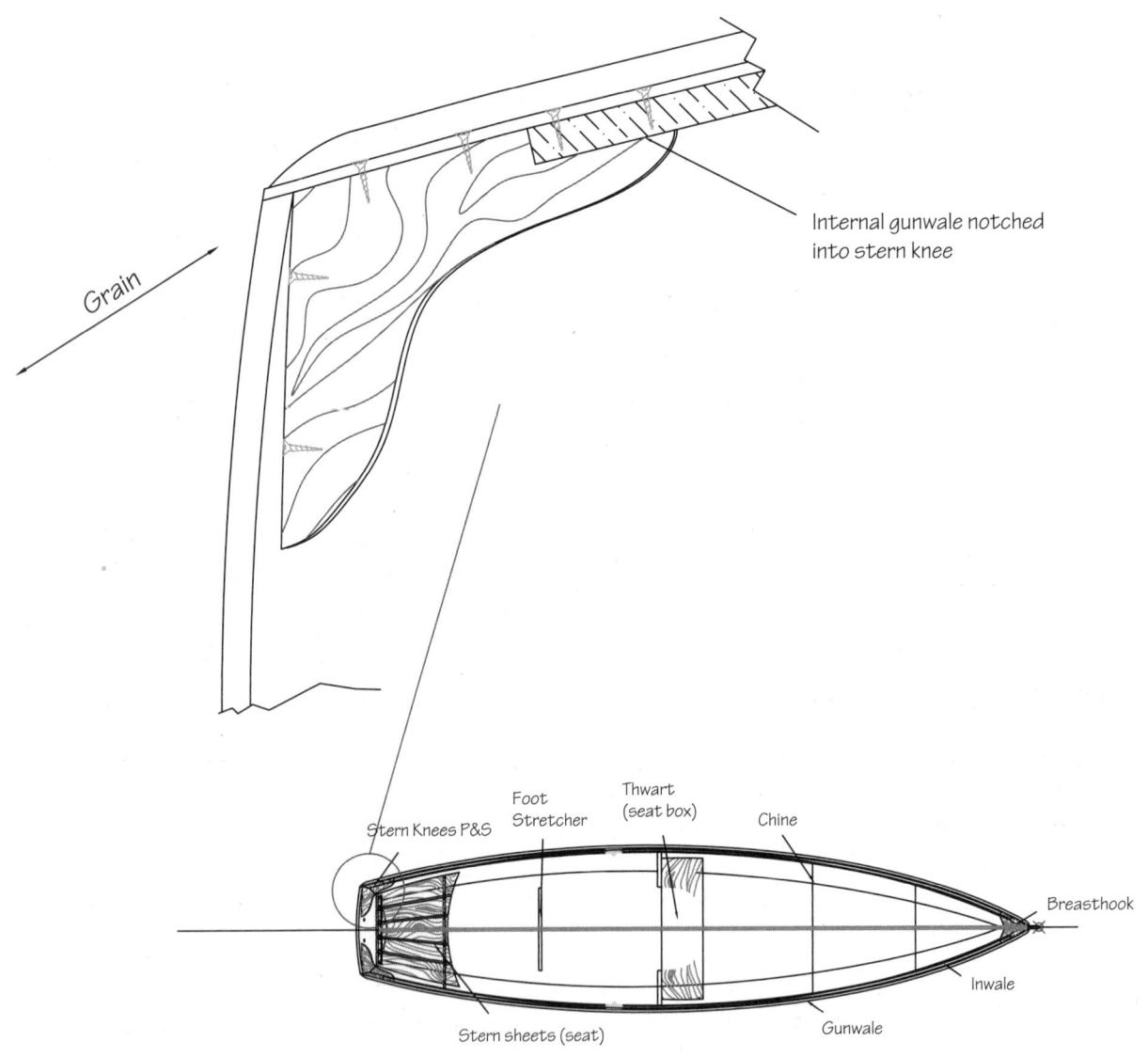

Figure 23-20. *Stern knees brace the transom of an open boat. Note the direction of the wood grain shown for greatest strength.*

CHAPTER 24

Carlins, Decks, and Cabins

If your Stitch-and-Glue boat is anything more than a small open daysailer or skiff such as the Candlefish 13 or 16, you have now arrived at the stage of constructing the deck and the framing necessary to support it (plus, importantly, to support people walking on it). Stepping up the ladder of ambition, your plans may also include a cabin trunk or pilothouse. All these structures will require some meticulous measuring and thoughtful planning, but they are easily approached. Take it one step at a time and you need not be intimidated.

Nearly every design that needs a deck, cabin, or pilothouse (let's call them DCPs for short) has structural bulkheads already in place in the hull. These will become the foundations of the DCPs. Assuming that at this point these bulkheads are in an untrimmed form that would poke through the plane of the future deck, the first task is to trim them down to the proper crown, or curvature, for the deck to land upon. This crown is essential for two reasons: It allows the deck to shed rain and splash, and it adds strength by holding the plywood deck form in tension. A dead-flat deck would need a lot of additional framing to make it stiff enough to walk on and handle the stresses of the boat in use. Framing adds weight, and every extra pound in the DCP structures raises the boat's center of gravity, which we don't want.

Most of my designs use a crown of ¼ or ⅜ inch per foot of beam (the width of the boat at the sheer), so for a boat with a maximum beam of 8 feet you would have either 2 or 3 inches of crown, respectively, at that widest point, depending on which ratio is specified. Where the beam narrows toward the bow or stern, the crown keeps the same proportion: at 4 feet of beam you would have 1 or 1½ inches of crown. If you live in metric land, ¼ inch per foot translates to 21 mm of crown per meter of width and ⅜ inch per foot to 31 mm per meter.

Marking the crown on the bulkhead tops is simple if you make a jig. Measure the widest part of the boat. Select a plank or plywood piece from your scrap pile long enough to span this distance, plus a couple of inches on either side, and tall enough to exceed the maximum crown by a couple of inches. Calculate the crown you will need to mark on the jig with the formula above. Let's say the maximum beam is 6 feet, so at ¼ inch per foot the crown will be 1½ inches. Mark the centerline and beam width on the pattern and measure down the distance from the top edge for the port and starboard corners. Set three small nails into the pattern at the peak and the corners. Take a relatively stiff batten stick that is longer than the pattern, bend it around the nails, and scribe this curve with a pencil. Cut along this line with a jigsaw or bandsaw, and true with your block plane to the marked line. This curved bottom edge can now be placed against each bulkhead as a pattern for marking the crown. Take care to align the pattern's centerline with the boat's centerline each time. If you haven't already dressed the tops of your sheer edges and sheer clamps, you will see that you need to do it now. Take a plane and bevel these hull edges to provide a fair and smooth landing for the deck along the entire length of the boat. When using a jig like this, I also recommend marking it as described above and then flipping the jig around so it is opposite from the side you originally marked from. You might be surprised to find the two scribed lines are not the same. Repeat this procedure for all the crowns you are marking on the bulkheads and average out the true crown line.

Once the top of each bulkhead has been marked (these lines indicate the underside of the deck to be installed), carefully trim them down to the lines with a jigsaw or circular saw, then do the final dressing with a block plane or small belt sander. Now is the time to consult your plans again and find the top-view (in architecture terms, the *plan view*) drawing that shows the half-breadths of the inside edge of the deck. This line would likely near-parallel the curve of the sheer as seen from above the boat and would lie several inches inboard of the actual sheer itself. It indicates the position of the carlins, longitudinal wooden members that we will let into the bulkheads. Mark their positions carefully, measuring symmetrically from the centerline out toward the edges, and not from the sheer in. Once these marks have been transferred to each of the bulkhead tops, I like to set a small wire brad at each mark and lay a long, stiff batten beside it on top of the bulkheads from the forwardmost bulkhead to the aftermost point (usually the transom). The idea of using the batten is to fair the marks for one single, sweet curve from bow to stern or the length of the DCP. Your faired marks might differ slightly from the plan, but if there's any dramatic deviation, you will need to go back and recheck your measurements. This is similar to your panel loftings that you did previously.

The carlins will add support for the deck and provide platforms for us to attach the coaming, cabin sides, or a pilothouse structure. The size or scantlings of the carlins will be called out on your plans. Depending on the design, they may be laid flat and would be wider than tall, or they might be set vertical and would be taller than wide. They should be made of a softwood such as vertical grain Douglas fir or Port Orford or Alaska cedar. Take care to use wood that is straight-grained so that it will bend fairly (read *smoothly*) in both profile and plan views. Notch into each of the bulkheads the whole depth and width of the carlin and dry-fit for that smooth bend. Once the carlins have been let into the tops of the bulkheads, you will have your decks well-defined by the edge of the sheer on one side, by the inside edge of the carlin on the other, and by the tops of the trimmed bulkheads. More framing might be necessary as shown on the plans for your particular design, but the basic structure is no more complicated than the above description.

Now is the time to lay out your deck material, which normally will be marine plywood. On a small dinghy, you might use a single layer of ¼-inch ply, but on larger boats you will laminate two or more ¼–⅜-inch layers to enable the wood to bend enough to conform to the curves of the crown and the sheer, making a smooth and graceful curvature. I occasionally scarf the deck pieces, particularly when there are long, sweeping side decks. Butt-joining plywood for side decks would be difficult and awkward. However you do it, stagger any joints in each layer so that when the laminations are overlaid there are no common joints on top of each other. This will ensure a strong and stiff deck that will support crew weight.

I like to dry-fit the decks and screw them temporarily into the framing of the carlins, bulkheads, and any other deck or cabin beam reinforcements. You will want to climb inside the hull with a flashlight to check that the deck is making good contact everywhere with the sheer clamps, carlins, and all other frame pieces. Mark where you need to shave or possibly shim the supporting members.

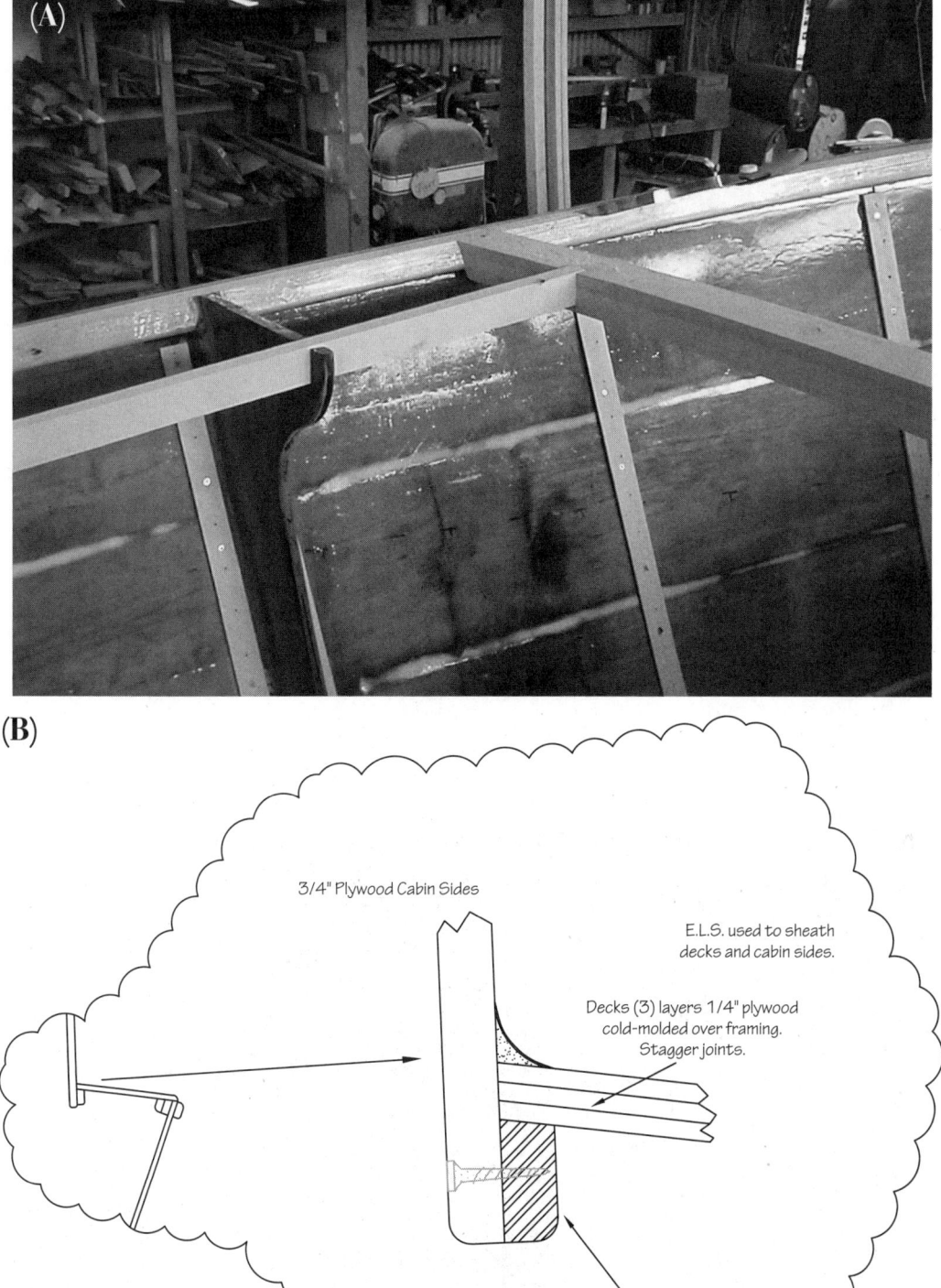

(A)

(B)

3/4" Plywood Cabin Sides

E.L.S. used to sheath decks and cabin sides.

Decks (3) layers 1/4" plywood cold-molded over framing. Stagger joints.

Cabin Side Carlins 3/4" x 1-1/2" cleat. Fasten 6" O.C. #8 x 1" SMS over epoxy.

Figures 24-1A–C. *Carlins provide landings for side decks, cabins, or pilothouses, and tie the deck-support structures together.*

Figures 24-1A–C. (Continued)

Figures 24-2A–C. *Side decks are glued to the sheer clamp and carlins. Epoxy squeeze-out should be cleaned up thoroughly while still wet.*

Figures 24-2A–C. (Continued)

If you have very small air gaps in some places, you can use extra screws to clamp down the deck as you glue it, but any significant gap—let's say more than $\frac{1}{16}$–$\frac{1}{8}$ inch—must be addressed or your screw clamping will distort the deck. One builder of my Zephyr design ended up with what looked like the power bulge of a Corvette's hood, suggesting the only sailing dinghy in existence that might be sporting a giant V8 under its forward deck. Be certain you have a symmetrical, fair, beautiful curve to the deck before proceeding to fasten it down.

While you have the deck pieces temporarily fastened for the dry-fitting, consider whether you will want to paint the underside of the deck. If you do, pencil off all the areas where the clamps, carlins, and beams will contact the deck. Remove the deck and thoroughly seal the underside with two coats of epoxy. Then, if you're painting, mask the contact areas and take care of the painting off the boat before the permanent deck installation. Sealing or painting overhead in the cramped space of the hull would be neither neat nor fun.

Now you can fasten the lower lamination of the deck very securely to the sheer clamps, carlins, and bulkheads, and any other framing with thickened epoxy and flat head screws. Don't skimp on the fastenings; a screw every 6 to 8 inches is about right. These should all be stainless steel or bronze screws as they will stay in the structure. Be sure to clean all the squeeze-out everywhere underneath with a sharpened tongue depressor and finally an alcohol-and-rag wipedown. After all the laminations are in place and the glue has cured, trim flush any overlap or overhangs at the carlins and the sheer. Fasteners in the final layer of the deck either need to be set deep and covered up with epoxy wood/flour bungs, or you can use temporary fasteners and pull them out after the epoxy sets and then methodically fill the holes with epoxy.

At the sheer where the deck joins the topsides of the boat, the exposed end grain at the deck edges should be sealed with several coats of epoxy and then fortified with several layers of fiberglass, Dynel, or Xynole cloth lapped over the edge. This is a critical step because this joint is subject to a great deal of stress and chafe, and any break in the sheathing is an avenue for the plywood to eagerly wick moisture into its exposed end grain. Dynel or Xynole are preferable because they will stretch before they fracture, while fiberglass cloth will not. In my shop we always sheathe these edges with at least two layers of Dynel (Figures 24-3A and 24-3B).

At this point, you are ready to define any structures that need to land on or clamp to the carlins and the inside edges of the deck. It is worth spending some time to make mockups before committing to permanent installations. The cockpit coaming, for example, serves as the backrest for everyone sitting in the cockpit, so its angle and height are important to test for comfort. You can experiment by temporarily clamping scrap pieces of plywood in place. A cabinside or pilothouse side mockup can be quickly and cheaply built with lauan door skins and cardboard so you can test the style and placement of portlights or windows. These patterns can then be transferred to your marine plywood for the final assembly.

I should address a delicate topic here. Some boatbuilders have been known to be tempted to raise cabin heights for more headroom, alter angles of cabin sides, and otherwise tinker with the original design. Proceed with great caution here. The design was done with great care and consideration

Figures 24-3A–B. *Dynel or glass cloth sheathing over the hull-deck joint is essential.*

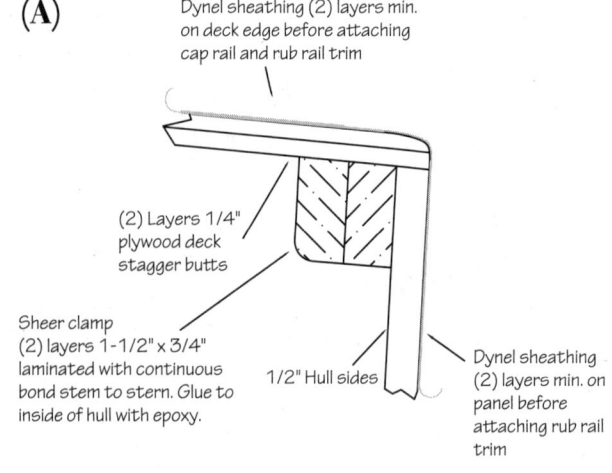

(A)

Dynel sheathing (2) layers min. on deck edge before attaching cap rail and rub rail trim

(2) Layers 1/4" plywood deck stagger butts

Sheer clamp (2) layers 1-1/2" x 3/4" laminated with continuous bond stem to stern. Glue to inside of hull with epoxy.

1/2" Hull sides

Dynel sheathing (2) layers min. on panel before attaching rub rail trim

(B)

Figures 24-3A–B. (Continued)

for aesthetics and function. A taller cabin adds windage and almost always ruins the proportions of the design. There were good reasons for the designer to specify the original proportions, so tread warily. This isn't a designer's ego speaking here; it's experience—and a concern for the builder to end up with a good boat.

In most cases, you will start with the dimensions for the coaming and cabin or pilothouse sides on the building plans, making small adjustments as necessary after experimenting with a mockup. These pieces will also be made with plywood. If the plan specifies curved cabin sides, they can be laminated in place on the boat with multiple layers of ¼-inch plywood, or they can be sprung into place using thicker plywood panels, typically following the curvature of the carlins. But most often the curves in top view are gentle enough that

full-thickness marine plywood can be used and the most you might need to do is to scarf up longer sheets to cut the parts from. Just as with the hull panels, the longer and smoother the bend of the panel, the more fair and smooth the line will look from all angles.

Most often the cabin and pilothouse sides will drop into slots cut into the tops of the bulkheads along the inside edge of the fitted carlins. I cut these new slots after gluing in the carlins, not before, as I want the fit of the carlins to be as accurate as possible and not sloppy. Dry-fit the parts, making sure their angles of lean on the port and starboard sides are identical. (Coaming sides normally lean outward and cabin sides inward.) To ensure symmetry, it is a good idea, to clamp temporary props or battens to these pieces as you glue them to the carlins and to leave them in place until the glue cures. Use plenty of epoxy to seal that critical deck edge where it meets the new part, and again, clean the excess meticulously with an alcohol-wet rag. These are very difficult places to sand later if you have worked sloppily.

Your sailboat cabin will normally have either portlights (which open for ventilation) or deadlights (which do not). A much taller pilothouse on a motor cruiser or motorsailer will enjoy airy windows that welcome the daylight and provide visibility for the skipper and crew. In either case, I find it better to cut the openings with a jigsaw after the cabin sides are in place because the plywood sides will bend more uniformly without these cutouts.

After the epoxy has cured, you are ready to build the rest of the structure of your cabin or pilothouse. At the upper edge of each side and the front and back bulkheads, I like to make a cleat, typically ¾ × 1½-inch hardwood, to serve as a secure landing for fastening the roof. This can be on the outside where it will look like a molding, or on the inside if you would rather add a decorative batten later to the exterior (Figures 24-4A, 24-4B, and 24-4C).

There can be a bit of ticklish business at the junctions of the cabin sides with the forward and aft bulkheads, especially if those bulkheads actually form the front or back of the cabin. The corners must be strong enough to withstand torsional stresses, but they are also important architectural details. You can make hardwood rabbeted corner blocks for the sides and bulkheads to slot into, which are very strong and good-looking but may take a considerable bit of cut-and-try fitting. Or you can just allow the two structures to meet, epoxy-gluing both faying surfaces, and then you can strengthen the joint using several layers of fiberglass tape over a heavy fillet of wood flour and epoxy. This composite joint, honestly, works best here for the same reason that it performs critical service in the chines and keel junctions in the hull. Obviously, the hardwood corner blocks are essential if you are going to varnish your cabin inside or out and show off the wood. My inclination these days is to paint and leave the fine woodwork for a few trim and accent pieces (Figures 24-5A, 24-5B).

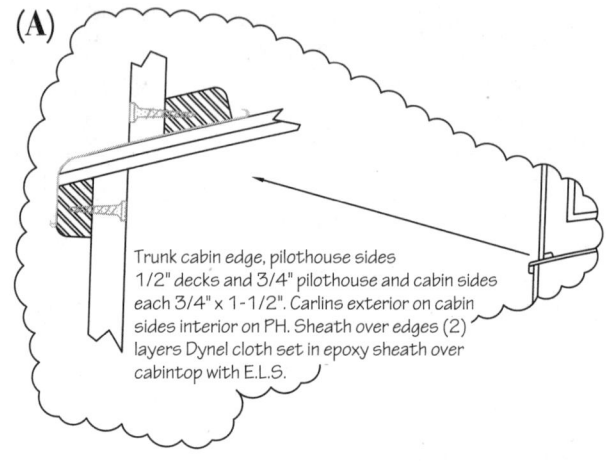

(A) Trunk cabin edge, pilothouse sides 1/2" decks and 3/4" pilothouse and cabin sides each 3/4" x 1-1/2". Carlins exterior on cabin sides interior on PH. Sheath over edges (2) layers Dynel cloth set in epoxy sheath over cabintop with E.L.S.

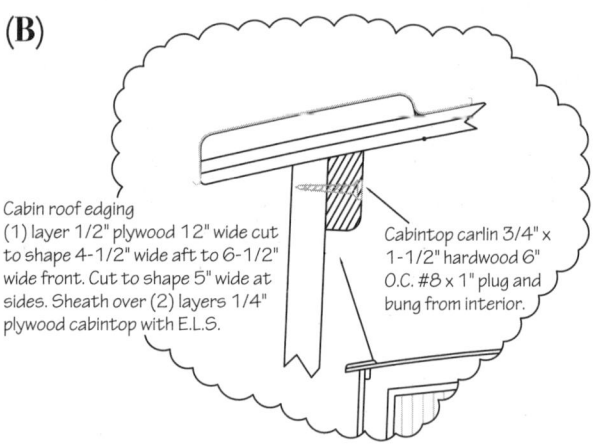

(B) Cabin roof edging (1) layer 1/2" plywood 12" wide cut to shape 4-1/2" wide aft to 6-1/2" wide front. Cut to shape 5" wide at sides. Sheath over (2) layers 1/4" plywood cabintop with E.L.S.

Cabintop carlin 3/4" x 1-1/2" hardwood 6" O.C. #8 x 1" plug and bung from interior.

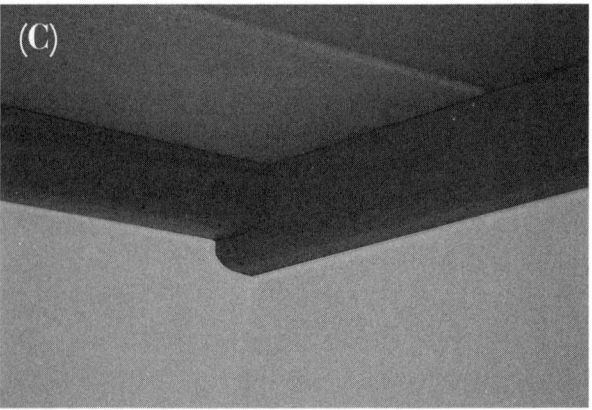

Figures 24-4A–C. *Details of cabin side attachments, roof landings, and interior frame joinery.*

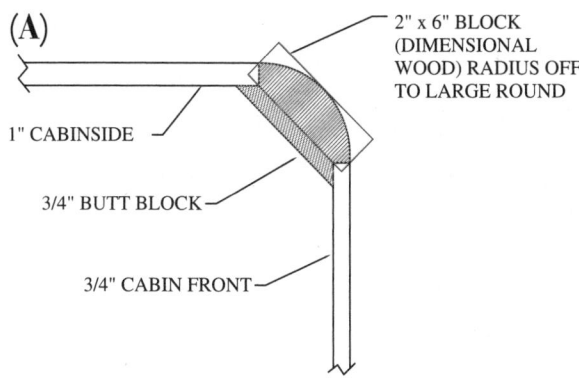

(A)

2" x 6" BLOCK (DIMENSIONAL WOOD) RADIUS OFF TO LARGE ROUND

1" CABINSIDE

3/4" BUTT BLOCK

3/4" CABIN FRONT

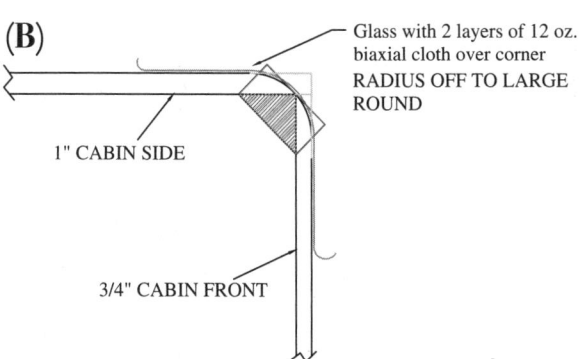

(B)

Glass with 2 layers of 12 oz. biaxial cloth over corner RADIUS OFF TO LARGE ROUND

1" CABIN SIDE

3/4" CABIN FRONT

Figures 24-5A–B. *Here are two methods of executing cabin corner joints.*

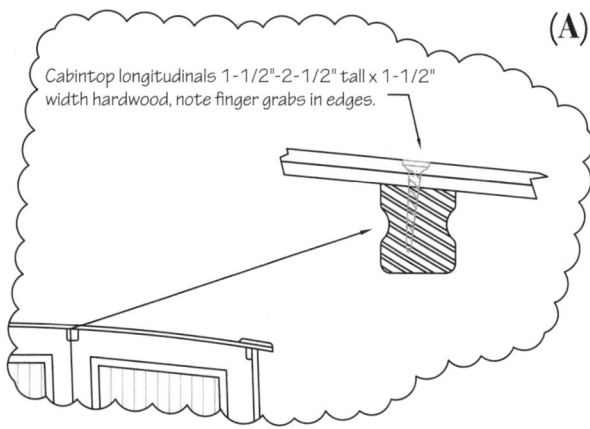

(A)

Cabintop longitudinals 1-1/2"-2-1/2" tall x 1-1/2" width hardwood, note finger grabs in edges.

Figure 24-6A. *Cabin or pilothouse longitudinals provide strong roof support and define the roof's curvature.*

(B)

(C)

Figures 24-6B–C. *A roof can be constructed without athwartships beams to reduce the potential for head-banging.*

The cabin or pilothouse will also have either athwartships (sideways) or longitudinal (running lengthwise) framing to support the roof. These frames are typically let into the cleats you have attached at the upper edges of the sides. Longitudinal framing is usually a bit heavier in scantling than athwartships framing as usually fewer of the longitudinals are fitted. A typical scantling for longitudinal framing would be hardwood, 1½ × 2½-inch high. Athwartships framing must bend to an arc, so you will laminate it on a mold that mimics the crown of the cabin top. Consult your plans for scantlings of these frames (Figures 24-6A, 24-6B, and 24-6C).

After the roof framing is completed, the cabintop or pilothouse roof can be laminated from multiple layers of marine plywood. If scarfing is necessary, the overlaps or joints should be staggered, just as we did with the deck. For pilothouses

I like the roof to overlap the sides by at least a couple of inches and as much as 12 inches in front and back, like a hat brim. To give this "brim" an elegant scale in keeping with the overall design of the boat, I will often laminate on the top an extra layer of ½-inch plywood about 5 to 8 inches in width all around the perimeter. This also serves as a valuable stiffening of the edge and makes the roof edge look heavier than its actual thickness. Attach the roof to the cabin sides and supports just as you did the deck, taking the same care to avoid unglued gaps and to clean up errant epoxy. Wrap the edges with a couple of layers of Dynel or Xynole as you did the deck.

If you are building a pilothouse with standing headroom, your basic structure is now complete. If you are building a sailboat's cabin, you now have the task of constructing a sliding hatch. This can be a tricky operation. Sliding hatches vary widely in design and hardware, and I am not aware of any

Figures 24-7A–C. *The "Devlin-style" pilothouse brow adds attractive thickness to the pilothouse edge without much extra weight.*

(A)

Perspective View
Pilothouse Roof

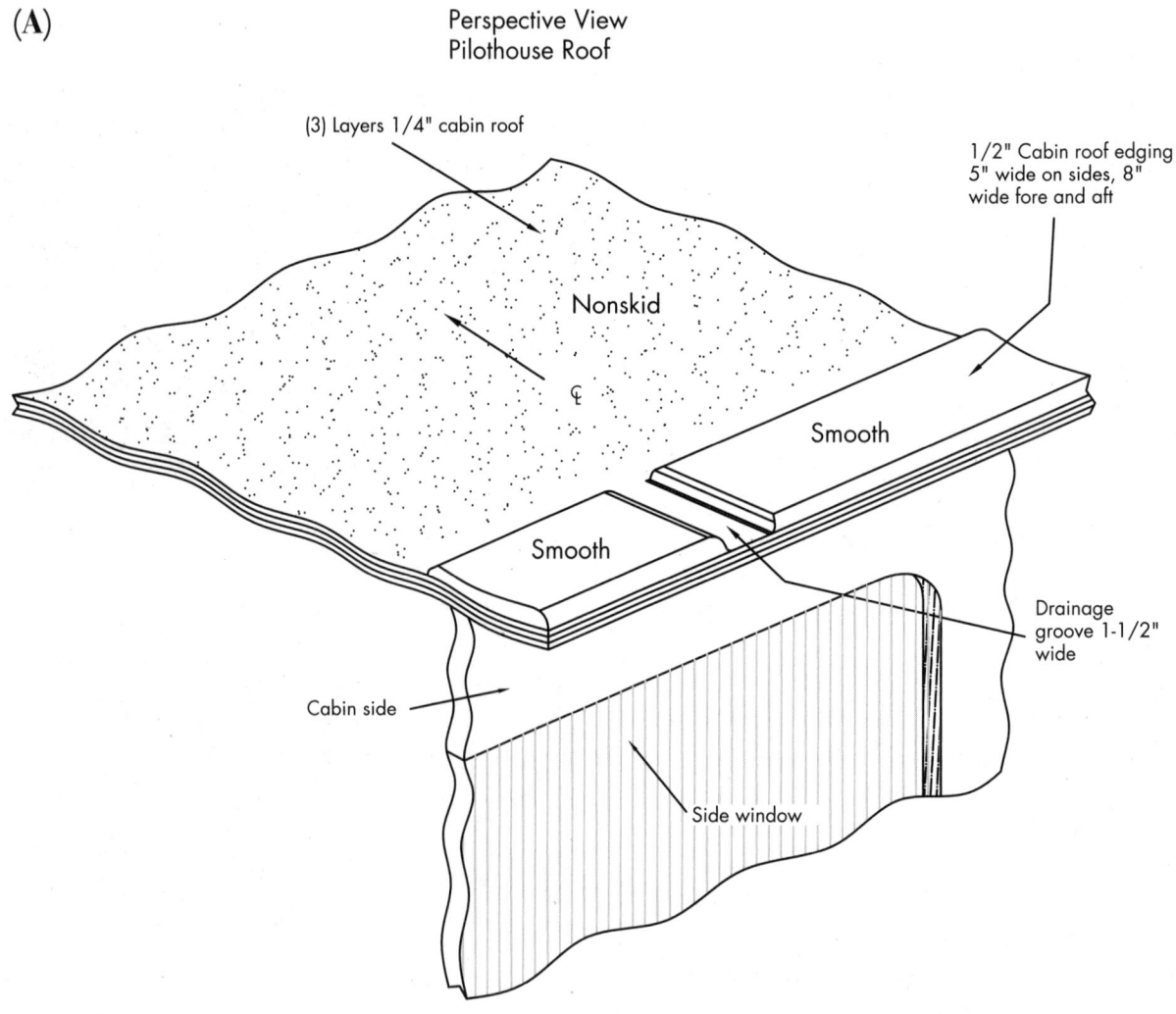

(3) Layers 1/4" cabin roof

1/2" Cabin roof edging 5" wide on sides, 8" wide fore and aft

Nonskid

Smooth

Smooth

Drainage groove 1-1/2" wide

Cabin side

Side window

(B)

(C)

Figures 24-7A–C. (Continued)

ready-made commercial sliding hatches available to the builder. I will outline only the basics of construction here.

The width of the hatch opening will match the width of the companionway opening at the top, and its length will typically extend 2 to 3 feet forward toward the cabin front. Since its lid will have to slide forward on rails mounted on the cabin roof, the length of the opening must be less than one-half the length of the roof. Consider any large accessories, such as an inboard motor or daggerboard, that may some day have to migrate in or out through this opening. I believe these hatch sliders mate best with longitudinal beam framing, and if you have athwartships cabin roof framing, it is important that you not cut out too much of it and weaken the roof structure. Once you are absolutely certain of the dimensions of the opening, you may take a deep breath, maybe mutter a hopeful prayer, and cut it out with a jigsaw.

Aligning the rails alongside the opening is the most exacting part of the operation. They must be perfectly straight, lie absolutely parallel to each

Figure 24-7D. *The pilothouse finishes off very attractively.*

other, stand plumb on their inner faces, and fit the crown of the roof. They are best made of hardwood, a bit more than twice the length of the hatch opening, and joined just ahead of the hatch opening by an athwartships wooden piece that will serve as a dam to keep water from splashing into the cabin. This dam piece will be shaped to follow the crown of the roof and must be just low enough to allow the hatch to slide over it.

Depending on how you make the channels that enable the hatch to slide, you might be able to use the section of the roof laminate that you cut out for the top of the hatch itself, or you might have to laminate a new, overlapping one that matches the roof crown. Some builders use this as an opportunity for a crafty piece of furniture, making a carapace with varnished tongue-and-groove boards. If I were tempted to do this, I would use stable laminations of plywood as a base; a cover that might swell or shrink is certain disaster. There are any number of ways to make the sliding components themselves with hardwood, aluminum, brass, UHMW plastic, or StarBoard®. At this writing, you can download an inexpensive electronic back issue of *WoodenBoat* magazine (#67, November/December 1985) that contains helpful instructions and drawings. The two most critical considerations: The hatch must not bind and it must not leak (take a look back to pages 245 and 246 and the illustrations Figures 23-14A–C for detailed drawings of these hatches and slides).

CHAPTER 25

Painting

Boat painting, along with its equally fusspot cousin, varnishing, could easily be the subject of an entire book. It's a complicated process with endless swirls of controversy. But in this chapter I will try to uncomplicate it as much as possible and distill it to the best practical advice you can use as an amateur builder.

You will first need to decide: What to paint, what to finish bright (i.e., with varnish)? The first consideration here is protecting the epoxy sealing and sheathing of the exterior of your boat from sunlight's ultraviolet degradation. Epoxy has no built-in protection against UV exposure. According to one epoxy manufacturer (System Three), an epoxy sheathing exposed to direct sunlight will first dull, then turn chalky, and finally crack and delaminate. In the worst-case latitudes (the tropics), total breakdown of unprotected epoxy could occur at about 15 months. A coat of opaque paint provides this critical epoxy sheathing with the best protection. Several coats of varnish with ultraviolet filters (check the label!) will provide adequate protection but must be renewed annually. This sounds like a lot of work, which it is, and that brings me to the second consideration. How much time and will do you have available to invest in annual maintenance of your boat? Make a realistic and honest assessment before you begin to haunt the chandlery's varnish aisle. My predilection, when building a boat for my own use or for sale on spec, is to paint nearly everything so as to keep the maintenance as undemanding as possible. In the first 18-foot Pelicano bassboat we built a few years ago, we left only the windscreen frame and companionway slides bright. Still, I understand the inner urge to deviate from my advice. I do it myself.

Another decision: what color (or colors)? Again, I have a strong preference for light colors. They are less susceptible to fiberglass cloth print-through (as explained in Chapter 6), and they also protect the epoxy from heat damage. This is critical in hot, sunny climates, but it can be an issue even in our cool Pacific Northwest. Some epoxy resins have a heat deflection temperature as low as 145°F. This is the temperature at which the resin begins to move, or deflect, under stress. It can also have a glass transition temperature—the stage in which it begins to behave more like a rubber compound than hard plastic—of about 155°F and if your boat is painted a dark color you can easily reach these surface temperatures for weeks at a time in the summer months.

One of the boats I used to own, a 22-foot 8-inch Arctic Tern sailboat, was originally painted a light gray. After four years, the original paint was in good condition, but my curiosity got the better of me and I painted her a dark blue. She looked very smart.

A year later, during her annual haulout, I closely inspected the hull surface. In her gray seasons, the hull had been fair and smooth. Now I could see the pattern of the fiberglass cloth sheathing telegraphing through the paint. The only difference was the color, and one year of being warmed by the sun. And this was the benign Pacific Northwest sun.

I waited for an unusually warm day. When the outside temperature hit a shocking 92°F, I placed a thermometer on the dark blue topsides and got a reading of 142°F. I then placed the thermometer on a nearby surface painted off-white. It registered 113°F. Quite a difference.

If you insist on a dark hull (and here too, I have been caught occasionally violating my own good advice), try a simple test first. Prepare two identical, fiberglass/epoxy-sheathed plywood panels matching your hull surface. Build them exactly as you have built your boat, using the identical procedure for prepping the surface. Paint one with your proposed dark paint and the other with a light color, preferably white. Set these freshly painted panels in the sun, and using an ordinary meat thermometer or a handheld heat-sensing gun, test the surfaces. You may well see a 20-plus-degree difference. If the dark panel rises to 125°F or above, you may be risking not only the fairness of the surface finish but also the integrity of the structure.

THE RELIGION OF PRIMER

No matter what type or brand of paint I intend to use, I religiously follow one rule with Stitch-and-Glue boats: the first step for applying paint to my hull both inside and out is always a conversion primer.

A conversion primer offers more than simple assurance that the paints to follow will adhere better. Most single-part paints have a formulation that depends on part of their chemicals penetrating into the substrate of the surface. The rest of the active solvents that make them thin enough to allow application by brush or roller have to evaporate. These aromatic solvents are present in all painting systems except for powder-coating paints, which are 100% solids and could not be used for a Stitch-and-Glue boat. If the surface about to be painted has been

sealed with epoxy, it can't be penetrated, and the result is paint that might never cure properly. The conversion primer acts as a chemical bridge that will allow all types of paint (at least in my experience) to cure, and it prevents any unwanted and potentially catastrophic reactions with the epoxy. There are several brands available that work well; our shop uses Ditzler DP40 and Awlgrip 545 (both are epoxy primers) primers with amazing success rates. While the manufacturers recommend spraying, both can be successfully hand-applied with foam rollers. Two coats are generally recommended. After the first coat cures, we carefully go over the entire hull, filling any remaining blemishes and divots with an epoxy fairing putty. When we are finally satisfied with our filling and fairing, we sand the entire hull smooth with 150-grit paper. Then we apply the second and final coat of DP40 or 545. If any bumpy geography still appears after this, we can transition to using a polyester resin-based Bondo-type filler (these use a catalyst to trigger the curing process). The final preparation before applying topcoat paint is to make sure that no bare epoxy spots or bare filler putty patches appear without a coat of conversion primer sealing any porosity they might present. You must have a uniform and complete primer surface ready to accept the final coats of enamel paint.

At this point, it already sounds like the painting process is plenty involved and time-consuming. It is indeed, and this is the step in boatbuilding where you will be faced most directly with a personal decision: how (nearly) perfect must it be? There is no answer that fits every boat or every boat owner. Some people will be very satisfied with a 20-foot boat—that is, one that looks fine from 20 feet away but probably should not bear closer scrutiny. Our shop puts a lot of time into surface preparation and we get a very fine finish in return, but this level of effort isn't a moral standard that everyone should observe. And as you'll learn in the following sections, you can use certain tricks to disguise a slightly imperfect surface. Just don't stray from the religion of conversion primer.

Figure 25-1A. *This hull has been sheathed, sanded, and faired, with two coats of sealing epoxy applied.*

Figure 25-1B. *Epoxy primer is rolled on before the finish painting.*

Figure 25-1C. *Spraying a second coat of conversion primer.*

WHAT TYPE OF PAINT?

Following is a discussion of four paint systems in order of durability (from softest to hardest) and ease of application (easiest to hardest). All will work on a Stitch-and-Glue boat when applied over a conversion primer. Which of these four types of paints you choose will depend on what level of finish you want, how much durability you are aiming for, and whether you plan to hand-apply or spray. There really is no reason for an amateur boatbuilder to fear spraying; you can rent a compressor and spray gun to avoid the expense of buying one, and practice on some inconsequential pieces before you begin painting your boat. However, roll-and-brush painting is less hazardous if you're working in a space shared with other things stored in it (such as a car or your house utilities), and it's much easier to repair and touch up after the dock rash and dings that will inevitably pock your boat.

The first category is the *alkyd marine enamels*. These are single-part paints (you do not need to mix a hardener or catalyst for them to cure) that are in part oil-based. They dry extremely slowly through evaporative action of the active solvents. They are fairly easy to apply with good results from brushing (rolling and tipping) and fair results from spraying. Their drawbacks are that they are only moderate in durability, and they dry so slowly that they invite inevitable problems from collected dust and bugs.

Oil-modified polyurethanes are a single-part paint system with about the same ease of application as the alkyds. They dry a bit faster and are a little more durable. Their pigments are more finely ground than alkyds, thus making them a bit glossier. They might not work as well on traditional carvel-planked wooden boats because they can't expand and contract with the seasonal swelling and shrinking of the wood as well as the alkyd enamels, but this is not an issue with the much more stable Stitch-and-Glue construction. They may be sprayed or brushed.

Acrylic urethanes are next up on the scale. These are two-part paints, needing to have the components mixed in precise ratios just prior to painting. They offer good durability but limited versatility in that they need to be sprayed for best results. They are essentially automotive paints. They work very well on Stitch-and-Glue boats and are used primarily in professional shops that need to paint quickly and efficiently. One option for the home builder is to complete the project through the primer stage, then hire an automotive painter for a house call and a few hours' labor for the final spraying. His or her years of experience may well be worth the cost if you're aiming for the finest possible finish. Another advantage of these paint systems is the ease of repair. Run or sags from the original application can be sanded out with fine grit sandpaper, and a final buffing can be applied. Patches might be sprayed on scuffed or damaged areas and likewise buffed to match the surrounding paint areas with good results.

Top of the line in marine paint systems are the *linear polyurethanes*, or LPUs. These are two-part paints and are durable beyond comparison to all others in this discussion. They can be brushed or sprayed, but either way they must be applied in thin coats; several coats are usually necessary for proper coverage. Frankly, they are very fussy in application and are so extravagantly glossy that they'll show every little imperfection in whatever surface they're covering. The pigments in true LPUs are very finely ground, and the resulting gloss can be almost mirror-like in quality, so surface preparation must therefore be meticulous. They may not be worth considering if you're building your first boat and are just beginning to acquire experience, but for the perfectionists among us, they are at the top of the heap.

LPUs contain irritating and possibly carcinogenic isocyanates, so they must be handled with great care. It's very important to avoid any contact with the paint and its vapors. We use disposable Tyvek suits, disposable gloves, and full face-mask respirators with an outside fresh air system. Home builders should at least use a respirator with fresh organic vapor cartridges installed and should take great care to keep any of these chemicals off their skin.

A few sentences above, I used the words "extravagantly glossy." This characteristic is not necessarily a virtue on boat finishes. Many owners prefer a matte finish because it is much more forgiving of imperfect surface preparation, easier to touch up and maintain, and in line with wooden boat tradition. Most marine paints are formulated as high gloss, but their manufacturers sell a liquid flattening agent designed to be used with their paint. Depending on the mixing ratio, it will transform the high-gloss sheen to a satin, matte, or flat finish. It will take some experimenting to determine which is to your taste. Since the flattener contains no pigment, it will increase the number of coats required to cover, something to consider, especially if you are painting your boat a dark color.

When you think you finally have prepared the boat's surfaces for painting, you still have to prepare the shop area around it. Any dust that is stirred up while the paint is being applied or drying will magically migrate to the painted surface. A very thorough vacuuming of the floor and any other surface that has accumulated dust is essential. Give the hull a final tack cloth wipedown *after* vacuuming the environment around it. (Prepackaged tack cloths contain a slightly sticky compound, and if wiped smoothly and without too much pressure over all the surfaces, they will remove most of the persistent surface dust.) Just prior to painting, we also wet down the shop floor with a light sprinkling of water, which helps to pull dust out of the air and aids in keeping human activity from launching more dust during the painting process. Even after all this preparation and even considering that I am painting in a buttoned-up shop, I still try to avoid painting on windy days. I have learned the hard way that painting in the spring with lots of pollen in the air can be like painting a boat with a nonskid paint!

If you are forgoing spray application, the best way to apply paint is with a combination of disposable foam rollers and foam brushes. Ideally, two people work together. One rolls on the paint in sections of no more than 300 square inches (about 24 × 12 inches). The other follows immediately with a good-quality foam brush to tip off and smooth out the fresh paint with long, even, horizontal strokes. The idea is that the roller lays down the paint very evenly and without runs, and the brush smooths out the roller stipple pattern that might be left on the surface of the boat. Use the yellow foam rollers (with about ³⁄₁₆-inch depth of foam) and the disposable charcoal gray brushes of 3- or 4-inch width. Keep plenty of the rollers and brushes available, and at the first sign of any degradation of either, discard and replace.

Allow this first coat to cure at least overnight (the time will depend on the paint and ambient temperature). The paint should preserve no fingerprint when a finger is pressed into the surface. If a light fingerprint shows, wait another day before doing the test again. Once the surface is cured enough to not show a finger imprint, sand very lightly with 220- or 320-grit sandpaper. I advise not using any finer grit than 320 because it will not provide adequate "tooth" for the next coat to grip. Carefully clean the surface again with a vacuum for loose dust; wipe down with a clean lint-free cloth and then a tack cloth before applying the next coat. (Don't forget to wet the floor down again.) Two coats may be all you need, but if you've added a flattening agent or Penetrol (which some painters like because it makes paint flow on more smoothly), you may need a third or fourth.

Here are a few more tips that will help ensure painting success, acquired through more than four decades of (sometimes painful) experience. These are guidelines, though, and you choose your own path:

- Don't rush to paint the boat after finishing the glass/epoxy sheathing. The epoxy should cure for at least a month, ideally two, before painting. Use the waiting time to build or finish some other parts of the boat, such as spars or seating. This is also another good reason to prime the topsides before rolling the boat right-side up; the conversion primer also needs time to cure and settle.

- You cannot overstir the paint or paint/catalyst mixture in the case of two-part systems. After stirring adequately, as instructed by the manufacturer, let the mixture sit, or induct, then stir it again after several minutes. Don't let the instructions on the can mislead you into thinking that no induction time is needed. It is, and patience is the key to success.

- Always strain the mixed and inducted paint before using it. You'll be surprised at what turns up in it. Disposable cone strainers typically cost a quarter or 30 cents apiece, so they're very cheap insurance against frustration. A lot of paint suppliers will give you cone strainers as a perk for buying paint from them.
- All foam brushes are not made equal. The cheap, floppy ones on sale at the hardware store are so disposable that they won't last through a 30-minute painting session. The chandlery where you buy your marine paint will probably have brushes of appropriate quality. They will feel moderately firm and have a tight, fine cell structure in the foam.

When prepping the hull for paint, I mentioned using polyester-based Bondo-type fillers on top of the conversion primer. These work very well if you observe a few basic rules. First, mix only very small amounts at a time when applying the fillers. As soon as they start to curdle and application is anything less than creamy smooth, it is time to set aside that batch and mix another. Second, apply patches to only those areas that will be sanded smooth during that body-work session. Leaving these fillers overnight, or worse, for several days, results in a lot more labor being needed to sand them; they harden progressively as they cure. And third, these fillers always need to be overprimed to seal up their porosity. Trying to apply a topsides paint over unsealed fillers is just asking for the application of extra coats of paint to cover up an uneven smoothness and gloss.

Figure 25-2. The hull has been primed, body-worked, sealer-primed, and sanded smooth in preparation for final topsides paint. The floor is wet down to minimize dust kick-up, and everything that might be oversprayed is masked off. Finally, we're ready to paint.

Figures 25-3A–B. *The forward cabin interior on this boat was pre-painted before installing decking. It will now require extra care when gluing the decks in place, but we avoided the even more difficult task of painting inside a small cabin's confines.*

Figure 25-4A. *Long-board sanding of the primer will expose high and low spots that need to be filled or sanded further.*

Figure 25-4B. *These low spots need attention.*

Figure 25-4C. *The long-board sanding has exposed different layers of primer, indicating highs and lows.*

Figure 25-4D. *Note the cloud of polyester filler used next to the stem to fill out a low spot.*

Figure 25-4E. *Just about ready for a final coat of primer.*

Figures 25-5A–B. *Rolling and tipping works best with a two-person team: One rolls and the second follows right behind with foam brushes while the paint is at its most pliable to smooth out the stipple pattern in the roller's wake.*

Figure 25-5C. *Not bad for a rolled-on paint job.*

PAINTING THE DECK

So far we have just been discussing painting the hull exterior. The deck, if you have one, is another issue. It will likely be another color—maybe an off-white, cream, or other light color that will contrast to some degree with the hull color and even more importantly with the cabin side paint color. In my opinion, a monochromatic boat will soon seem boring. The deck will also be walked on, so a non-skid surface where crew can walk or stand safely is essential.

Prepare the deck with conversion primer just as you did the hull. Remove or mask any structures that won't be painted, such as hatches. Then spray or roll on two or more coats of conventional paint in the selected color over the entire surface before worrying about the nonskid application areas. After curing (use the fingerprint method to test; imperfectly cured paint will retain a tape pattern), mask off those areas where you *don't* want a nonskid surface. I like to keep any area that will need to be masked later as smooth, such as against toe rails or cap rails at the sheer, or areas adjacent to companionway hatches, handrails, or deck hardware. Then roll or spray on the nonskid paint or compound. Most marine paint companies make a prepared nonskid deck paint or a nonskid compound that can be added to their own topsides paints. A very inexpensive alternative is a jar of silica sand or ground pumice, which you will sprinkle evenly onto fresh wet paint. After it cures, simply roll on a second coat of thinned paint over the sand to secure it in place. Some of the pumice nonskid additives have minute particles of iron or magnetite in them, which will likely give slightly rust-colored specks in the painted surface later after much contact with saltwater. Painter beware.

Over the last 20 years, I have become very enthusiastic about a radical alternative to deck paint: spray-on polyurethane truck bed liner. It's easy to apply, tintable to just about any color, provides a textured surface, gives the underlying epoxy more protection than paint ever could, and is as close to zero maintenance as anything used on a boat can be. Still more advantages: these coatings protect against salt corrosion, as well as damp and extreme temperatures, and are waterproof and flexible. They even help to deaden sounds and vibrations. Coverage is about 125 square feet per gallon, and if you spray them, your job will be done in short order. One warning: Thoroughly mask any surface you don't want coated with the liner. Once it's sprayed onto anything, it sticks like cement.

Figures 25-6A–D. *The cabintop has a rough-textured nonskid surface, but the areas where hardware will be bedded are kept smooth for easy cleanup of bedding compound squeeze-out.*

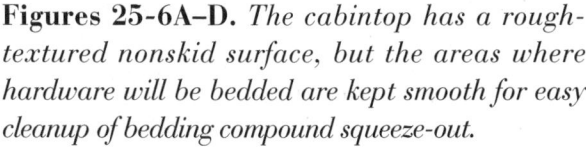

(D)

Figures 25-6A–D. (Continued)

Figure 25-8. *The cabin roof is complete with skylights, handrails, radar dome, and Charley Noble (a chimney for galley ventilation).*

Figure 25-7. *Cockpit seats of the Surf Scoter 23 contrast white interior hull paint, teak flooring, and a light beige deck paint.*

Figure 25-9. *Note the scupper in the roof to allow rain runoff—a detail you don't want to forget.*

INTERIOR FINISHES

My preference for interiors used to be an all-bright finish for sides and furniture with a painted overhead. As with many other issues, my thinking has gradually changed—in fact, reversed. A lot of boat cabins have smallish portlights or windows that are relatively stingy with the amount of daylight they admit, so it is all too easy to make a cabin feel dark and oppressive with too much natural wood finish—especially if it's a dark wood such as mahogany. My preference now is to paint a lot of the interior with a soft white paint, setting off the clean painted areas with bright dimensional wood framing and trim. This scheme seems to gather and reflect the light more efficiently. And it can look elegant as well as practical. However, varnished wood is actually more forgiving of imperfection than white-painted wood because the eye is drawn to the wood grain instead of to slight flaws in the curves or surface. White-painted surfaces must be smooth and finished very fairly to look good and not highlight defects.

This is a matter of individual taste, but I prefer a matte or satin finish below on both painted and bright surfaces. Gloss seems garish and out of place, especially in a small cabin. One Arctic Tern builder who originally finished his cabin with high-gloss varnish sanded it all off and started over. "I had pictured a warm, subtle, embracing kind of feeling," he wrote in his builder's blog. "Once the glossy overhead was fitted, it was like a funhouse in there; reflections everywhere. You no longer really saw the wood, just a bunch of reflections."

Whether I'm planning to paint or varnish, I seal all the interior panels and surfaces with epoxy and sand with 220-grit paper before applying the finish. For the underside of the deck and cabintop, I save myself a lot of messy drip and body contortions by prefinishing these panels if possible before I fasten them in place. It's a good idea to mask the areas where these pieces will land on their support structures so that you will not be gluing painted or varnished surfaces.

A final thought on painting: As with so many aspects of boatbuilding, good preparation is vital. Make sure all surfaces are clean and dust-free; going through a stack of disposable tack rags is nothing compared to the expense and effort of sanding and rolling on a whole extra coat of paint. Use clean, fresh rollers, brushes, paint-mixing tubs, and strainers. The patience you will learn from this effort will build character, and as no small bonus you'll end up with a fine-looking boat.

Figure 25-10. *It's a great day when the masking comes off and we start to see what all the hard work of painting has yielded.*

Figure 25-11. *With some of the hardware now installed, she really starts to look good.*

Figure 25-12. *More hardware has been installed and the boat is nearly done.*

Figure 25-13. *The boat finally rolls out of the shop and into the daylight, and we can finally see her from more than a few feet away. This is the time the boatbuilder either begins to celebrate or begins writing out the list of things that need to be redone.*

CHAPTER 26

Exterior Trim and Hardware

EXTERIOR TRIM

Toerails, rubrails, caprails, and other trim can make or break the appearance and finish of a boat. If the trim is out of scale, the boat will appear out of proportion. If the trim work is sloppy, the cosmetic effect will cheapen the overall quality of the boat.

All the trim on my boats share one characteristic: It is added after the basic boat is completed and painted. There are two reasons for delaying the trim until after the final painting: One is to achieve a uniform and fully protective paint job; the second is to minimize the tedious job of masking required before painting.

Each design has its own trim details and scantlings, but the basic procedure is always the same: make the parts, dry-fit them to check for accuracy, and finish the final sanding and prefinish (coatings) before installing them. I recommend sealing both the front and back and all the surfaces of these trim pieces before mounting to the boat so that there is an even coating of protection all around. You do not want moisture to work its way under an unsealed bit of toerail or trim and swell the piece or migrate into the wood enough to cause the final varnish or paint coating to fail. My goal is to only minimally touch the trim after I've installed the pieces with screws or bolts, bunged the fastener holes, and chiseled the bungs off flush with the surface. Most

of that final finish work is to simply apply another couple of coats of oil, Cetol, or varnish to seal the bungs. This procedure generally applies to bright-finished trim pieces. If a piece were to be painted, then I would apply it as a structural piece in the normal building sequence, then paint it along with the rest of the boat.

It is becoming increasingly difficult to buy long lengths of mahogany and teak to use for exterior trim, so the alternative is to scarf the stock to appropriate lengths from shorter stock. When scarfing, always consider the run and appearance of the grain so the pieces will match well, and align the scarfs so they shingle past obstructions instead of catching them.

Rainwater and splash will run across decks, course over the top edge of a rubrail, and sheet down the topsides, discoloring the topsides paint. To prevent this damage, cut a drip groove about ⅛ inch wide and ¼ inch deep in the lower edge of a rubrail (Figure 26-2).

Place mechanical fasteners at least every 8 to 10 inches along the length of the trim, and always bed the trim in a flexible polyurethane caulking compound or a bedding compound. For exterior trim, the woods of choice are teak, mahogany, purpleheart, locust, and oak. These are all hardwoods that will stand up well to the elements and are

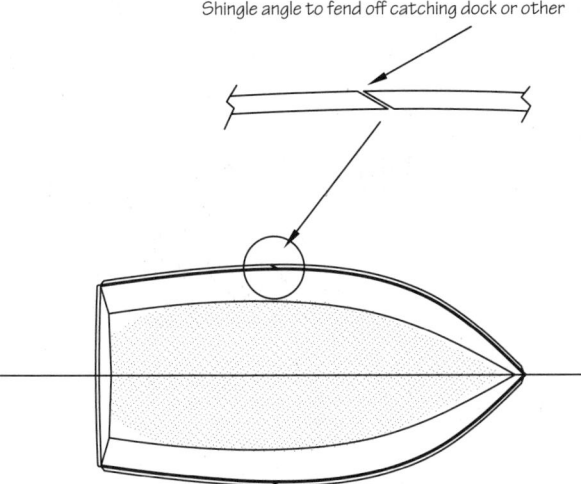

Shingle angle to fend off catching dock or other

Figure 26-1. *Align your scarfs in the gunwales and rail guards to prevent an obstacle from lifting or hooking the joint.*

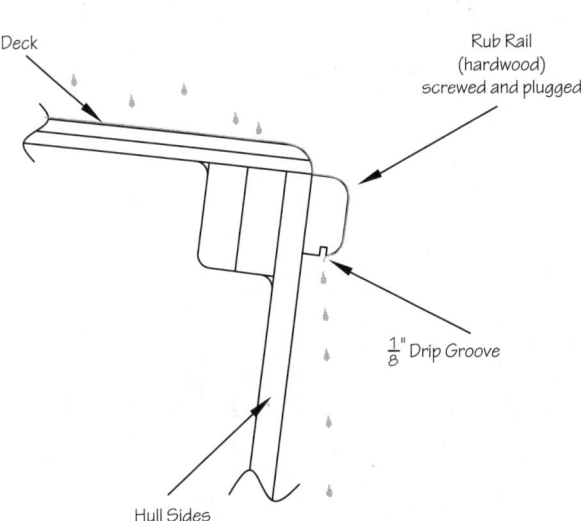

Deck

Rub Rail
(hardwood)
screwed and plugged

$\frac{1}{8}$" Drip Groove

Hull Sides

Figure 26-2. *A small ⅛-inch saw cut in the bottom of the rubrails will prevent water from sheeting down the topsides of your boat and eventually discoloring the paint.*

considered stable, so they won't move excessively in a marine environment. Whatever wood you choose, bear in mind that the first purpose of trim is to bear chafe, so consider how the wood will stand up over the long run: Will it stain dark when damaged or scuffed? Will those blemishes affect the overall appearance of the boat? I generally shy away from

the light blond oak or locust woods for that reason, opting for the darker mahogany, teak, or purpleheart. But there is no hard and fast rule. All the woods above are good choices for exterior trim.

Bedding compound is an important subject, and boatbuilders have a great variety of opinions, matching the great variety of compounds on the market. The two most important principles are (1) bed *every* piece of trim or hardware so as to prevent water from sneaking into the structure and rotting either the trim piece, or worse, the hull or deck from inside, where it may go unnoticed until it is too late; (2) make all these pieces removable. Sooner or later you will want to repair them, replace them, or change their locations. Why make it difficult (or impossible)?

A bedding compound forms a water-resistant gasket between a piece of trim or hardware and the hull or deck (Figure 26-4). It should remain slightly rubbery and flexible throughout its working life so that vibration or impact will not open up a gap in the seal. For items I think are more likely to be removed, I like Dolfinite, which is easy to apply; compatible with wood, fiberglass, metal, and plastic; and can be obtained in white, gray, or a mahogany-like color. The polyurethane bedding compounds come in tubes and should be thought of as adhesives first and sealants second. Sikaflex 291 and 3M 4200 are moderately tenacious. Sikaflex 292 and 3M 5200 should be reserved for when you want a marriage to last forever; they will ferociously resist anything being taken apart. For bedding hardware, we have also had very good results using butyl tape, which we obtain from RV suppliers. It never hardens completely and can conform to the expansion/contraction of the hardware over time. Windows and portlights are fine examples of hardware that can best be bedded with butyl tape.

Whichever compound you're using, make sure the mating surfaces are clean, dry, and free of dirt and grease. After the component is installed, carefully clean up any squeeze-out with a knife, sharpened tongue depressor, and a rag wetted with mineral spirits. Avoid the most aggressive solvents, as they can eat into young and not-fully-cured paints.

Figure 26-3. *Clean polysulphide bedding compound from around a portlight trim ring with paint thinner and disposable rag.*

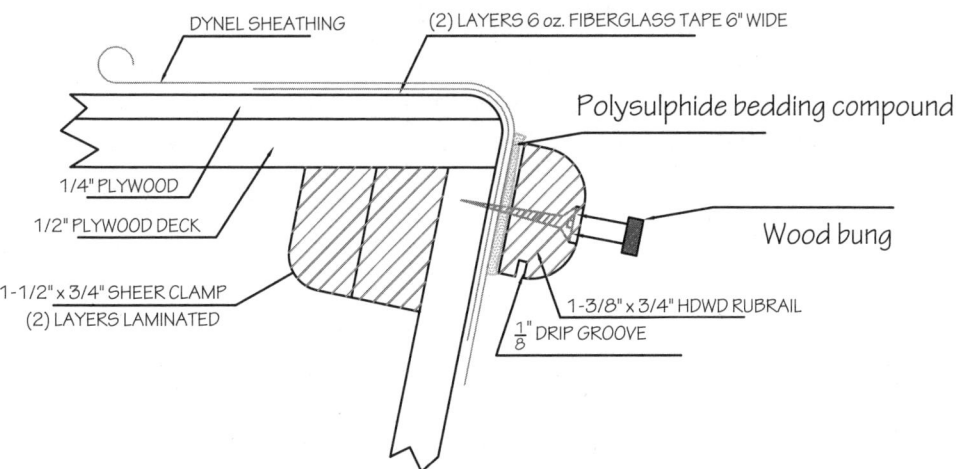

DYNEL SHEATHING

(2) LAYERS 6 oz. FIBERGLASS TAPE 6" WIDE

Polysulphide bedding compound

1/4" PLYWOOD

1/2" PLYWOOD DECK

Wood bung

1-1/2" x 3/4" SHEER CLAMP
(2) LAYERS LAMINATED

1-3/8" x 3/4" HDWD RUBRAIL

$\frac{1}{8}$" DRIP GROOVE

Figure 26-4. *Cross section of the hull-deck joint with sheer clamp, fiberglass reinforcements, rubrail, and bedding compound.*

HARDWARE

The hardware you install on your boat will have to appeal to your own sense of aesthetics above all. Beyond that, it should be practical and durable. I prefer to use plain bronze whenever possible, and I don't polish it because I prefer the light-green patina of oxidized bronze. However, some things are simply not available in bronze these days, and in such cases I defer to stainless steel.

When you use stainless steel fasteners, padeyes, and other hardware, get the highest stainless grade available. Type 304 (sometimes referred to as 18-8) is good, but type 316 is much more resistant to crevice corrosion. Type 304 will appear to rust and look less than shipshape after a while. If you buy stainless hardware, look for smooth and polished surfaces to reduce the problem of crevice corrosion. When used underwater, stainless steel is particularly vulnerable to crevice corrosion; better to confine its use to above the waterline or inside the boat. For fasteners or any hardware that will be continuously exposed to water, such as the rudder's pintles and gudgeons, bronze will endure much longer.

All marine hardware is frighteningly expensive, especially bronze, but hardware intended for home

use just doesn't hold up. You've come this far in the building project, and you owe it to yourself to finish your boat off with the best hardware available.

For secondary insurance beyond your bedding compound, coat and seal the edges of all hardware-fastening pilot holes with epoxy. I keep a supply of pipe cleaners to coat the interiors of the smaller holes. Any breach in the epoxy sealing will later come back to haunt you, so be thorough. Some of my customers go the extra mile when installing hardware on deck by first drilling an oversize pilot hole, filling it with epoxy, then re-drilling a smaller hole after the resin has cured. This is the ultimate insurance policy against water wicking its way into the interior grain of a plywood deck or cabin roof.

Use through-bolts to mount hardware whenever possible. Wherever a component will have to respond to high loads, such as a bow eye or deck cleat, provide it with a backing plate. One-eighth-inch silicon bronze plate stock works well for customized backing plates. It drills much easier than stainless steel and can be cut with a jigsaw or bandsaw with a dull, fine-toothed wood-cutting blade. Other possibilities are G-10 (fiberglass composite board) and King StarBoard® (a marine plastic board). For some backing plates, I simply use an extra layer of marine plywood—cut to shape, epoxy-sealed, and bedded in place.

I use stainless steel bolts with nylon-insert, aircraft-type nuts for attaching most hardware. These nuts don't need lock washers and won't back off the threads as the boat ages. For those instances where you don't need a backing plate, I still recommend a large, flat stainless steel fender washer to help spread the load over as large a surface as possible, and bed the hardware meticulously.

CHAPTER 27

Propulsion

How you power a boat has everything to do with the ultimate enjoyment of the craft. Gone are the days when boats could only be powered by sail or human muscle. Today we have a variety of engine options—inboard, outboard, gasoline, diesel, hybrid, and pure electric systems. Electric systems divide into battery and solar power. Even a sailboat will likely utilize one of these options as backup propulsion. If you are a purist who insists on sails alone, you will still use oars as backup (and they will serve a small sailing dinghy perfectly well), but in that case I will allow you to skip over this chapter. As for the rest of us, we are going to talk motors.

The matter of the form of propulsion is a complicated one, and our options and choices are now changing rapidly. But it always begins with a straightforward exercise: a complete review of how you expect to use the boat you're building, wherein you ask yourself the hard questions about your real goals (as opposed to fantasies) for the use of your boat, and how these uses will interact with the needs of the boat's hull design (more on this later).

As I have mentioned, I live near the Northwest coast of North America, and I have a unique opportunity for using a boat. I can walk down the street from my home about one-eighth of a mile, board a boat, and cruise almost 1,000 miles north to Alaska. Almost that entire journey will be in waters sheltered by myriad islands that protect my boat and me from the open Pacific. But realistically, now: Will I ever actually have the time to take this ambitious and wonderful cruise? If the answer is yes, then I should be thinking of powering my dream boat with an engine that would suit this use. A few of the questions I should then be addressing are: What are the power requirements necessary to propel it at the designed cruising speed? What is my desired cruising range, and can I install fuel capacity to provide that range? What about the availability of parts if something goes wrong with the engine, and what are my skills in maintaining and repairing it? What is my (and my crew's) tolerance for engine noise, and how can I minimize excessive racket? The answers to these questions will help determine the size and type of engine, the size and location of fuel tanks, and whether my personal proposition for using the boat was realistic in the first place—or must be modified.

Even the auxiliary power for a sailboat must be assessed in light of the boat's use because it will vary with different environments. Regarding the Alaska cruise above, many sailors who have made that excursion come back reporting that they had their sails up for only a couple of days out of the month or more that the cruise took to complete, with the vast majority of those days spent motoring. Summer winds in Puget Sound and the Inside Passage are notoriously fickle and often very light. The engine for a cruising sailboat in our region is, in reality, more a primary than a secondary propulsion system. In other environments, this order may be reversed, but these are the waters that I live in and I must remain realistic about them.

When I wrote this book the first time, my personal boat was one of my 23-foot Arctic Tern sloops, and she had a 10-hp diesel inboard. With the sails, the diesel, and a good anchor, I could deal with most problems very nicely. With a 24-gallon fuel tank and a .4 gallon/hour fuel consumption at a 5-knot cruise, I could expect a useful range of about 250 miles. (On a boat, there are no fuel tanks that will allow you to use all the fuel held in them; I estimated I could use about 20 gallons of that capacity—so dividing 20 gallons by 0.4 yields a runtime of 50 hours.) Although I never undertook that Alaska cruise in the Arctic Tern—I simply didn't have the time in a 5-knot boat—this engine and fuel capacity made her as capable a cruising sailboat as a 23-footer could be.

This same boat could be powered with a much less expensive 6- to 8-horsepower gasoline outboard, which would change her capabilities. With the same fuel tank the range would be only slightly less, but it might not have enough power to push through some of the Northwest's more extreme tidal currents—hence some six-hour delays waiting for an opposing current to change. The outboard would certainly be noisier in the cockpit at sustained cruising speeds, and very possibly she would not have been able to motor as fast as the inboard diesel. Finally, transom-mounted outboards may become a problem in very rough seas, with the prop levering out of the water on wave crests. With an outboard, that Alaska destination would clearly be unrealistic. But for daysailing or short-distance cruises of three or four days in protected waters, it might make perfect sense.

We now have the third option of electric auxiliary power. If I'm using my 23-foot sloop as a daysailer 90 percent of the time and I'm occasionally taking an overnight cruise to a destination 10 or 15 miles distant, and I know I'll have a marina slip there and can plug into shore power for recharging, then an electric outboard makes very good sense. The higher cost (as opposed to a gasoline outboard) may be offset by the much lower motor noise and the opportunity for motorsailing with virtually silent assistance in very light winds. If I need more motoring range—say, 40 or 50 miles—it may still be possible to outfit this boat with additional battery banks. But conventional absorbent glass mat (AGM) marine batteries would consume too much space, and at this writing the cost of additional lithium-ion batteries engineered for safe marine use will likely total more than the motor itself.

This is what I mean by a complete review of your intentions for your boat.

Whatever kind of power plant you choose, back it up with adequate ground tackle and a VHF radio to call for help if the worst comes to pass. And maintain the engine religiously. If you neglect engine maintenance, expect bad karma. It will indeed let you down when you need it the most.

OUTBOARD POWER

If you choose an outboard for power, you have a couple of options for how to hang it on the boat. Your design might have been done around the idea of placing an outboard in a motorwell, which is (sometimes) a nice way to deal with the outboard. But there are some problems. I did some motorwell designs back in the old days of two-cycle outboards, and back then it was easy to get the templates or outlines of the motors to overlay on our working drawings. With four-stroke motors now nearly universal, the powerheads are larger, and it is almost impossible for designers or builders to obtain these templates, and you don't want to design a motorwell that proves too small for the motor the owner will need. In other words it is very difficult to design a motorwell that could accommodate a wide variety of power options on the outboards. On sailboats, another disadvantage is that it may be difficult to design a well so that the shaft and propeller can be tilted up to reduce drag.

If you're building a powerboat with a semidisplacement or planing hull,[*] an outboard motorwell is probably not that appropriate for your design anyway. The open space of the motorwell removes lift area from the bottom of the boat right where it is most critical and will degrade potential performance. Your boat will be faster and more fuel-efficient if you leave the outboard on the stern, not in some cutout forward of the stern. For most applications, I have gotten away from motorwells and have gone to the conventional bolting of the outboard onto the transom.

Outboard mounting on a small sailboat raises special issues. If the rudder is hung on the transom on the centerline, as I tend to do in many of my designs, the motor must be offset to one side—and then it's crucially important that the propeller and rudder not make contact when either is swung during docking maneuvers. Some small sailboat owners cringe at the counterweight effect of levering 60 or more pounds of motor out behind the transom on a conventional mount, so they build a vertical mounting slot into the transom itself to keep the weight within the boat envelope. This works as long as the slot is robustly reinforced and

---

* A **displacement** hull pushes the water aside as it makes its way forward; its maximum or "hull speed" in knots is the square root of the waterline length in feet multiplied by 1.34. A displacement boat with a 16-foot waterline will have a hull speed of 5.34 knots. An increase in power will only cause the stern to squat and make a larger wake—but go no faster. A **semidisplacement** hull does not quite plane, but its hull is designed to provide some dynamic lift with water pressure on the underwater sections of the hull; with adequate power it can top out at 2 to 2.5 times the speed of a pure displacement hull of the same waterline. A **planing** hull typically has a flatter deadrise (the "V" angle of the hull bottom) and can skim over water at more than 2.5 times displacement speeds. This type of hull uses the water pressure to help reduce the effective weight of the boat and move more rapidly through the water.

the motor has room to swivel, but be careful as the geometry can be very tricky and you might end up with no advantage at all for the whole exercise.

Among the best advice on outboards I can give you is to buy the motor early in the boat project. You can either buy a commercial mount for it and carefully mock up the parameters of the mounting, or make your own homebuilt bracket that approximates the ultimate mount to your hull. In either case you want to assure that the motor will have enough room to swing from stop to stop, tilt, and not come into contact with motorwell walls, tiller, or rudder.

INBOARD MOUNTING

For an inboard engine in a Stitch-and-Glue boat, I have developed some methods that make installation a little easier. Engine beds and the angle of the propeller shaft are the first issue. Generally, modern marine engines require their motor mounts to be in line (parallel) with the shaft. You will need to do a bit of layout work to accommodate these angles.

You will need to begin by cutting an oblong hole in the keel area of the boat at the expected position of the shaft log (the alleyway for the propeller shaft to enter the boat) and its attached packing box; this hole will help locate the shaft line. (The packing box is the gland that allows water to cool the bearing where the shaft transitions to the outside of the boat, and that keeps this water outside the boat.) Stretch a string from the position of the aft-most strut or shaft bearing (this would be a bearing or a support very near the propeller) to the bulkhead located at the front of the engine box or compartment to help measure the shaft angle. Your designer or the engine manufacturer will indicate the maximum allowable shaft angle, which is usually not more than 15 degrees tipping aft from the level waterline. Inside the engine room, attach the string along the centerline of the shaft to the front of the engine room. It used to be difficult to get a good reading on the angle of the string, but with the digital levels and laser sights available today that indicate degrees, the job is much easier.

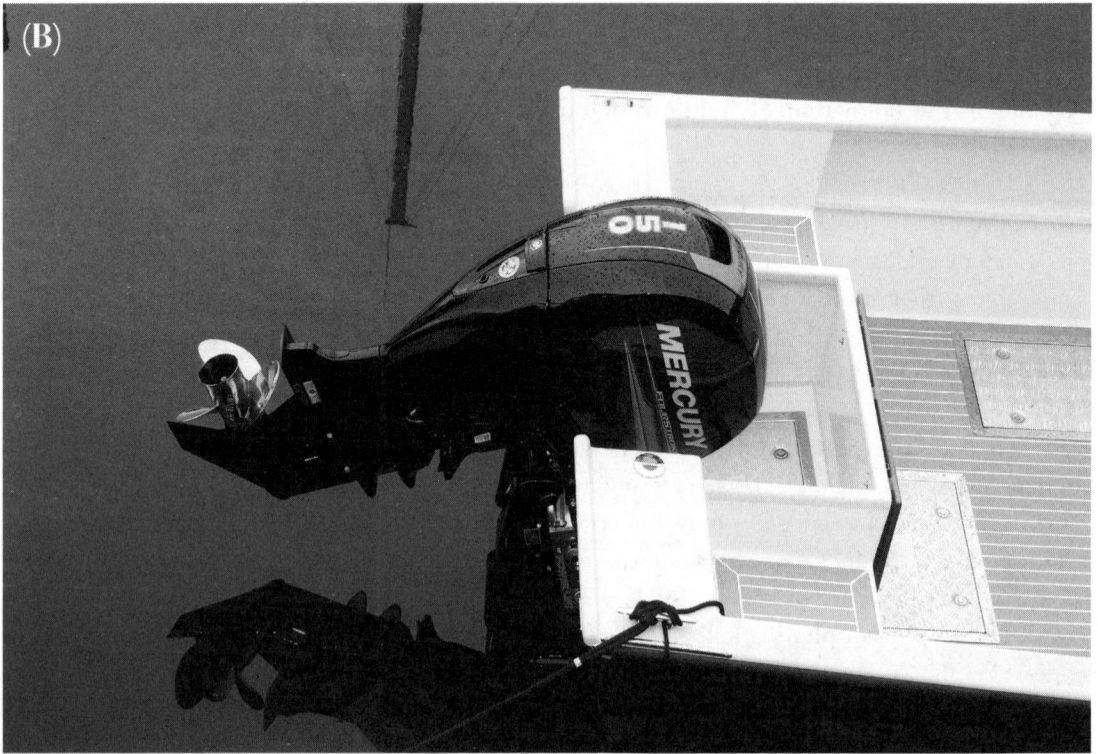

Figures 27-1A–B. *A Surf Scoter 23 with outboard power.*

Rent the tools if you need to; it is very important to get this angle right. If your string cannot stretch straight, you'll need to re-drill or adjust the hole cut in the bilge. Be careful not to overbore this hole. As soon as you get the shaft log into position, it will be glassed solidly into place, and you don't want to create more work for yourself at this point in the project (Figures 27-2A and 27-2B).

Centek Industries of Thomasville, Georgia, manufactures an excellent fiberglass shaft tube stock under the trade name Vernatube™. Made by a filament winding process, these strong fiberglass

Figure 27-2A. *A taut string will show the alignment for drilling the propeller shaft tube installation.*

Figure 27-2B. *Shaft tube and rudder post tube are installed.*

tubes can easily be bonded into a Stitch-and-Glue boat. After determining the appropriate tube size for your engine installation, enlarge the hole in the bilge so that you can slip the tube through the bottom of the boat at the proper shaft angle, all the while keeping the string or wire dead center in both ends of the log-tube. Bond the tube in place with high-density epoxy fillets, backed up by plenty of layers of biaxial fiberglass cloth, making sure the shaft is centered at each end and aligned with the string. This permanently fuses in the shaft log-tube. On a fin-keeled sailboat, you'll want the tube to extend to just slightly past the actual exterior bottom of the boat. Glass the interior and exterior surfaces adjacent to the tube with epoxy and several layers of biaxial tape. If your boat has a wooden full keel, like a displacement power boat, install the shaft tube by glassing where it exits into the inside of the boat only.

Figure 27-3. *This photo shows the alignment of the string for the centerline of the shaft and the installed shaft log.*

The outside of the tube will extend through the keel in a notch cutout.

Once the shaft tube has been glassed in place, the engine beds can be installed. These are the members to which the attachment feet of the engine will fasten. For small diesel engines, the engine mounts are usually parallel to the shaft, although on certain models they may be as much as 1 inch above or below the shaft line. Check the actual engine to be used to ascertain this offset if necessary. Using ¼-inch plywood template stock, make a pattern for the engine mounts. Then, taking careful measurements of the angle from the boat's bottom, laminate the engine beds from layers of marine plywood. Depending on the shape of the hull and configuration of the motor box, the bed may decrease in height until it runs out aft or, in some cases, runs up to the cockpit longitudinals once clear of the engine. I often like to have a bridge or top to the engine beds that extends over to the sides of the engine box or to the cockpit longitudinal supports. This design allows several parts to serve double duty and creates an extremely strong engine bed grid, with the top forming a walkway along each side of the engine (Figures 27-4A, 27-4B, 27-4C, and 27-4D).

You have a couple of options for mounting the feet of the engine to the bed: Either fasten them directly, or fabricate two long beds of stainless steel angle iron or strong aluminum angle iron that will allow you to through-bolt the flexible engine mounts. The engine mount bolts will bolt

Figure 27-4A. *Cross section of all the necessary reinforcements for the engine bed.*

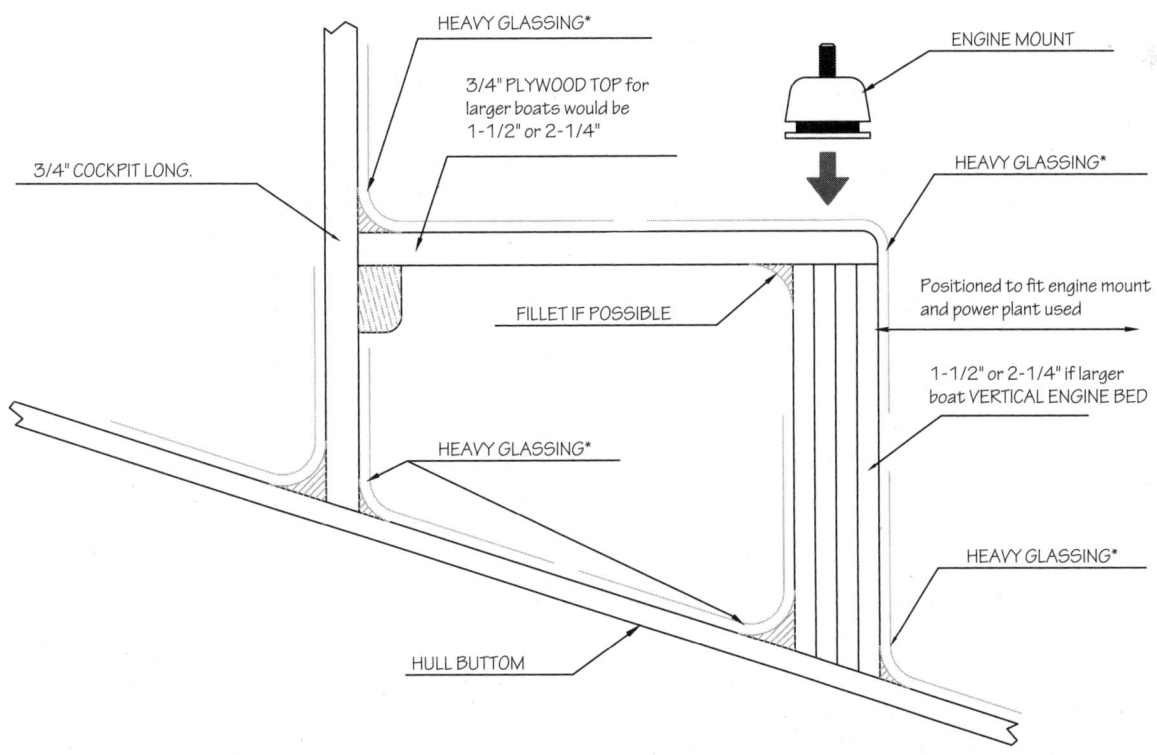

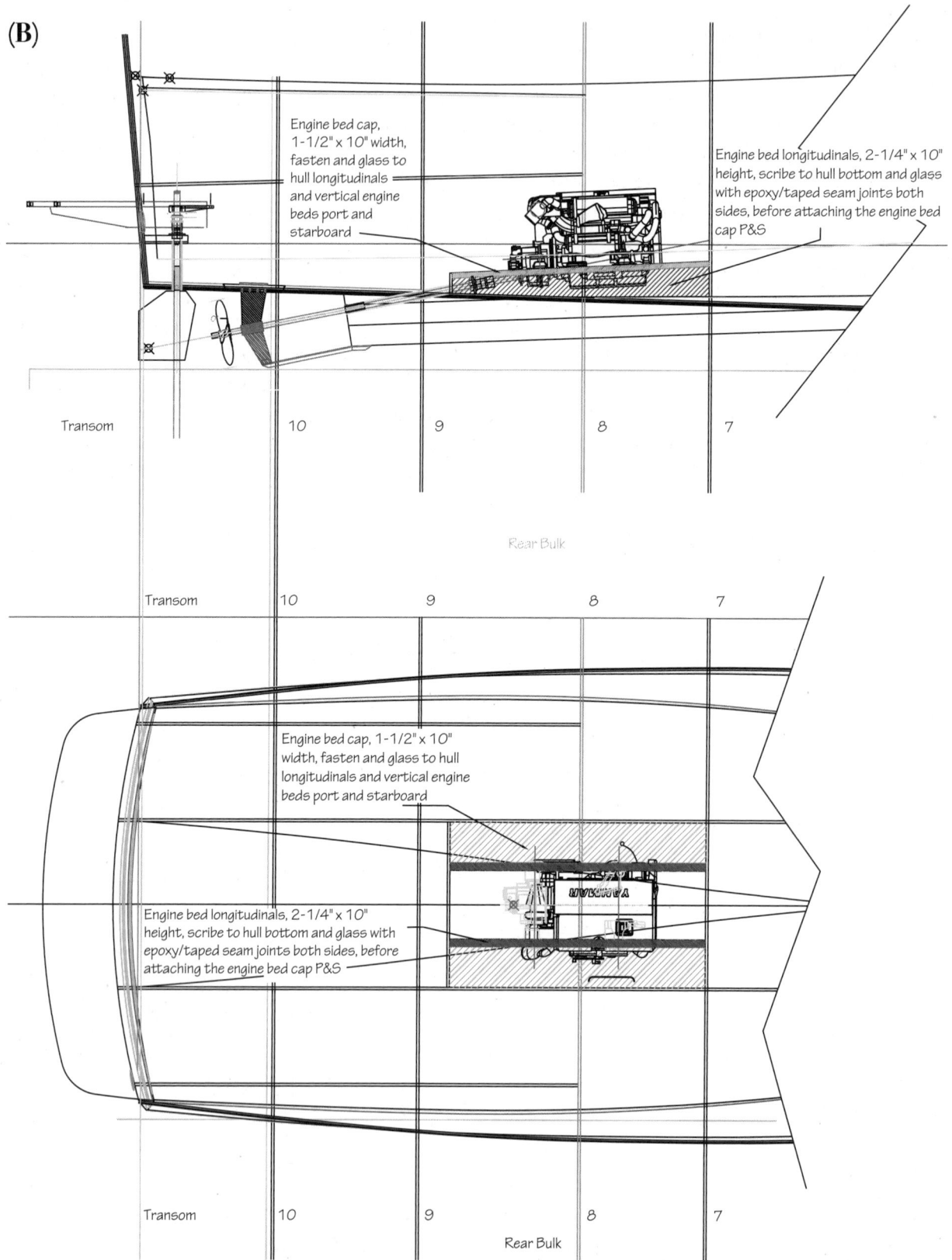

(B)

Engine bed cap,
1-1/2" x 10" width,
fasten and glass to
hull longitudinals
and vertical engine
beds port and
starboard

Engine bed longitudinals, 2-1/4" x 10"
height, scribe to hull bottom and glass
with epoxy/taped seam joints both
sides, before attaching the engine bed
cap P&S

Transom   10   9   8   7

Rear Bulk

Transom   10   9   8   7

Engine bed cap, 1-1/2" x 10"
width, fasten and glass to hull
longitudinals and vertical engine
beds port and starboard

Engine bed longitudinals, 2-1/4" x 10"
height, scribe to hull bottom and glass with
epoxy/taped seam joints both sides, before
attaching the engine bed cap P&S

Transom   10   9   8   7

Rear Bulk

Figure 27-4B. *Engine installation details.*

(C)

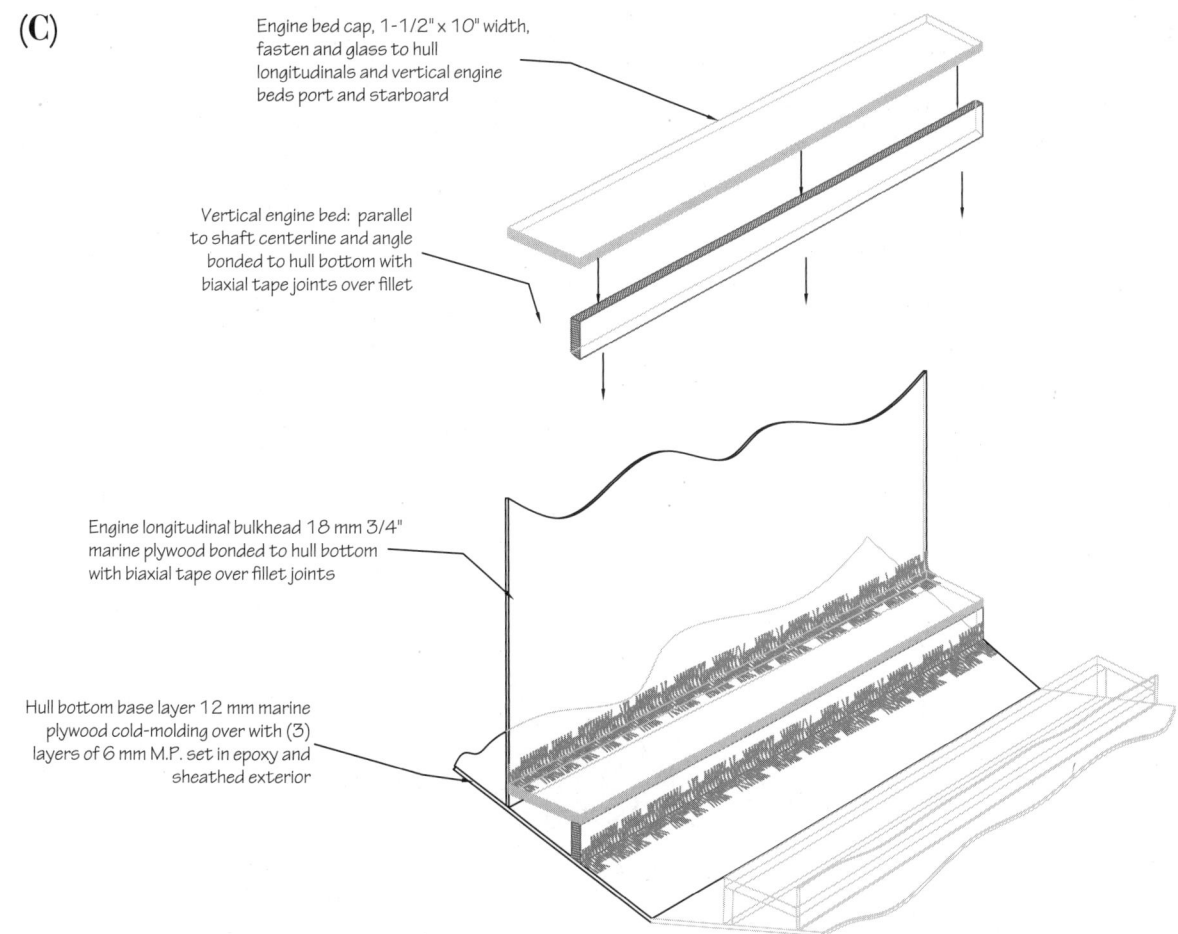

Engine bed cap, 1-1/2" x 10" width, fasten and glass to hull longitudinals and vertical engine beds port and starboard

Vertical engine bed: parallel to shaft centerline and angle bonded to hull bottom with biaxial tape joints over fillet

Engine longitudinal bulkhead 18 mm 3/4" marine plywood bonded to hull bottom with biaxial tape over fillet joints

Hull bottom base layer 12 mm marine plywood cold-molding over with (3) layers of 6 mm M.P. set in epoxy and sheathed exterior

Figure 27-4C. *Engine bedding details.*

down on these metal bed logs, and the bed logs themselves can be bolted to the wooden beds you have fabricated. Using this method, you can adjust slightly the angle of the metal beds and refine the engine mounting angles to a very precise degree. The flexible engine mounts now used throughout the boating industry themselves have extra adjustment capabilities to further fine-tune this mounting angle. You will want the shaft to mount to the mating coupler on the engine's reduction gear to an accuracy of just a few thousandths of an inch so the installation will run with the least possible noise and vibration.

Three guidelines: Make the whole structure as solid and as strong as possible. Be sure to eliminate crevices for dirt and oil to stand in, and allow easy cleaning of the engine compartment. Provide for as much adjustability as possible to set up the proper angles and to make a strong installation.

Another helpful item is a flexible shaft log (if you recall, this is the bearing and gland that will keep the exterior water from climbing up into the shaft log and getting into the inside of the boat). These packing boxes can be fastened to the end of the fiberglass stern tube with sturdy stainless steel hose clamps and are somewhat self-aligning

(D)

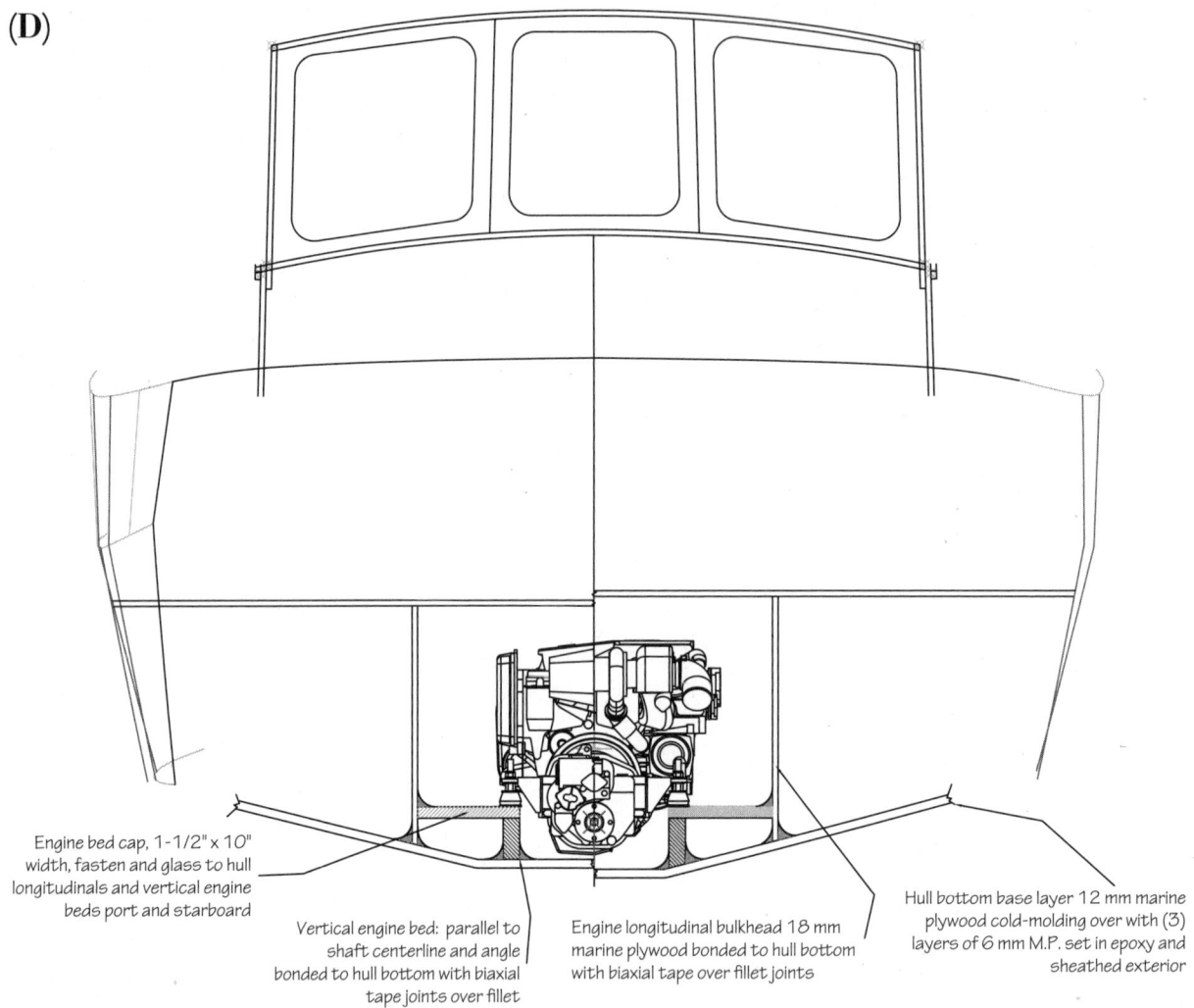

Engine bed cap, 1-1/2" x 10" width, fasten and glass to hull longitudinals and vertical engine beds port and starboard

Vertical engine bed: parallel to shaft centerline and angle bonded to hull bottom with biaxial tape joints over fillet

Engine longitudinal bulkhead 18 mm marine plywood bonded to hull bottom with biaxial tape over fillet joints

Hull bottom base layer 12 mm marine plywood cold-molding over with (3) layers of 6 mm M.P. set in epoxy and sheathed exterior

Figure 27-4D. *A stern view of the engine bed.*

because of the flexible nature of the thick radiator-type hose. There are old-school models that require you to periodically replace a flax cord that is squeezed by the nut to help seal the shaft, or you can opt for one of the leakless and low-maintenance models now available.

The rest of the installation is a matter of hooking up the wire harness of the engine to the gauge panel, plumbing the fuel supply into the engine, and in the case of a diesel, plumbing the return line (a small amount of fuel is run through the engine and back into the fuel tank as a way of cooling the engine's injectors). Exhaust will need to be run and then plumbed through the hull sides or transom. Install the propeller shaft and propeller, and the installation is ready to go.

ELECTRIC OPTIONS

We've already discussed electric outboards, but there are a couple of other choices: inboard and pod. And you have the possibility of augmenting or even

Figure 27-5. *The whole engine bed area has received several coats of epoxy to seal it.*

Figures 27-6A–B. *The engine is now in place with sound-deadening insulation lining the compartment.*

replacing their batteries with solar power. One of my recent projects was a 27-foot picnic boat named *Wayward Sun*, which despite being designed simply to take passengers for evening cruises on quiet local lakes, undertook a 1,200-mile cruise from Washington State to Alaska (Figure 27-10). And they made it—on solar power alone, not once recharging the batteries (see *WoodenBoat* magazine No. 285, March/April 2022). Make no mistake: it was a slow trip, averaging 27 miles a day at an average speed of 3.7 knots. An electric motor's power consumption curve rises very rapidly in relation to small gains in speed.

An inboard electric motor is mounted much like an inboard diesel, though its much lighter weight and compact size provide more flexibility in placement. A pod motor is mounted on a fin under the hull, or sometimes integrated into the rudder.

Figure 27-7. *An 850-pound engine needs strong support both when it's lifted up and when it's settled into its permanent home.*

If you're considering any of these possibilities, you need to have some level of comfort in designing and handling electric systems. For solar electric, I recommend hiring a professional consultant at least to design the system.

The question a lot of people have about marine electric power is this: Is it reliable enough at this point? It's a legitimate question because some early adopters have had unhappy experiences. My feeling is that the industry is maturing,

Figure 27-8. *A center bay between a twin-engine setup before installation. Note how all surfaces are smooth, painted, and finished to make engine maintenance cleaner and easier.*

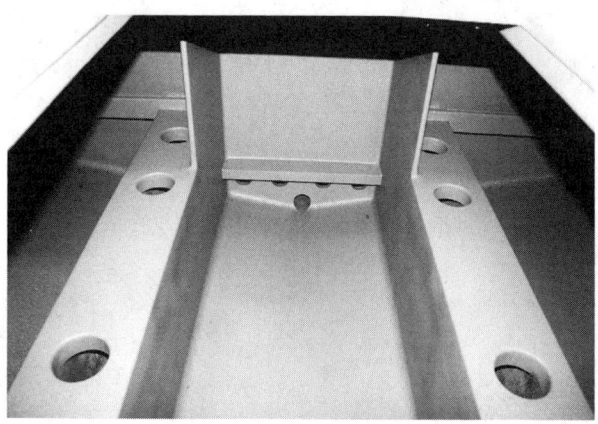

Figure 27-10. *My Solar Sal 27 was the first electric boat to cruise from Puget Sound in Washington State to Alaska without recharging from shore power or generator.*

Figure 27-9. *The completed work area between the two engines with fuel supply, water strainer, and other equipment.*

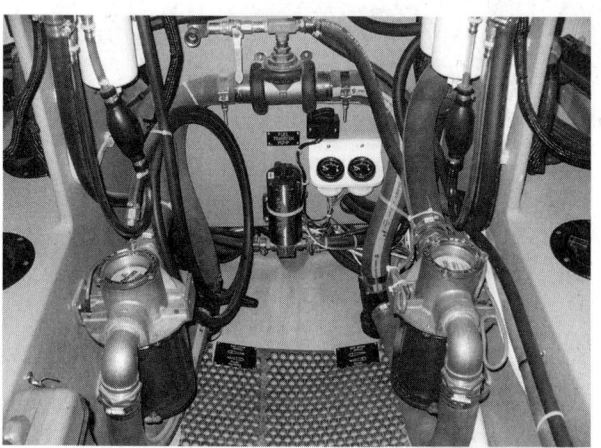

Figure 27-11. *The Torqeedo electric pod drive on the Solar Sal.*

and reliability is now on a par with petroleum-fueled engines. This is not to say you can blithely purr into the sunset with never a worry because you can't do so with any boat propulsion system. Even 21st-century gas outboards and diesels now incorporate electronic engine management, and any time you have electric devices operating in a marine environment you have the potential for trouble. This isn't pessimism; it's just physics and chemistry.

Figure 27-12. *As a catamaran, Electric Philosophy maximizes rooftop space (about 15 by 35 feet) for solar cells while minimizing drag from wetted surface area. It also has lithium battery banks storing enough power for nearly two days' cruising even without help from the sun.*

CHAPTER 28

Launching

Much like childbirth, launch day mixes pain with joy and arrives with agonizing slowness. The delays and projects just don't want to conclude. It might seem that this boat in which you've invested so much time and money is unwilling to start its life off the building cradle. I've seen many a launch in my life, and my most vivid mental image is that of a builder chasing the boat with a brush as it is eased into the water to touch up that last coat of paint on the hull. No matter how hard I try to get all the pre-launch jobs done, a few always seem to slip through the cracks and present themselves on launch day. Even scheduling the launch day is fraught with complications and delays. I know this frustrates my customers and friends, but it is incredibly hard to pin down the balance between the punch list of items still to be finished and my emotions, and commit to the final day. In my experience with literally hundreds of launches, it may be because I know acutely how much effort will be involved in the actual launching, and I want to delay this expenditure of energy as long as possible. At least one of my customers has confessed that he put it off because Launch Day looked like Judgment Day, where his character, through his workmanship on the boat, would receive its ultimate grade. He is right in that assessment. Once the boat is outside the shop and floating in the water it is subject to endless scrutiny and critique, not least from the builder's own conscience.

But at some point, the builder must declare, "No more paint! No more varnish! No more woodwork!" Otherwise, even a dinghy could be worked on forever. I've observed that I make my resolution to finally shove the boat through the shop door just after my mood has bottomed out and is on its way back up again. We've got a saying in our shop: "You've got to hate 'em to finish 'em." That is, you need motivation to finish the boat, and that motivation may be a momentary flash of loathing. Building a boat is a gigantic creative endeavor, and the complicated emotions woven through it themselves need some personal attention.

Launch celebrations run the gamut from large parties with hired bands to low-key splashes with a few bare words of dedication. I like the story about how some African boatbuilders used to launch their dugout canoes. The builder would not actually attend the launch, but instead would conceal himself nearby, within earshot. There he would wait for the crowd's response to the launch. If he heard cries of joy he would go and join the party, but if he heard groans and curses of derision he would bolt for a good head start on his angry pursuers. He knew that after a few weeks he could return to less impassioned feelings and start the next boat.

That story could be apocryphal, but here is some real advice. If you are going to have a formal party, launch the boat twice: first as a dress rehearsal (a "builder's launch") without spectators. Crank the engine, rig the sails, and work through the boat's systems. Satisfy yourself that everything's in working order. Do this on a day with a forecast of mild weather, and if launching in ocean water, for heaven's sake consult the tide table to make sure it will be possible to get the boat back out of the water for the later formal launching. There's plenty of stress at a launch party; you don't need to add to it by having to troubleshoot at the last minute and worry about the feelings and emotions of your customer or your friends and loved ones as they simultaneously worry about you. These people, remember, have also been through a giant journey and one that may not have been all joy to them either. They deserve a happy occasion. When you're satisfied she is not going to embarrass you (the boat, I mean), proceed with the formal launch. This one's for the owner, family, friends, and finally you the builder. Give it the full attention and energy that it deserves; this is indeed a celebration of creation and perseverance.

Some people want long dedications, others don't. I think it is always appropriate at least to offer a few of your own words about the boatbuilding journey. But no matter what anyone else might have to say, my own personal remark for every boat is, "Over the land and into the drink,

Figure 28-1A. *It's always a late night before launch day with many details left to wrap up.*

Figure 28-1B. *The boat's name may be one of the final chores.*

Figure 28-1C. *Name boards have been carved, painted, varnished and are ready to install just before launch day.*

Figure 28-2. *Rolling out of the shop on launch day—not yet time to relax, but the end is near.*

please, God, don't let her sink!" I enter that small petition for every boat even over the top of some long-winded owner's declaration, sometimes out loud and sometimes under my breath. It seems appropriate that the person most responsible for the project, this transaction of turning plywood, timber, and sticky goo into a boat with life and spirit, should have the final word just before the keel hits the water. The point is, the work ends only when the launch is successfully over, the boat's tied up securely to a mooring or dock, and I'm back home with my feet up, hoisting a few to the memories of the vessels that preceded her and dreams of those that might follow.

The launch is a celebration of both the journey and the arrival at the destination. If it takes on somewhat mystical or even spiritual overtones, this is nothing that even the most ardent nonbeliever can run away from. I wrote at the outset many chapters back that anything that has a person sweat over it, bleed on it, curse at it, and experience joy with it ends up with the spirit of all that labor and emotion built into it. This spirit deserves a rousing welcome into the world.

Figure 28-3. *A few words from the new boat's owner just before the keel touches the water tend to be rather emotional.*

Figure 28-4. *Let the celebration begin!*

Figure 28-5. *The first time she shows her head as you throttle up might be the most emotional moment of the whole process.*

CHAPTER 29

Inspecting, Maintaining, and Repairing

Stitch-and-Glue boats are remarkably tough and resilient. When I wrote the first edition of this book way back in the 1990s, I had seen only one vessel return for a major repair. With some of the fleet of Devlin boats now beyond the age of 40, I have of course encountered many more, and some of them have suffered adversities that were almost painful for me to see. There have been car-topped boats that blew off their roof racks and bounced down the highway at speed. There have been boats that were T-boned by other boats. And then there was a 19-foot Winter Wren sailboat that was left outdoors in a storage lot for 15 years with no cover over her; the cockpit and cabin were so full of rainwater that the tires on the trailer popped from the weight. But I have rarely if ever seen a boat that was so badly damaged it could not be brought back to life. These boats are not common and disposable; they tend to be loved, at least that is always our hope, and that makes the effort worthwhile.

Neglect is the ultimate enemy of every boat. Walk through any marina and you'll see some boats that seem essentially to have been abandoned, their decks covered with green slime, their brightwork peeling and cracking. At some point, a boat's deterioration will advance so far that any attempt at restoration will seem too overwhelming, and that will be the end. It's so much better to accept the fact that a boat needs a certain amount of regular maintenance and to budget the time to do it. If the boat is loved, the work won't feel like a burden. And if it's not loved, why are you keeping it?

I recommend creating a log book and owner's manual where you keep information about all your boat's systems and what you do to them. A three-ring binder notebook works well for this purpose. In it you will organize copies of all the essential manufacturers' information for the boat's systems, such as the motor maintenance specifications and the operating instructions for the depth sounder. You will preserve details on all the things *you* did in building the boat that you won't remember five years later, such as a complete wiring diagram for the electrical system. And there will be a maintenance log where you record the dates of the operations you perform—for example, when you last varnished the caprail and changed the motor oil.

The maintenance a particular boat will require varies according to several factors: type of boat, type of use, whether it lives in the water or on a trailer, its home climate, and its exposure to the elements. It's difficult to generalize, but there is one rule that must be followed: investing in protective coverings, such as a waterproof canvas cockpit cover, always pays off. This cover may seem expensive, but when you factor in the time you will spend in repainting, revarnishing, and repairing water damage, it pencils out very favorably. Remember that ventilation is vital: whatever the type of cover, the space underneath it needs to be able to breathe so as not to trap insidious moisture. In a cold, damp climate like our Pacific Northwest, it is a good idea to place a low-wattage marine dehumidifying heater in the cabin during those cold months when the boat is not used often. These heaters help stir the air by the natural effect of warm air rising, and the resulting ventilation will help to maintain a dry, clean interior.

INSPECTING

Safe, reliable, and long-lived boats are the consequence of their owners' good habits. One of those habits is to inspect your boat just before and just after every boating season. When you find something that needs attention, note it in your logbook and get after it in a timely manner. Don't allow issues to pile up until they present a punishing and discouragingly expensive mountain of work.

These are the main areas I would recommend paying close attention to in your regular inspections:

Hull bottom: Look for cracks or separation (which would invite water intrusion) in the sheathing or attachment of stem, keel, or skeg. In fact, close inspection of every part of the boat below the waterline is vital; you must seal any breaches in the watertight integrity of the hull as soon as they appear.

Hull bottom and sides: Inspect major scratches and dock dings for penetration of the fiberglass–epoxy sheathing. Even if the scratches are superficial, you might as well touch up the paint right away. Touch-ups will take a matter of minutes instead of waiting for the blemishes to

mount into a repainting task that demands days of labor.

Hardware: Check the tightness of all fasteners. Inspect screws and bolts for corrosion. If you can see any discoloration of the wood around the fastener head, water is likely getting into the fastener cavity in the wood, and it will rot both the wood and the screw or bolt. Stainless steel will corrode when in a tight fitting where it is deprived of oxygen. If something is loose, you are best advised to remove the hardware completely and then re-bed and tighten the fasteners, preferably with new ones. There was some reason that caused them to loosen, and completely redoing the installation should rectify the problem.

Rigging: On a sailboat, there are enormous and relentless stresses on both the standing and running rigging, and the entire rig is only as strong as its weakest link. A broken shroud can quickly lead to a dismasting, and it will occur at the most critical time—a big blow, where you most need the rig to stay intact and work for you. Inspect all the lines for chafing and all the hardware for signs of corrosion, fatigue, or loosening from a mounting point. If your boat lives in a warm climate and/or saltwater, a good freshwater showering after every brisk sail on all the stainless standing rigging will help prevent saltwater-related corrosion.

Brightwork: Inspect regularly for signs of varnish breakdown; check for peeling, separation at joints, and weathered or discolored wood where the protective finish has worn away. If you have used bedding compound on an exposed piece of trim such as a rubrail, inspect it annually inch by inch for small cracks that can allow rainwater to squirm behind the rail. These are really critical seize-the-moment issues; the time to address them is as soon as you notice trouble developing. Procrastination will only assure a compounded and much more time-consuming issue.

MAINTAINING

Cleaning: It may sound elementary, but thorough, periodic cleaning is essential to any boat's long-term health. Sand, grit, and salt will quickly wear away paint and varnish, corrupt moving parts, and work themselves into crevices where they will contribute to a general air of shabbiness. Frequent freshwater rinses, a collection of chamois rags, and a powerful vacuum are all good tools to keep these demons at bay.

Motor: Your outboard or inboard will come with an owner's manual that includes a detailed schedule of preventive maintenance. Read it and follow it. For small outboard-powered boats that live on trailers, preparations for the motor's winter off-season are particularly important. Drain fuel from the hoses, fuel pump, and carburetor. If the boat is used in saltwater, wash the exterior of the motor and shaft thoroughly, and flush the cooling system by running the motor for a few minutes in freshwater. This is the minimum; manufacturers may recommend further procedures. There are sacrificial zincs on all outboards of somewhat modern construction, and these need to be replaced annually on any outboard that is run in saltwater. The principle is that the less noble zinc will attract the corrosive electrical current present in saltwater before the more noble metals in props, shafts, and other critical hardware. Inboard engines and drives are particularly dependent on these zincs, and they should sometimes be replaced more often than annually if the local marina or mooring is known to have stray current corrosion issues.

Exterior: Any wooden parts exposed to water or sunlight must be protected, and that protection must be renewed from time to time. The best protection, as I have written, is opaque paint in light colors. For brightwork, it is much easier to sand lightly and revarnish a moderately deteriorated area than to wait until the disease advances to the point where you must strip everything down to bare wood and start over. Exterior brightwork will likely need a complete renewal every year with at least two coats. The first "heals" the old surface layer, and the second renews the UV protection for the wood.

Hardware: If you have bronze hardware, it will over time (a very short time!) oxidize and acquire a green patina called verdigris. Most traditionalists advise leaving it as is; the patina does not damage the bronze. If you insist, you can remove it by making a paste of white vinegar, salt, and flour, applying and leaving it on the bronze for 30 to 60 minutes, scrubbing with a brush, and then applying a polish to provide a little protection before the next round of oxidation starts up. You will repeat this tedious process several times a year. To my eye, weathered bronze is a beautiful thing and not something to fight. Bronze is the best of the alloys available to the boatbuilder for metal parts and fittings for its combination of durability and strength, so let us celebrate it for what it is.

Safety equipment and spares: Because accessories such as emergency lights and air horn are seldom used, they're easy to ignore and forget—until you really need them and then discover their batteries are dead, or the horn's air is no longer compressed. A simple procedure is to check the operation of all these accessories at monthly intervals through the boating season and schedule a battery replacement annually. The batteries you extract probably have some good life left in them and can be recommissioned for noncritical uses around the house.

It should go without saying to check the fire extinguishers and have them recharged if necessary. Note that I wrote "extinguishers," plural. A fire aboard a boat is a sobering event. I have experienced one myself, and after expending all three of my onboard extinguishers, I still wondered, not in idle curiosity, whether the fire was truly out. Fire extinguishers are classed according to a code indicating what type of fire they will deal with: Class A is for fires that leave an **ash,** such as paper, wood, or plastics. Class B is for fires that **boil,** meaning flammable liquids such as gasoline, diesel, or alcohol. Class C is useful for fires that involve a **charge,** meaning anything related to electrical systems and equipment. On my large boat I now keep five fire extinguishers of type A-B-C that will fight all the fire types. I use the extinguishers that have metal tops (not plastic) because these more expensive extinguishers can be refilled and checked annually by professionals for optimal performance.

REPAIRS

The winter of 1990–1991 was unusually tough in the Pacific Northwest, marked by more rain than anyone could remember. In just two days, storms battered the Puget Sound area with 16 inches of rain, accompanied by several huge wind storms. On one of those evenings after work, I was watching the news and saw a story about a marina under assault from a particularly fierce north wind. The wave action was so intense that it destroyed the marina's breakwater, which allowed steep wind-driven waves to batter the docks. Television cameras panned the watery chaos as boats bounced about, some even being thrown onto the beach by the ferocious surf. People were trying to hold their boats away from the rocks with their hands and feet at great risk to their own health. The eventual casualty toll would include some 30 sunken boats. In the midst of this, I noticed a flash of green bouncing in and out of sight and recognized it as one of my 22-foot Surf Scoters.

Its owner called me a couple of days later to report the sad story. His boat had been tied to the main stem of the dock leading from shore. Off the main dock were a number of smaller finger piers. When the breakwater started to fail, the finger piers broke away from the main trunk, taking boats and dock remains with them. Our poor Surf Scoter remained tied to the main dock, where broken

concrete with jagged edges and steel rebar projecting at waterline level battered and chewed into her hull. A fishing boat also drifted into her and ground away at the topsides. The damage toll included a severely chafed and gouged stern, holes punched into the waterline on the port side, and a large, new, impromptu porthole through the topsides just above the galley. The stem was badly damaged just above the waterline where the boat had surged against the concrete dock, chafing through a $\frac{3}{16}$-inch-thick, $\frac{1}{2}$-inch-wide brass half-oval on the stem and a 2×6-inch timber chafe strip on the dock.

The day after the storm, the owner's wife ran the boat 10 miles across a bay to the nearest haulout facility. This was no small feat itself as she could stand at the helm and see the water from at least four holes in the hull. A couple of weeks later, I looked the boat over to make an estimate of the repair costs, and not long afterward towed the boat back to my shop by trailer for repairs. I am detailing them here to serve as a basic guide to how one goes about repairing impact damage to a Stitch-and-Glue boat.

The stem was the easiest to repair. We cut away the damaged material down to good wood, and then we glued a Dutchman in place with epoxy. A Dutchman, in boat-renovation circles, is a wooden patch piece shaped to fit as precisely as possible into a gap of hull or structure that has been cut away to the same shape.

Turning to the new portside "porthole," I cut out the damaged area with a jigsaw to enlarge the hole into sound, undamaged wood, then routed from the exterior of the boat a step around the perimeter about 2 inches larger in diameter from the original hole, cutting about halfway through the $\frac{1}{2}$-inch-thick hull side. This created a "stepped" opening. Two $\frac{1}{4}$-inch plywood patches were cut: one to fit into the hole and a second, slightly larger one to fit into the routed shelf around the hole. I glued the largest patch in first, fastening through the narrow flange around the hole. After the glue dried, I glued and fastened the smaller patch into place, bringing the repair flush with the hull's interior and exterior surfaces. To hide the repair in the varnished interior, I cut a long butt-block panel of plywood to cover it. The galley countertop was reinstalled in its original position, and with the butt block in place, the patch looked like part of the original construction.

On the exterior, the repair and surrounding painted area of the hull were sanded until the glass cloth layers were visible. The patch was ground to ¹⁄₁₆-inch below the surrounding surface level of the original hull. Two pieces of 6-ounce fiberglass cloth, one covering just the patch and the other cut larger to overlap the hull 6 inches on all sides of the patch, were laminated in place with epoxy and a final peel ply layer on top. A light sanding and a coat of microballoon/epoxy fairing compound leveled the repair surfaces. To save a little time and effort, I covered the fairing compound with 2-inch-wide clear cellophane packaging tape, and then I smoothed the tape with a squeegee. When the tape was later removed, it revealed an extremely smooth patch. (The tape also holds the filler material firmly in place on vertical surfaces, preventing sags.) All that remained was to prime and paint the repair. The Surf Scoter was repaired with about 90 hours of labor, much of which was taken up in the complete stem-to-stern exterior repainting of the boat. The patches were easy and involved no more skill than any part of the Stitch-and-Glue construction process.

Figure 29-1. *This boat was damaged as she approached a dock that had a spike sticking out of it. The owner quickly surveyed the damage and applied duct tape over the scrapes while waiting for repair.*

As any boat owner who has ever commissioned a boatyard to undertake a repair job knows, professional repair (or routine maintenance, for that matter) can be shockingly expensive. One great advantage of Stitch-and-Glue construction is that home builders can undertake their own repairs, even in cases of catastrophic damage such as the above. If you built it, you most certainly will have the skills to repair it. Through more than four decades as a designer and builder, I have also acquired enormous confidence in the durability of these boats, even in extreme conditions.

Surveying all the possible materials for boats, I have found that steel is resilient and very strong, though prone to corrosion issues. Welding patches of steel does not appeal to me. Aluminum allows freedom from paint and concern with some types of corrosion, but it can present issues when other metals are used in concert with it. And frankly, I detest the cold gray color of unpainted aluminum. Fiberglass

Figure 29-2. *After applying a wood flour/epoxy paste over the repair area, I am sanding the patch smooth.*

is strong and relatively maintenance-free, but it is prone to have issues with gelcoat blisters (moisture migration into the polyester/fiberglass laminate). And fiberglass boat interiors still need to be paneled with a wood or fabric ceiling or covering to avoid condensation problems in a cold environment.

A wooden boat doesn't sweat, so if you want, you can have a hull in which all interior surfaces are visible and easily accessible. Even if you're cruising in a remote area and suffer hull damage,

precut plywood patches with waterproof polyure-thane caulking at the edges and some bronze ring-shanked boat nails or stainless steel sheet metal screws are all you'll need to quickly fasten a temporary patch over most any hole. And with wood's relatively light-weight-to-high-strength ratio, a Stitch-and-Glue hull rates higher than most in efficiency and durability in extreme conditions.

I cannot overstate how valuable it is that the Stitch-and-Glue boat can be built by the amateur

Figure 29-3. *Body-working the repair area prior to repainting.*

Figure 29-4. *Later, with the boat hauled out of the water, it is now time for a more complete repair. The patch area had a grid cut into it that allowed the damaged wood to be removed.*

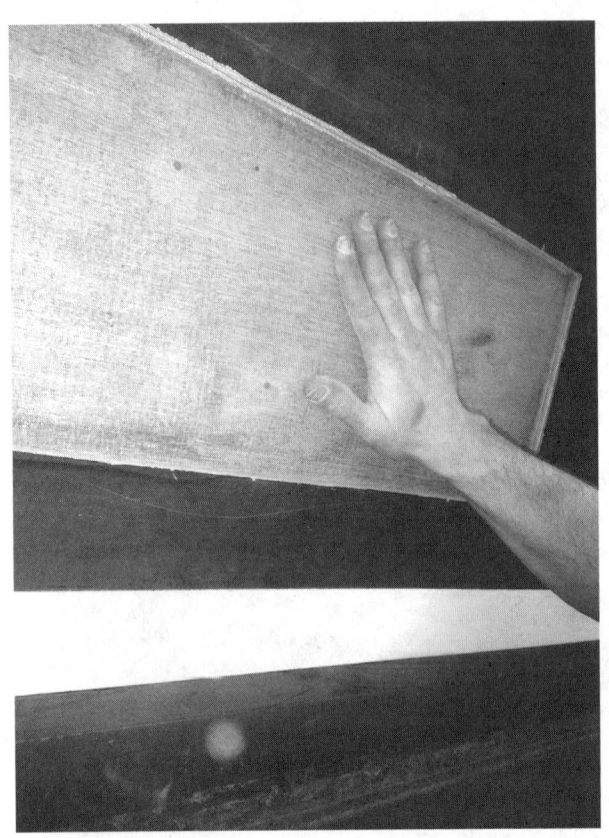

Figure 29-5. *A new butt block of marine plywood will be glued behind the cutout with epoxy.*

and maintained by the same amateur. This provides a sense of control and satisfaction that is almost vanishingly rare these days, when we need specialists to cope with almost every other commodity in our lives. Wood is inviting to work with, forgiving, and in many respects superior to its synthetic alternatives. When combined with the strength of modern epoxy and fiberglass, you have the most nearly perfect combination of approachability, maintainability, and durability.

Figure 29-6. *I have sanded the repair area to a slightly lower level, leaving a depression to contain several layers of glass cloth and epoxy over the patch.*

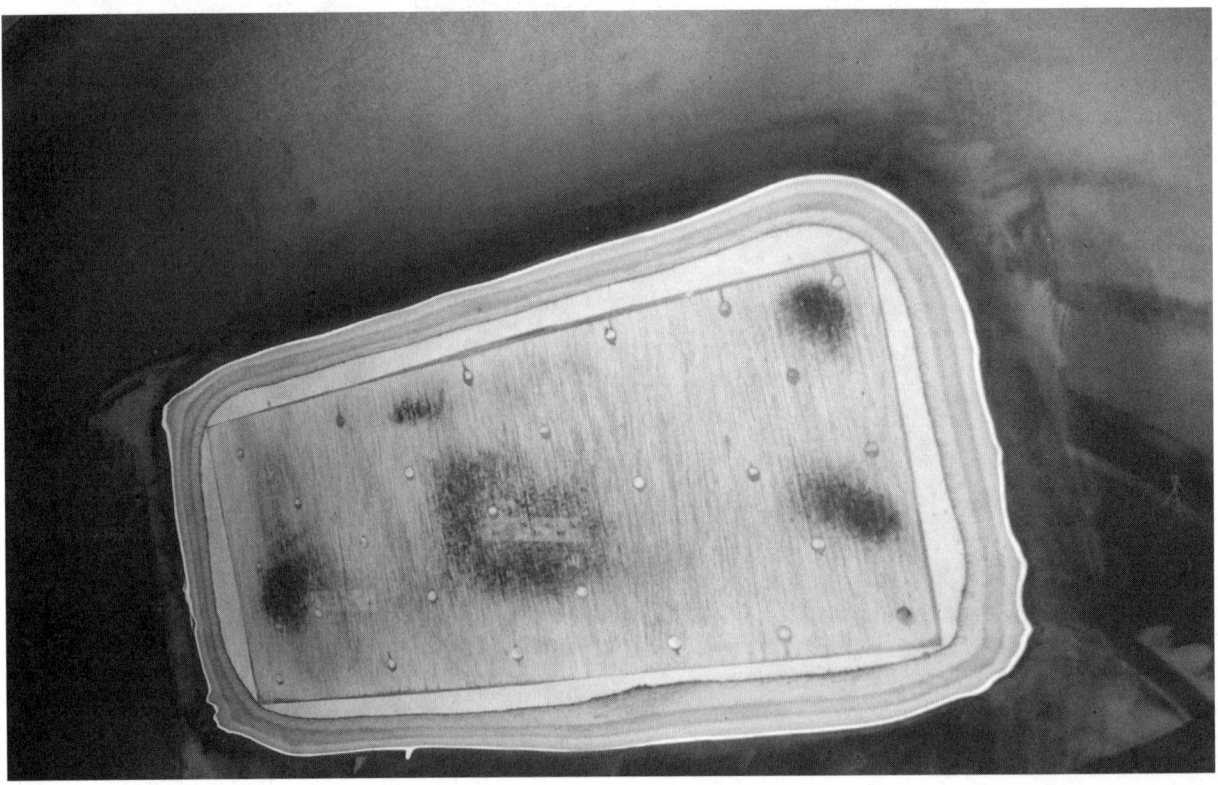

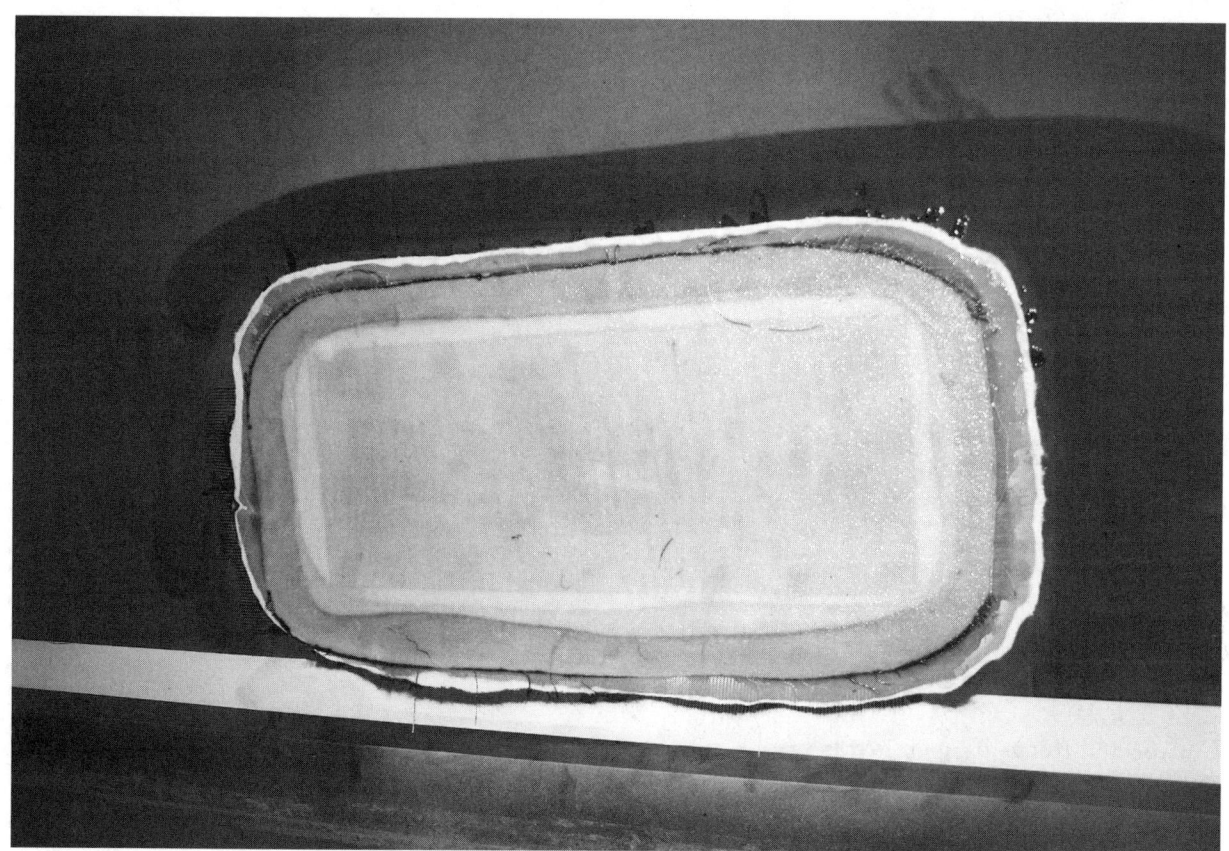

Figure 29-7. *Several layers of cloth have now been applied, and it is time to refinish the surface.*

CHAPTER 30

Building Devlin Boats: The Amateur's View

By Lawrence W. Cheek

Unexpected things happen inside your head and heart when you build a boat. You aren't fabricating a complicated toy. You're undertaking a life-changing experience.

You surf a tidal cycle ranging from elation to despair, and you learn to manage it, building a reservoir of perseverance that surprises you by seeping into all the other corners of your life. You learn that perfectionism is not your friend, so somewhere along the way you let go of the shimmering vision of uncompromised beauty you've held since you first gazed at the plans for your boat, and oddly, this surrender makes you happier. You gain confidence in yourself at the same time that you plumb new depths of humility, and you're startled to find that these two qualities don't contradict each other.

You've taken a plunge into the unknown, attempted something you really didn't know whether you could handle, and through sticking with it, built up your character along with your boat. Of all the reasons (some of them dubious) for building a boat, this is the one that's unassailable and that will stand up against all argument.

I have built three Devlin boats. The first, a 13 foot 6 inch Zephyr sailing dinghy, I chronicled in my memoir *The Year of the Boat* (Sasquatch Books, now out of print). I was underqualified for the job on every score, but I persevered. Halfway through, I named her Far From Perfect, which wasn't so much a self-deprecating poke as it was a way of accepting the imperfection I was building into her, even though I was doing the best I knew how. John Vigor, a respected sailing writer, wrote a kind cover blurb for the book and then sent me a private e-mail. "When I read the manuscript, I thought you were exaggerating your ineptitude," he said. "Then I saw the photo of the boat, showing all those screws in the stem without bungs. You really *didn't* know what you were doing."

My smartest decision, following Sam Devlin's advice, was to build this relatively uncomplicated boat first. A sailing dinghy wasn't my ultimate goal; it was a way of testing the water and gaining experience. Far From Perfect proved to be a delightful little craft to sail, and she never failed to draw appreciative comments at launch sites (which was more credit to Sam's design than to my execution), but almost as soon as she was finished, I bought plans for the vastly more ambitious Winter Wren II.

The Winter Wren is a gaff-rigged pocket cruiser, remarkably traditional in appearance despite the modern Stitch-and-Glue construction. At the time,

Figure 30-1. *Far from Perfect on her first sail on a very calm suburban Seattle lake.*

Figure 30-2. *The Winter Wren II Nil Desperandum rests at a dock in a peaceful cove on Puget Sound.*

it was the largest boat I felt I could reasonably build without help, and the longest (18 foot 8 inch on deck, 22 foot 7 inch, including bowsprit and rudder) that the garage could swallow. (I live in the Pacific Northwest, where one does not build boats outdoors.) The design beautifully fit all my requirements: trailerable behind a Subaru Outback rather than a gas-guzzling truck, simple cabin accommodations for weekend cruises, and safe sailing in the protected waters of Puget Sound. And most of all, foxy good looks. If you're going to invest 3,000 working hours and $20,000 in parts and materials, you need to be in stupid, helpless love. I loved the Winter Wren from first sight of the plans. Many others, boat-knowledgeable and not, have concurred. When I took the plans to an Office Depot to make a backup on a blueprint photocopier, a young clerk looked over my shoulder and gasped, in an almost reverent whisper, "Holy cow!"

Over the three years it took for the Winter Wren to assume form in my garage, the promise of beauty did not grow stale. Sam is more than a technically capable designer; he's an artist. He's taken plywood further than anyone suspected it could go. This boat (along with many others) belongs more to the sculpture garden than the lumberyard. When I started to build the cabin trunk, I thought I'd raise the roof an inch to carve out a morsel of extra headroom. I built a cardboard-and-duct-tape mockup before ripping into my $100-a-sheet okoume plywood, which I studied in place on the hull for two days and finally concluded, Damn, Sam's eye had it right, immutably. A change of *one lousy inch* ruined the proportions. My mockup was a cartoon cow with an oversized head. I tore off the cardboard and built the cabin according to plan. It felt like I had dodged the temptation of sin.

My third Devlin boat is the 21 foot 3 inch Song Wren, which finally, absolutely, *is* the largest I can shoehorn into my present shop (and I blew out all of Chapter 2's sensible suggestions for working space to do so). This was major boatbuilding. I had to repeatedly impose on friends to help move, hold, or persuade big pieces into place, and the ballasted keel, bowsprit, and rudder installation all had to wait until the final six weeks when I moved the boat outside. Its length overall comes to 26 feet 7 inches with a weight of 2,800 pounds. Although this is still a pipsqueak by contemporary yacht standards in America, it has the feel (to me) of a large and substantial boat, with sleeping accommodations for up to three that invite real cruises lasting several days. The substantial cabin volume provided an opportunity to build an inviting and comfortable living space with lots of varnished sapele, and the complexity of the sailing rig (a cutter, with gaff mainsail and two headsails) set me up for several 2 a.m. worry sessions (sometimes productive ones, actually). This boat consumed about 4,500 working hours to build, and the cost of parts and materials came to $36,000 in 2016–2019.

This sounds like a boatload of money, and it is. Few production sailboats in the 20- to 22-foot size are still being made, but two that are being manufactured carry list prices, fully outfitted, of less than $36,000. There are numerous cheaper ways to get out on the water. It takes a prodigious leap of faith to invest so much money and labor in an amateur venture. What I found is that building the first two boats gave me the courage to take the leap and that this courage and faith has pooled into an internal reservoir that has changed my life. I am not the timorous creature I once was.

Building that first boat, I felt I was always skating the rim of disaster, so near the edge that one more wrong move would be the last, the fatal tumble. Or maybe I'd already made the fatal mistake and was just too ignorant to know it. I still relished the work—or so I told myself. But when I laid it out in my book, the story was tinted with fear and the dark nag of pessimism. The truth, now that I have 15 years of perspective on it, is that it was a complicated time in my life, a whirlpool of contradictory emotions. It was overwhelming, frightening, and full of terrible self-doubt, at the same time that it was deeply engaging and incredibly fun.

On the second and third boats. my skills with tools and problem-solving dramatically improved, but even more importantly, my level of confidence rose higher and higher. I no longer felt there was any mistake that would amount to doomsday; whatever blunders I made could be redeemed. When I made the Winter Wren's spars, I remember thinking, *Hmm, the boom seems to flex more than I expected. Well, no big deal. I'll go sailing, and if it seems too flimsy in use, I'll simply rip it down the middle, fortify it with a strip of good, stiff mahogany, and put it back together.* When I realized I'd just hatched that contingency plan with neither fear nor dread, I knew I'd passed a milestone. And by the way, the original boom as Sam had specified it turned out to be just fine.

Another illustration of progress: On the Winter Wren, I cobbled together an electrical system that worked but looked, well, cobbled, with plastic conduits snaking around the cabin like tendrils of kudzu. Just being able to flip a switch and have electric light seemed like a stellar accomplishment. When it came time to wire the Song Wren, I knew enough to rout dados into the back sides of the cabin overhead support beams and hide the wiring in them. There's a complete switch and fuse panel for the lights, bilge pumps, and depth sounder. My craftsmanship still will not be mistaken for that of a professional, but I don't feel like an imposter any more. I am a boatbuilder.

I have, by the way, strayed occasionally from the Church of Stitch-and-Glue. Among the five sailboats and two kayaks I've built, four of the seven have been Stitch-and-Glue, two have been plywood lapstrake, and one was strip-planked. I've had no experience with cold-molding or traditional carvel planking. But I'm firmly convinced now that Stitch-and-Glue is the best option, at least for the amateur builder—the easiest, fastest, and least sleepless-night-inducing route to making a good boat. With the techniques described in this book, there are no limitations to the performance or beauty of these boats.

Building one of Sam's boats requires a lot of thinking, whether at 2 a.m. or a more reasonable

Figure 30-3. *The Song Wren 21 PattyB on her first sail.*

One advantage of building your own boat, of course, is that you can depart from plan and tailor it to your own likes and requirements. There are some things we amateurs can tinker with and others we should avoid touching. Most of my departures have been cosmetic, since I'm unqualified to tamper with Sam's engineering. I veneered the deck of Far From Perfect, which at first looked very cool. Dumb idea, as a sunny summer day loosened the contact cement underneath and blister-bubbles bloomed in the veneer. But other ventures have worked out nicely. On both Nil Desperandum and the Song Wren PattyB, I built lightweight ceilings of fir and sapele, respectively, inside the cabins, which lends them an air of refinement that fiberglass production boats of this size don't have. A more substantial alteration on the Winter Wren was to build 14 cubic feet of closed-cell foam and watertight air compartments into the bilge and assorted crannies. This works out to 874 pounds of positive flotation, a margin of safety over the boat's 685 pounds of lead ballast. In the worst-case scenario, a knockdown and total swamping, she will stay afloat. Most cruising sailboats will not.

There is one further life dividend that grows from building your own boat, and it is a profound one. The only term large enough to embrace it all is *character*, which encompasses more than self-confidence. In building a boat, you will reach a new level of integrity, which I define as the determination to do the right thing even when nobody is watching. People's lives may depend on the strength and seaworthiness of the boat you build, so you will not cut corners. You will do the best work you possibly can, and if you don't know how to do something right, you will gulp down your pride and yell for help. You may even realize that you have an obligation to something larger than yourself, something beautifully expressed by a character in the Richard Powers novel *The Overstory*: "When you cut down a tree, what you make from it should be at least as miraculous as what you cut down."

hour. As he'll readily tell you, these aren't paint-by-number plans. You get good drawings and measurements of all the critical pieces, but many details are left to the builder's imagination (or worry). Okay, I see from the plan how long and high the cabin sides need to be, but what happens at the corners where the sides meet the front and back and top? The plan is mum. And this is not a doghouse. The sides intersect at wicked angles and must of course be absolutely waterproof *and* strong enough to resist any torsional stresses the hull and deck might shoot up through them. Well, another of my fortunate decisions long ago was to dedicate a fat ring-binder notebook to articles and notes on boatbuilding I collected over several years. A *WoodenBoat* magazine article provided the guidance I needed. The corner posts took a ton of time, and I had to throw away several flawed attempts, but they worked.

To this point, I've been writing largely about the psychological and character issues you'll encounter

Figure 30-4. *A view from Nil Desperandum's cockpit under sail, showing her curved cabin sides and homemade sapele portlight frames.*

in building a boat, and I hope I haven't scared you. That's not my intent. It should be obvious, but I'll say it straight out: I've come to love building boats. More than likely you will also. It's a life-changing experience, and the changes will be entirely for the better. So suck it up and do it.

From here on, I offer practical advice. I've come to know Sam well, and I think he'll agree with most of it, but if we diverge, it's simply because we have different perspectives. I remain an amateur.

- What boat to build? I know neophytes who've leaped into the Ultimate Boat the first time out. One intrepid soul on an island near Seattle built a 43-foot schooner as his starter boat; it took him 33 years. But I know many more examples of boats-to-be abandoned because someone took on an overwhelming project and rightly got overwhelmed.

  Take a sheet of paper and on its left describe the boat you *know* you could build successfully—maybe an 8-foot dinghy. Then forget that one; it won't stretch your abilities. On the right, lay out your ultimate dream—the boat you'd build if you had all the time, money, space, and skill in the world. Then forget it, too, because you have none of these things. Now move to the middle and write out realistic specifications for the compromise. This boat will stretch the envelope of what you think you can do but not to the breaking point. It will suit the conditions in which you plan to row, motor, or sail. You will find it beautiful and it will give pleasure, even if it isn't as big or as sophisticated as you'd like. And it will fit in the space you have to build it.

- If you are a willful, flinty cuss who loves to do everything the hard way, you may disdain the idea of a Stitch-and-Glue boat kit, whose very mention emits the reek of beery Sunday afternoons and an IKEA box. You've got it wrong. A CNC-cut hull kit still provides all the challenges and creative exercise in boatbuilding you could want; it just erases the opportunity to make the irreversible Big Mistake—gluing

up a misshapen hull—and deprives you of the disagreeable work of scarfing plywood sheets. I've built boats both from scratch and kits and greatly prefer starting with a kit. It saves more in time, fear, and grief than the money it costs.

- Be prepared both emotionally and financially to build every part of the boat twice, particularly if it's your first. In practice it won't be that bad, but you will throw away a lot of misfitting, malfunctioning, or broken pieces, and things will go much better if you can toss them without regret or frustration. When you curse and kick yourself for making a dumb mistake, you carry the residue of that resentment forward into the next step and build it into the boat. Possibly the most important skill I've acquired in boatbuilding is to move from problem to solution without getting mired in emotional muck in between. I limit myself to a max of 30 seconds of umbrage and outrage, and then I flip an internal switch and simply go back to work. Few of us have this equanimity by nature, but it can be learned.

- The boat will take between two and four times as long to build as you think it will, so prepare accordingly. One of your greatest challenges will be to preserve momentum. There will be doldrums of tedium and discouragement as well as flights of ecstasy. What you do is ride them both and just keep going. If you expect this cycle, like the ocean's tides, it'll lose its power to stall you.

  On days when I don't want to work on the boat—when I'm feeling listless or pessimistic or the main task presenting itself is forlorn and tedious, like fairing the hull—I still take half an hour to do some small, innocuous job. This discipline keeps me in touch with the boat and represents forward progress, however small.

- You begin a boat with a vision of perfect beauty formed in your mind. If you have only an amateur's skills, as I do, somewhere you will have to reconcile that perfect vision with the imperfect that's possible. You can then choose to either feel like a failure or be satisfied with the equilibrium you've found. I like the advice Bill Withers, the Grammy-winning R&B singer/philosopher, says he gave his kids: "You know, it's okay to head out for wonderful, but on your way to wonderful, you're going to have to pass through all right, and when you get to all right, take a good look around and get used to it because that may be as far as you're going to get."

It's smart to establish a hierarchy of values that helps you decide how good to make various parts of the boat. Anything that affects the boat's integrity and safety is obviously of the highest importance, and there must be no fudging, no slacking, no excuses. Cosmetic issues are another story, and I think we amateurs should choose priorities. Some boatbuilders cannot abide the thought of even an unseen, out-of-the-way place like the bilge or anchor locker having anything less than perfect skin. I suspect such people are not a lot of fun to be around. My bilges look like hell, as bilges should. Among Nil Desperandum's visible pieces, I decided that since I didn't have the time or skill to make *everything* beautiful, I would lavish attention on a few distinctive details that would give pleasure. For example, instead of straight cabin sides that (to my eyes, at least) would visually contradict the graceful contours of the Winter Wren's sheer, I laminated them into a curve. This spawned a brood of complications that added a dozen hours of work, which is something I could not have done if I were in the business of building boats. The result, to me, is worth the effort.

- Regarding tools, I've done it both ways, minimalist and maximalist, and I recommend the best. Buy the tools you think you can't afford and don't deserve.

I built Far From Perfect with a cheap drill, cheap orbital sander, cheap jigsaw, and a cheap homeowner's smattering of cheap hand tools. I added a cheap bandsaw, the most useful stationary tool for a boatbuilder to have, if you can have only one. These tools were adequate but just barely. I always felt like I was fighting to make them do what I needed.

When I began Nil Desperandum, I knew I had grown serious about boatbuilding, that it was a passion worth an investment in quality. I won't lay out the complete list of tools here, but it cost more than $5,000 and included a table saw, jointer, drill press, and some German-made Festool power tools such as the extremely versatile orbital/rotary sander. I have since more than tripled that tool investment. There are Lie-Nielsen planes, objects of such grace and integrity that they almost qualify as sculpture itself. A table-mounted router makes interesting details possible, even easy. These tools were and still are better than I am, which creates both a sense of obligation to grow into them and feelings of pleasure and trust as I use them.

- You will need help. Problems will arise that you cannot or should not try to resolve by yourself. Happily, the Internet has become the boatbuilder's greatest tool since the power sander. There are online boatbuilders' forums where you can post a question and enjoy a shower of answers, sometimes salted with a sprinkle of abuse, within hours. A search term such as *small sailboat rigging* will turn up helpful technical papers or videos. Amateur builders' blogs will illustrate how others have solved your problem. The subscription video library Off Center Harbor (www.offcenterharbor.com) is an excellent resource of more than 500 how-to videos and visits to inspirational wooden boats.

Neighbors are important. Even on a small boat, periodically operations arise that require three or four hands, and you will need to borrow the extras. Assuming you're considerate—you take care not to annoy your neighbors with power-tool racket or noxious fumes, and you offer your growing skills to fix things for them—you'll probably find the neighborhood taking an avid interest and

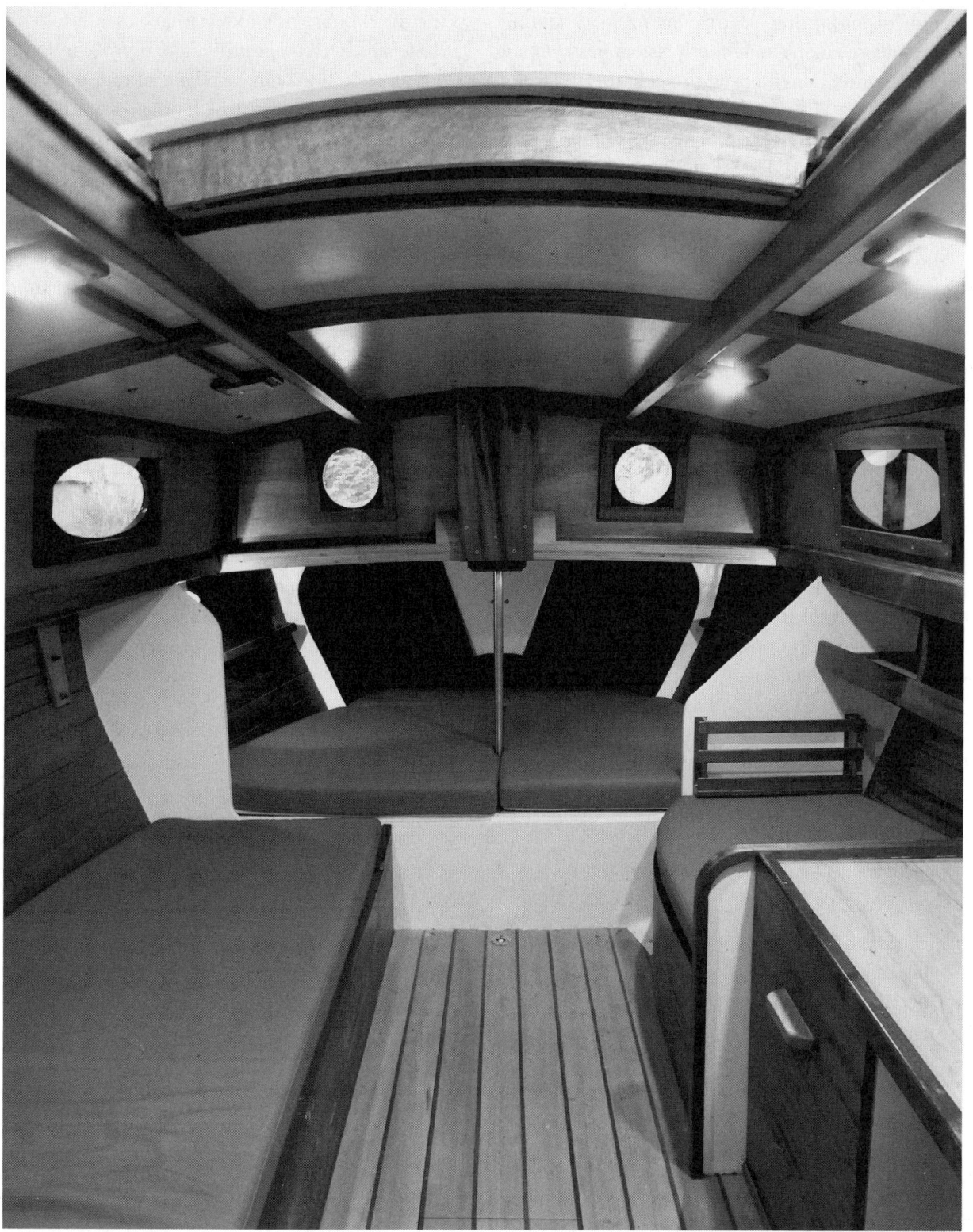

Figure 30-5. *Patty B's cabin was built to be inviting and comfortable even though its maximum headroom is only 50 inches.*

almost "adopting" your boat project. Among my most valuable neighborly assets has been an aeronautical engineer with substantial sailing experience. I called him in for consultations several times; in return I took him sailing. There was no downside to this relationship.

- Set up a separate checking account to fund the boat construction. This helps avoid anxiety and domestic disputes because purchases won't feel like they're gnawing into the family's essential living expenses. Be sure this account provides for cost overruns because they're all but inevitable. These overruns often result from your own mistakes, but this is not your moral failure fund—it's your learning opportunity account.

- Back in Chapter 25, Sam strongly recommends that we "paint nearly everything" on our boats' exteriors rather than varnishing. I have repeatedly and flagrantly flouted his advice, leaving vast acreages—spars, cabin sides, cockpit seats, and much more—as brightwork to celebrate Nil Desperandum's and PattyB's woodiness. I am now sorry. Varnished woodwork is wonderful, but spending great swaths of your life—in my case, up to two weeks a year—maintaining it is not. A wooden boat can be beautiful with little or no brightwork. Time is our one truly nonrenewable resource. Accept Sam's advice.

- Finally, on the relationship front, our family members can be affected by our projects either negatively or positively. Negatives may include draining the family savings, obsessing to the point of neglecting spouse or children, taking over the garage or carport, and crudding up the doorknobs with epoxy. I have committed all these sins. On the positive side, if you can build a boat, you need not be afraid of any domestic fix-it or build-it chore; you automatically become more useful. (I have now built most of the furniture in our home.) You may also become better company at least in the hours you are *not* working on the boat. The wife of a friend who built another Devlin Zephyr told me, "I've seen him smile more in the last eight months than any time over the last eight years. He was just glowing over it."

A spouse or partner who's on board with the project can be of incalculable help even if she or he never lifts a paintbrush. Once when I was in an emotional ebb tide on Far From Perfect, my wife Patty gave me a card with a painting of a grand square-rigger on the front and a note inside: *For the boatbuilder—love of my life*. When I wrote my book, I dedicated it to her. I reprise that dedication here because it has continued to apply to us through the building of seven boats now, and it also serves as a model for how our families, and maybe ourselves, might best embark on building a boat.

For Patty
who may have doubted
but never spoke of it.

Lawrence W. Cheek has written 15 books on travel, natural history, archaeology, architecture, and boatbuilding, and is a frequent contributor to WoodenBoat *magazine. He lives on Whidbey Island, Washington, with his wife, Patty, who shares his love of sailing and kayaking.*

In Appreciation

There is no road map a dreamer can follow to become a boat designer and wooden boat builder. You just have to keep your pilot light of inspiration shielded and recognize when the flames of creativity need to be fanned into bursts of endeavor and perseverance. This modern world can send many ill winds to blow out the light. Money is probably the fiercest one; figuring out how to make a living designing and building wooden boats remains one of the most difficult challenges. The necessary evils of taxes, licenses, bureaucracies, and all the required ecological permits constitute others. And finally, strange as it may seem, to make a living designing and building wooden boats, you need to sell your creations, marketing them and indeed standing by them for an inordinate amount of time. All these tasks must be done while carefully managing your own creative and physical energies. Many of your customers have only a sketchy understanding of what efforts it takes to design and build boats for a living, but you must work gracefully with them as they become your benefactors, your reason for being, and your passport to continue doing this thing with your hands, your mind, and your spirit.

With these pitfalls always lying in wait, I have found that I am only as good as the people who have helped me navigate them. So first I must give thanks to my parents for allowing me enough adversity in my early life to toughen my hide, and enough nurturing to inspire me to keep going.

Next, thanks to my first wife Liz, who shared my early dreams for this life, struggling along with me so that we could live the way we wanted. And today as I write this, I give thanks to my present wife Soitza, an ever-bright light in my life, truly an amazing person, largely sacrificing her own dreams to stand by my side and help with mine. I feel that with Soitza I was given the opportunity to pluck a diamond away from Mazatlán, Mexico (her original home). She has brought a new essence into a life that was already full of spice, but the flavors are different now, more poignant and intense than I might have expected. Our lives today are the results of our labors together, and it's been a strange and wonderful path.

Thanks to the workers who have stood by my side, lived with my many moods, and kept perspective on the dream. The list is extensive after so many years, but it includes especially Jim Henson, Randy Foster, and Joel Mill, Fred Fenske and my two sons Cooper and Mackenzie, who have chosen to stand at their dad's side and work hard with both their hands and their minds.

Thanks to my many customers, who with their hard-earned money have allowed me to follow my dream and who have had the patience to work with me as I helped them with their own journeys, in turn bringing their dreams to life.

Thanks to my friend Larry Cheek; his help and assistance in this book has been unlimited and his contribution in Chapter 30 shows a remarkable slant on this world of boatbuilding with Stitch-and-Glue construction. Larry's own journey in boatbuilding has been epic and shows to all what can be accomplished with some pluck and a lot of persistence.

And in no small part, thanks to a world of interesting and inspiring boats. To me, boats are like fine wine or music: They live in the hearts of their designers, builders, and owners. They elevate our spirits to a higher plane. The days when I feel most successful in this business are the days when I see that I have enhanced someone's life through one of my own creations.

Thank you all for the privilege of such moments.

Sam Devlin

Appendix:
Sam Devlin's Designs

At last count, the Devlin design catalog comprised a fleet of over 100 boats ranging from a 7½-foot dinghy to a 62-foot world cruiser. Nearly all the designs up to 35 feet are available as plans for the amateur builder, and some of the larger ones are also. You can browse the complete catalog online at www.devlinboat.com. For an appetizer, here is a gallery of some of the designs. We'll begin with power boats, proceed through rowboats and sailboats, and finally onto motorsailers.

POWER BOATS

Appx 1   Cackler

LOA	14'4"	4.38 m
Beam	5'9"	1.76 m
Draft	8"	20 cm with outboard in rotated-up position
Weight	375 lb	170 kg dry weight

This is a new-generation hunting/utility skiff designed for locales where seaworthiness is essential and for times of the year when long distances may need to be covered in a hurry to dodge incoming weather. I have been in one with three other hunters, two dogs, and six dozen decoys, and despite that load, her 35-hp outboard would still lift her up on a plane. For those wanting even more capacity, we offer models of her in 16- and 18-foot lengths. As utility boats these all excel in beachability, seaworthiness, and usability.

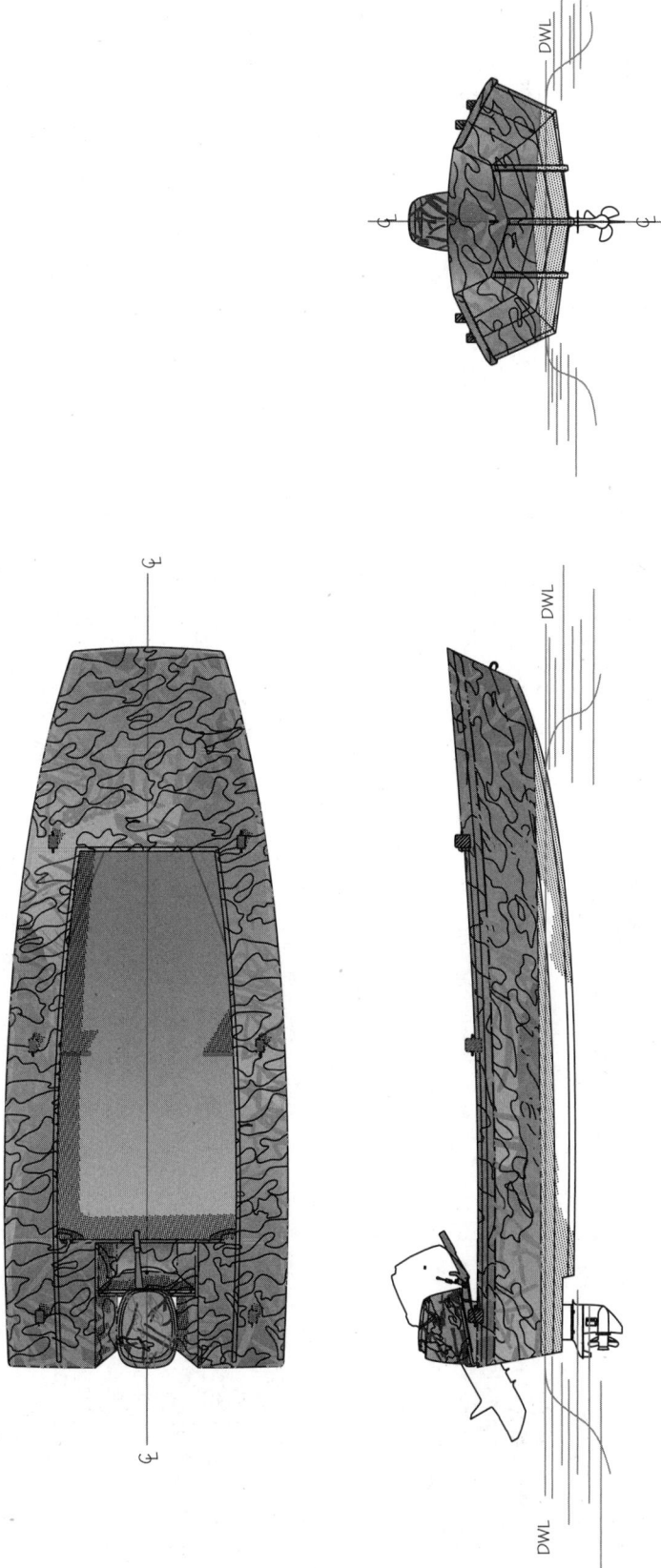

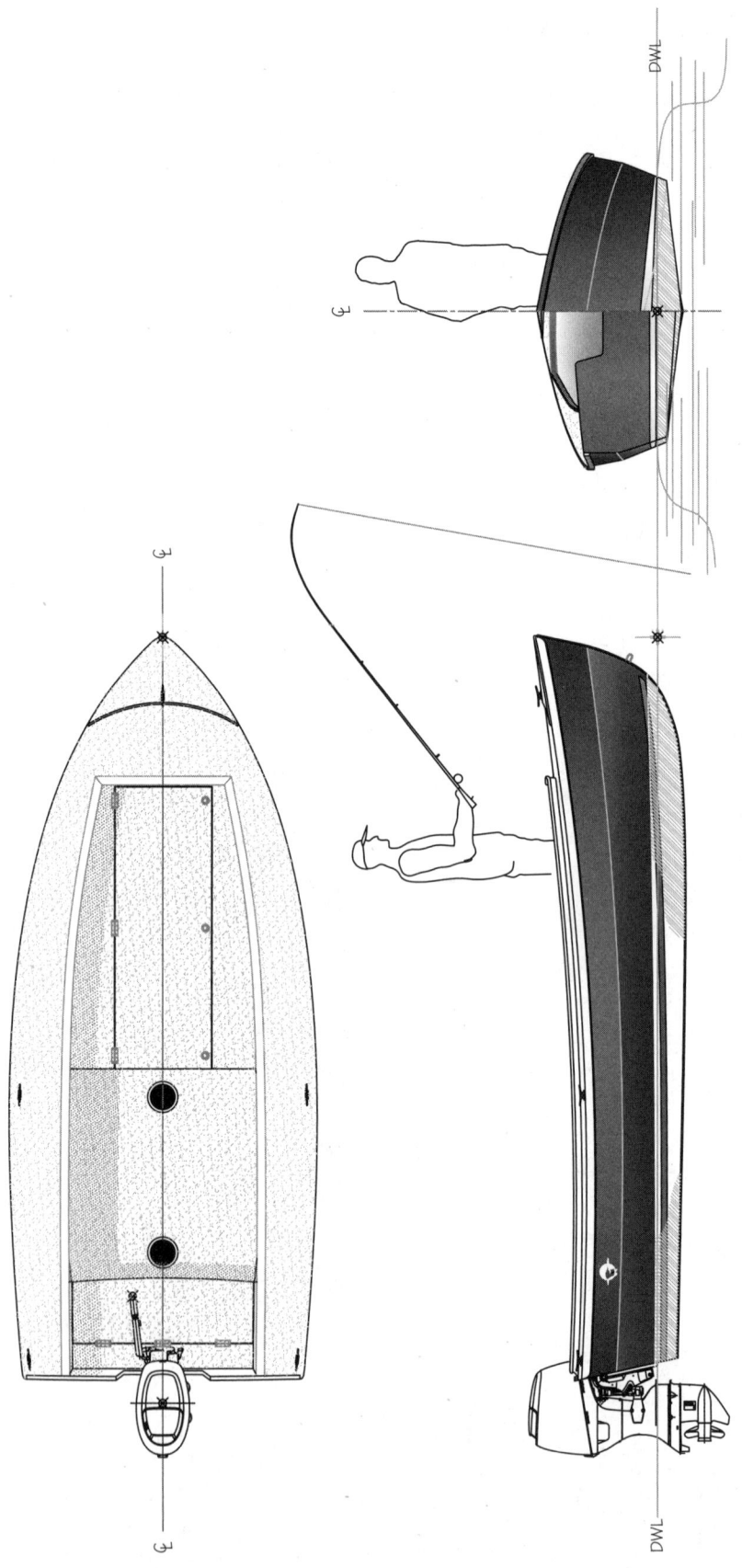

Appx 2   Candlefish 16

LOA	16ft	4.87 m
Beam	6'10"	2.09 m
Draft	7.5"	14 cm outboard up
Weight	512 lb	232 kg dry weight

I designed this skiff for an avid fisherman from the fine state of Maryland. She has a casting deck that allows the fly fisherman a stable platform for spending time with the gamefish of those waters. A 60-hp outboard is about the maximum that one would need for blasting about through shallow waters, but for my own uses in the Northwest something more like 40-hp would work fine. The Candlefish is also available in open models if you are looking for a simpler building project and the quintessential skiff. Also available in 13- and 18-foot models.

Godzilli 16 Mini-Tug

LOD	16'8"	5.07 m
Beam	6'2"	1.9 m
Draft	14"	34 cm outboard up
Weight	860 lb	390 kg dry weight

After the success of our Godzilla 22, I began thinking of a smaller Godzilla-type tug that we could use as a small yard tug for my boatbuilding company. A little bit of log salvage after winter storms and just a general fun boat on the water. As often happens, a prospective customer appeared with exactly that same vision, and soon a couple of Godzillis emerged from the shop. She has a high-thrust 20-hp outboard in a well at her stern, and with a proper tow bar can do an amazing amount of work. A miniature tug can hardly help looking "cute," but this one is also a real boat with the look and feel of a much larger vessel.

Also available in 22- and 25-foot lengths.

Flush Cargo Hatch

DWL

DWL

Appx 4   Dipper 19

LOD	18'-8"	5.6 m	
Beam	7'-3"	2.21 m	
Draft	1'3"	38 cm outboard up	
Weight	1,545 lb	1154 kg dry weight	

Why own a pleasure boat inspired by a tug? How about good looks, great accommodations in all sorts of weather, vast cockpit space, and the visual warmth and individuality of a wooden boat. With twin 10- to 15-hp outboards the little Dipper 19 cruises very economically at 7–8 knots. For more speed a single outboard of up to 90 hp would give a cruising speed at two to three times the speed of the smaller outboards and still have good fuel economy. Small as they are, these are real cruising boats that accommodate double berths, small galleys, and remarkable storage. We have shown the 19-foot model but a smaller sister at 16 feet would make for a fun option if you want something smaller and more trailerable.

Appx 5   Pelicano 20

LOA	20'-2"	6.14 m	
Beam	7'-5"	2.26 m	
Draft	12"	30 cm outboard up	
Weight	1420 lb	644 kg dry weight	

Pelicano 20 is a product of my frequent expeditions to Mazatlán, Mexico my wife's hometown, where I have had plenty of opportunity to observe the fishing skiffs called *pangas* venturing out to the Pacific before the afternoon trades blow in. The Pelicano 20 is my interpretation of the *panga*, adapted for easy trailering and multi-chined Stitch-and-Glue construction. It's a tough, fast, and versatile boat that can even be beached. Plans are available in assorted configurations and sizes from 18 to 23 feet. I have chosen to show you the middle of the pack, the P-20 bass boat version. With a 115-hp outboard, you will enjoy an easy 30 knots.

Available in 18-, 20-, and 23-foot lengths.

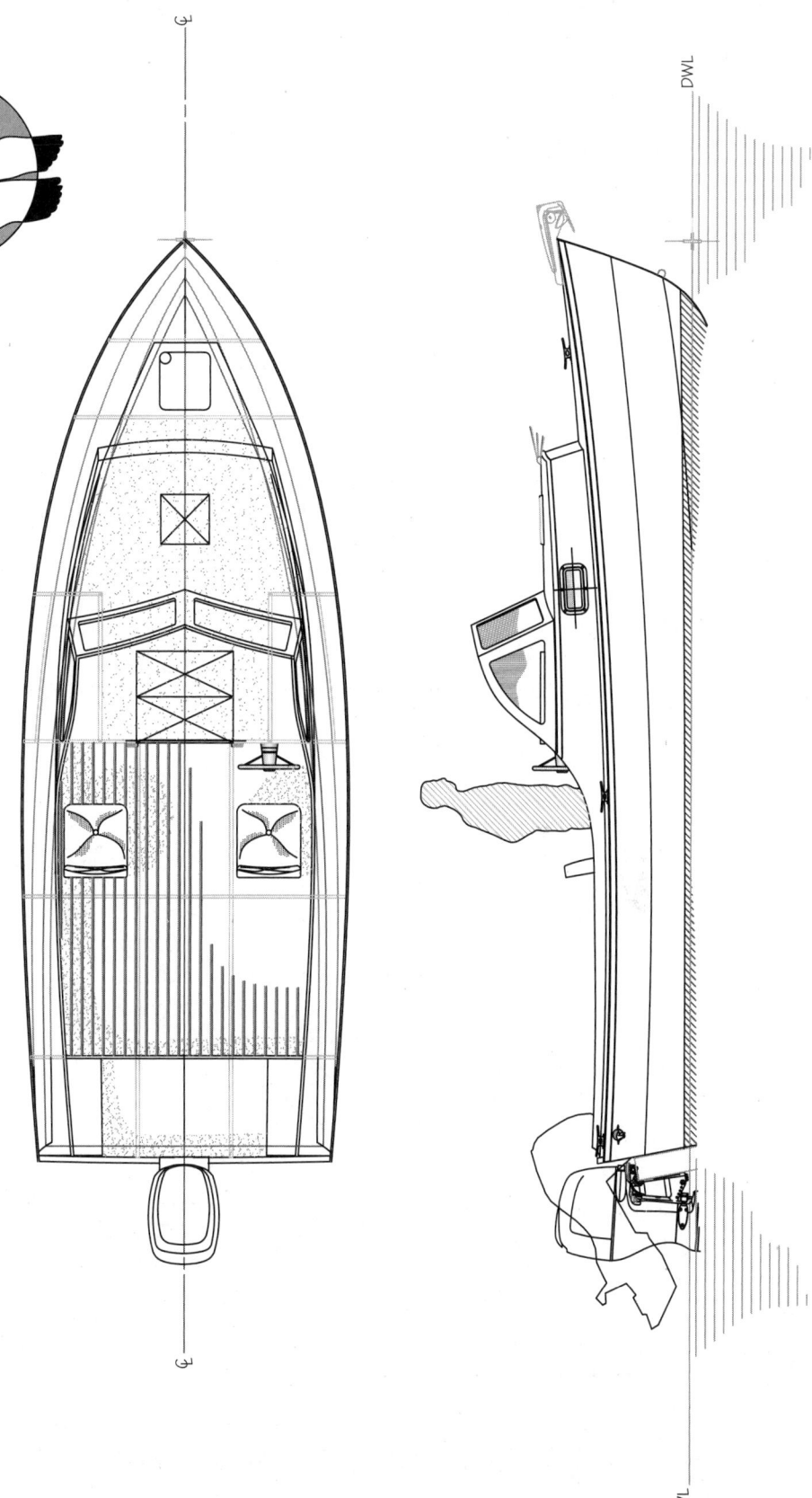

Appx 6   Honker 20 H.B.

LOA	20'-1"	6.1 m
Beam	7'-2"	2.19 m
Draft	13"	32 cm outboard up
Weight	2100 lb	952 kg dry weight
Headroom	6'-2"	1.88 m

The Honker 20 H.B. is a combination of one of our Garvey-hulled designs and the venerable Millie Hill houseboat. The result would be a seaworthy and fast (if given enough power) moveable Shanty boat that would excel at either being trailered to each launching and use or living on the river or waterway more full time. There is comfortable living space inside and plenty of cockpit space for lounging in the sun or enjoying an evening cocktail and appetizers at sunset. She is very trailerable and would be an amazing boat to own for many different uses and purposes.

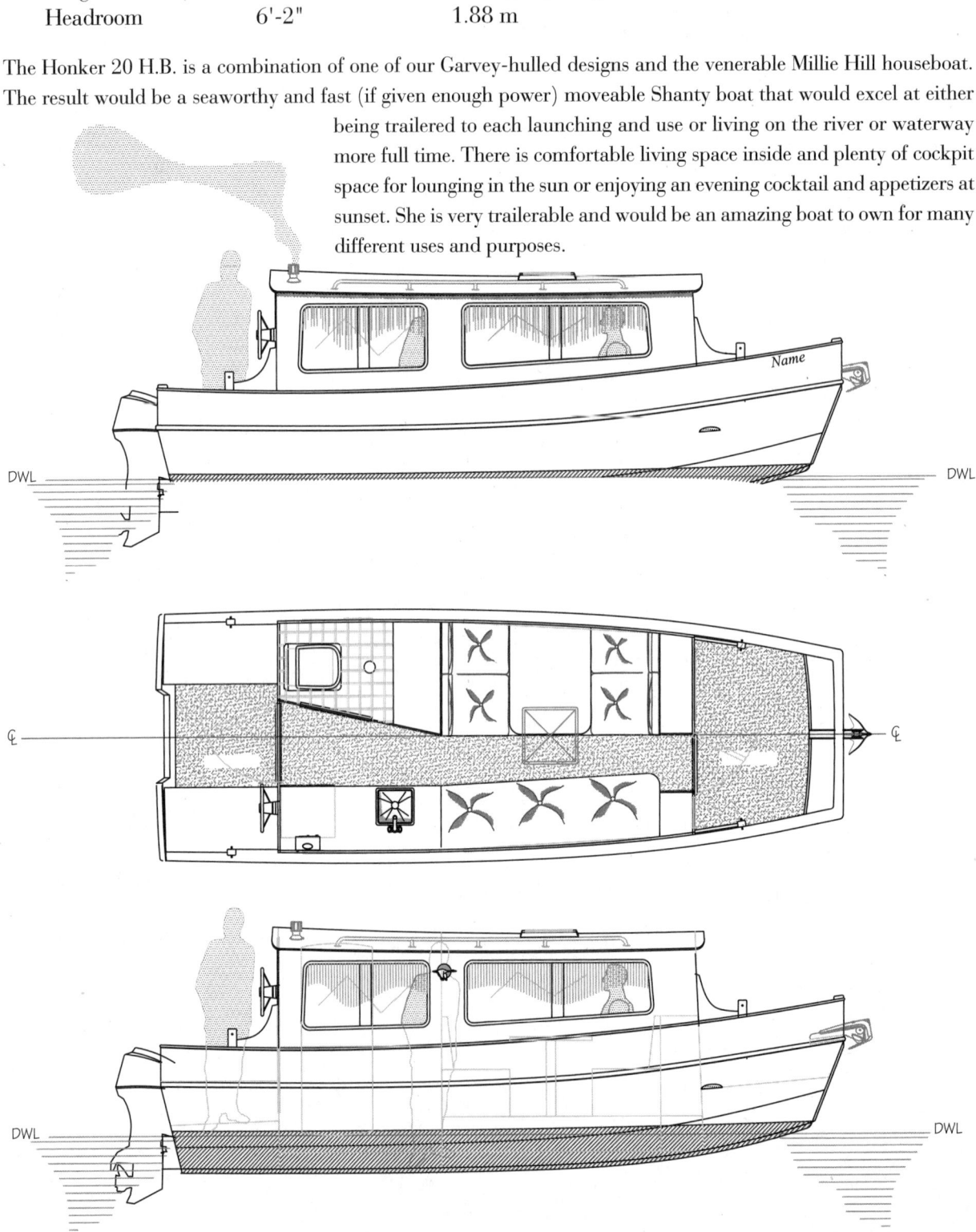

Appx 7   Dunlin 22

LOA	21'7"	6.59 m
Beam	7'10"	2.39 m
Draft	1'5"	43 cm outboard up
Weight	2,400 lb	1,088 kg dry weight

The Oregon scientist who commissioned me to build the first "Dunlin 22 pilot" asked a lot of her. He wanted a spacious cockpit to serve as a blind for observing bird and marine life, and enough utility to serve as a fishing and crabbing boat. He demanded good fuel economy and seaworthiness to punch confidently over a river bar into the open Pacific for a day's fishing. The design filled all these requirements, but I am still tinkering: There is now a stretched pilothouse option called the Dunlin 22 cruiser, tilted more toward cruising than fishing. Recommended power for either is a 90-hp high-thrust outboard, giving a top speed of about 20 knots and a good cruising speed of 13 knots.

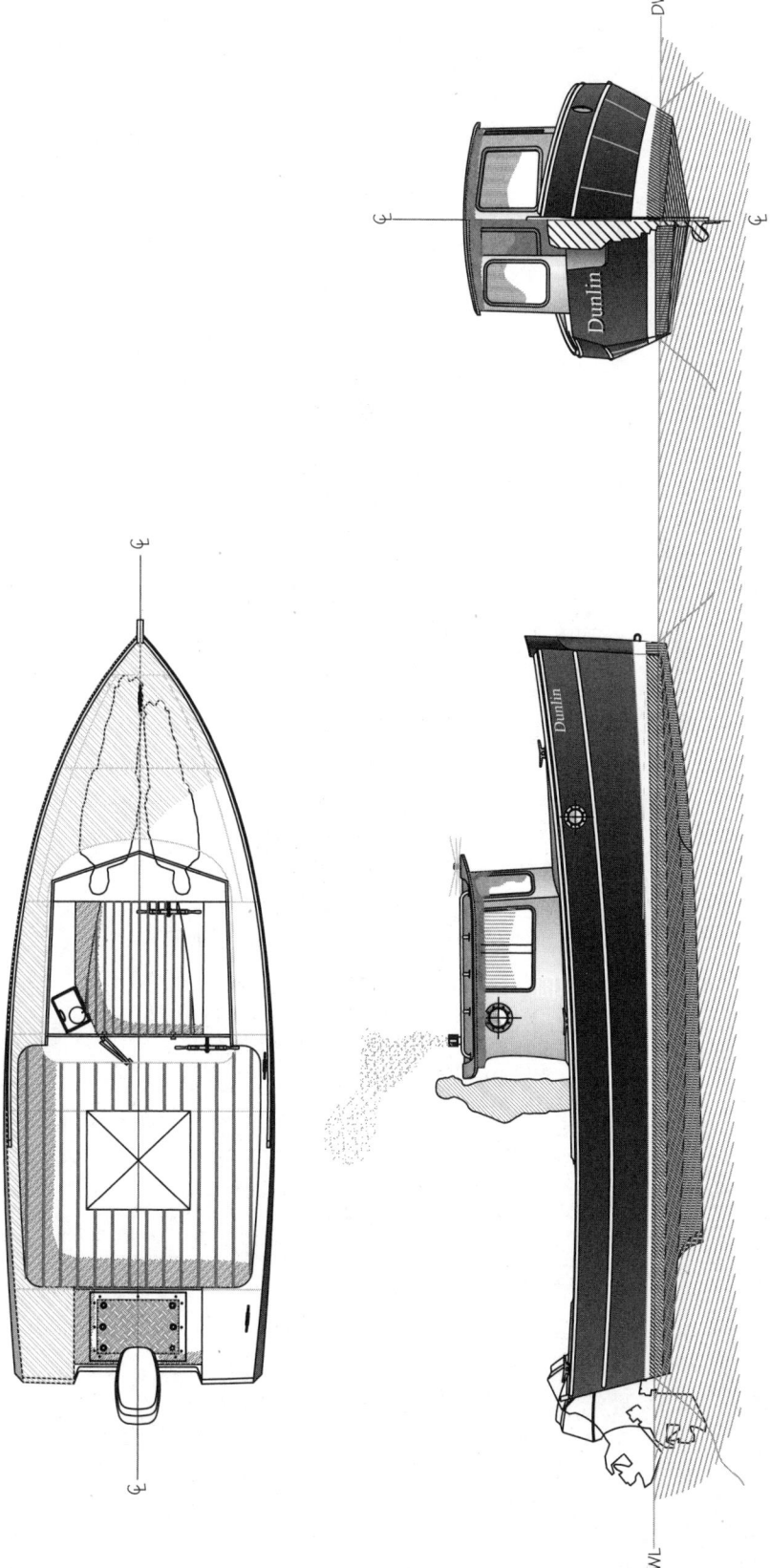

Appx 8   MugWump 22

LOA	22'-4.5"	6.8 m	
Beam	8'-5.5"	2.58 m	
Draft	1'-9.5"	54 cm	outboard up
Weight	3,400 lb	1,542 kg	dry weight

I think of the MugWump 22 as the water equivalent of an Airstream trailer, with amazing livability on land. This is an amazing combination of a capable boat coupled with the utility of a travel trailer on land. There is space galore for the cruising couple and the true utility to match your cruising dreams in the smallest, most convenient package possible. She's even trailerable to reach cross-country. Power is a high-thrust 60- to 70-hp outboard off the stern, and with covered cockpit and a large cabin she provides plenty of options of space and use.

Also available in a 24-foot, solar–electric-powered version.

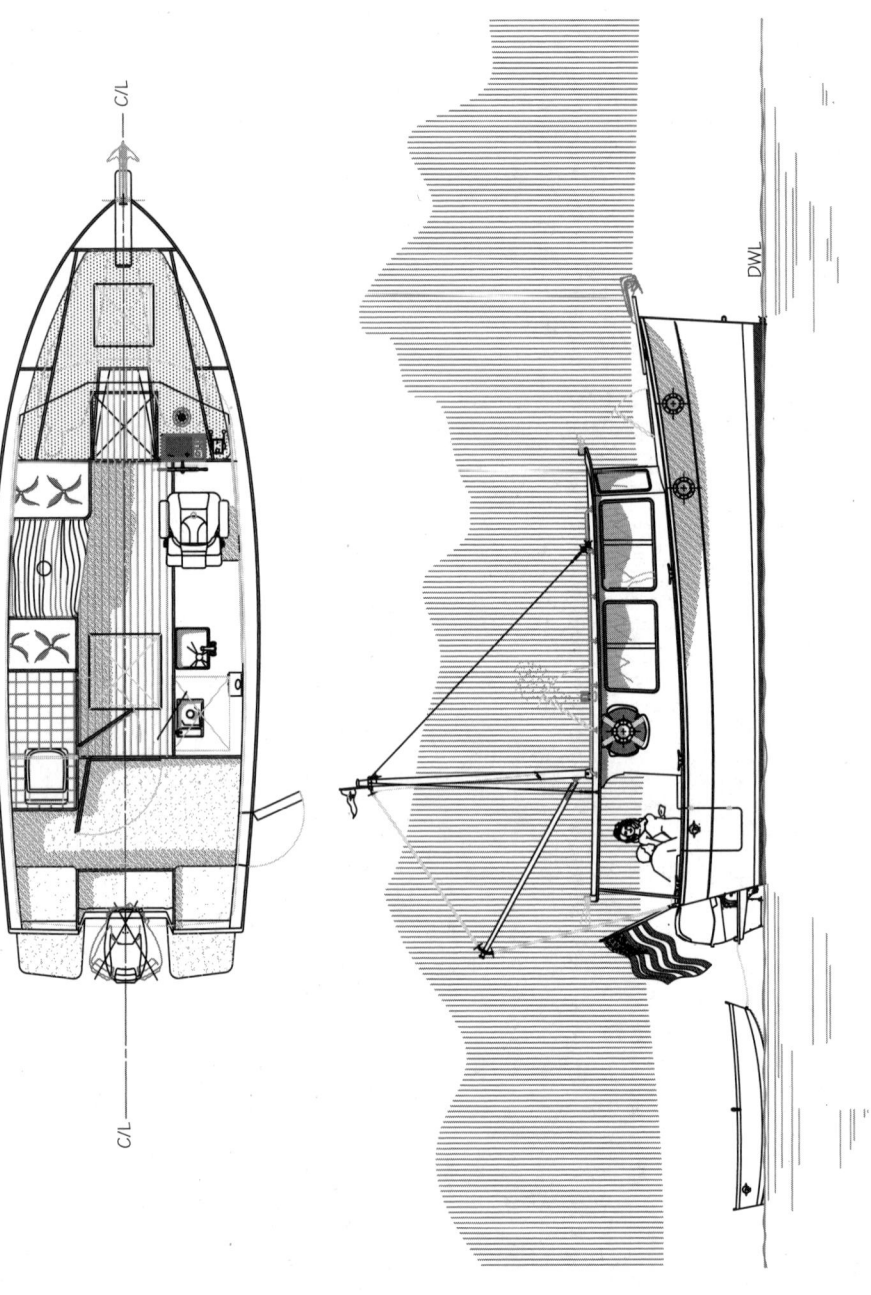

334

Appx 9 Mud Dauber

LOD	23'2"	7.05 m
Beam	8'7"	2.62 m
Draft	12"	30 cm outboard up
Weight	3,000 lb	1,360 kg dry weight

The Jeep of wooden boats: that's the idea here. Imagine you're building a cabin in a remote cove and need to ferry in supplies, tools, even a small tractor, and land on a beach. The Mud Dauber will do it. And after the cabin is completed and you need an all-purpose boat for exploring or hunting, or commuting back to civilization in threatening conditions, she's up to all this, too. Her hull bottom is a full 1 inch thick, and there are strong guards at the sheer. I can't claim she's pretty, but she is one tough boat. Imagine this also: a small box cabin that could be built and slid into position on her deck to allow you shanty-type utility to your cruising dreams and desires.

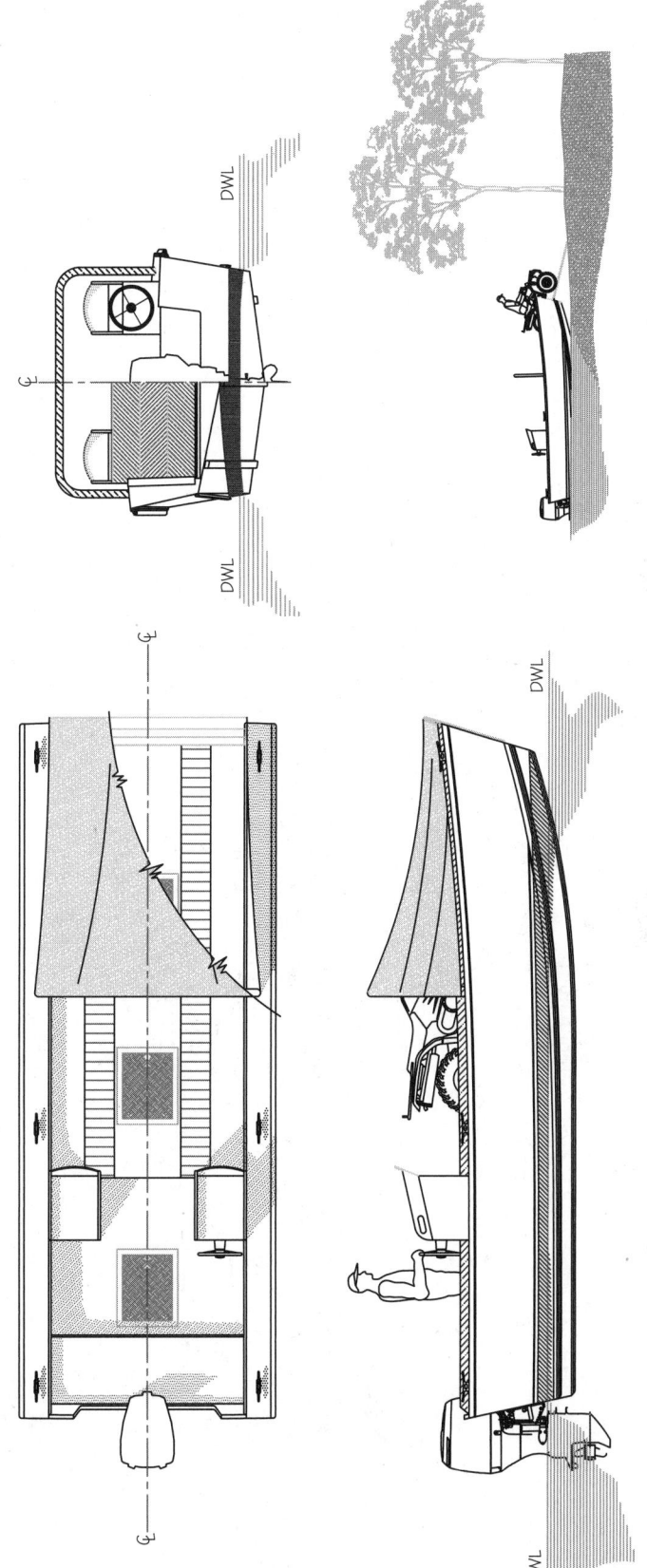

Appx 10 Surf Scoter 26

LOA	26'-8"	8.1 m	Draft	1'-9"	53 cm outboard up
Beam	8'-6"	2.6 m	Weight	4,600 lb	2,086 kg dry weight

Ignoring the ancient advice to not tinker with success, I have gone back to our venerable Surf Scoter design numerous times to tweak and spin off variations. The newest branch of the family is a serious, full-sized 26 foot 8 inch cruising boat with outboard power and enclosed head and galley. Her smaller sisters at 23- and 24-foot lengths sport smaller cabins and accommodations but still are very fun and capable boats on the water. This larger 26-foot boat will cruise with twin 90-hp outboards at 18 knots and burns just under 4.5 gph. You can also rig her with a single engine outboard if desired, but I like the redundancy of having the twins to rely on. Inboard and sterndrive versions of this design are also available. Surf Scoters are available in 22-, 24-, and 26-foot lengths.

Appx 11   Tugzilla 26

LOD	26'6"	8.08 m	Draft	3'7"	109 cm
Beam	10'1"	3.07 m	Weight	10,600 lb	4,818 kg

A customer originally commissioned this design as a tuglike cruising boat for the Columbia River. He began building it himself but in midstream changed his mind about what he wanted. I took over and completed it for myself as a 'for-real' tugboat equipped with a 1,200-pound lifting crane, 110-hp diesel spinning a big 24 × 18-inch prop, and a single napping berth in the fo'c'sle for a tired crew member. A stretched 33-foot version has been built in Turkey and commissioned to tow the 50-odd vessels in the Rahmi M. Koc Museum in Istanbul. But for our northwest waters the TugZilla has proven to be an amazing boat, allowing us to pick out disabled boats from marinas and tow them over to the local boatyard for work or hauling an amazing array of items and boats around south Puget Sound. You might be interested in her only as a serious cruising boat, but she truly has so much more potential in her and good looks to boot.

Appx 12   North Haven 26

LOD	26'8"		Weight	4,900 lb	1905 kg
Beam	8'-6"	8.1 m	Power	90- to 225-hp	outboard
Draft	18"	2.6 m			
		47 cm			

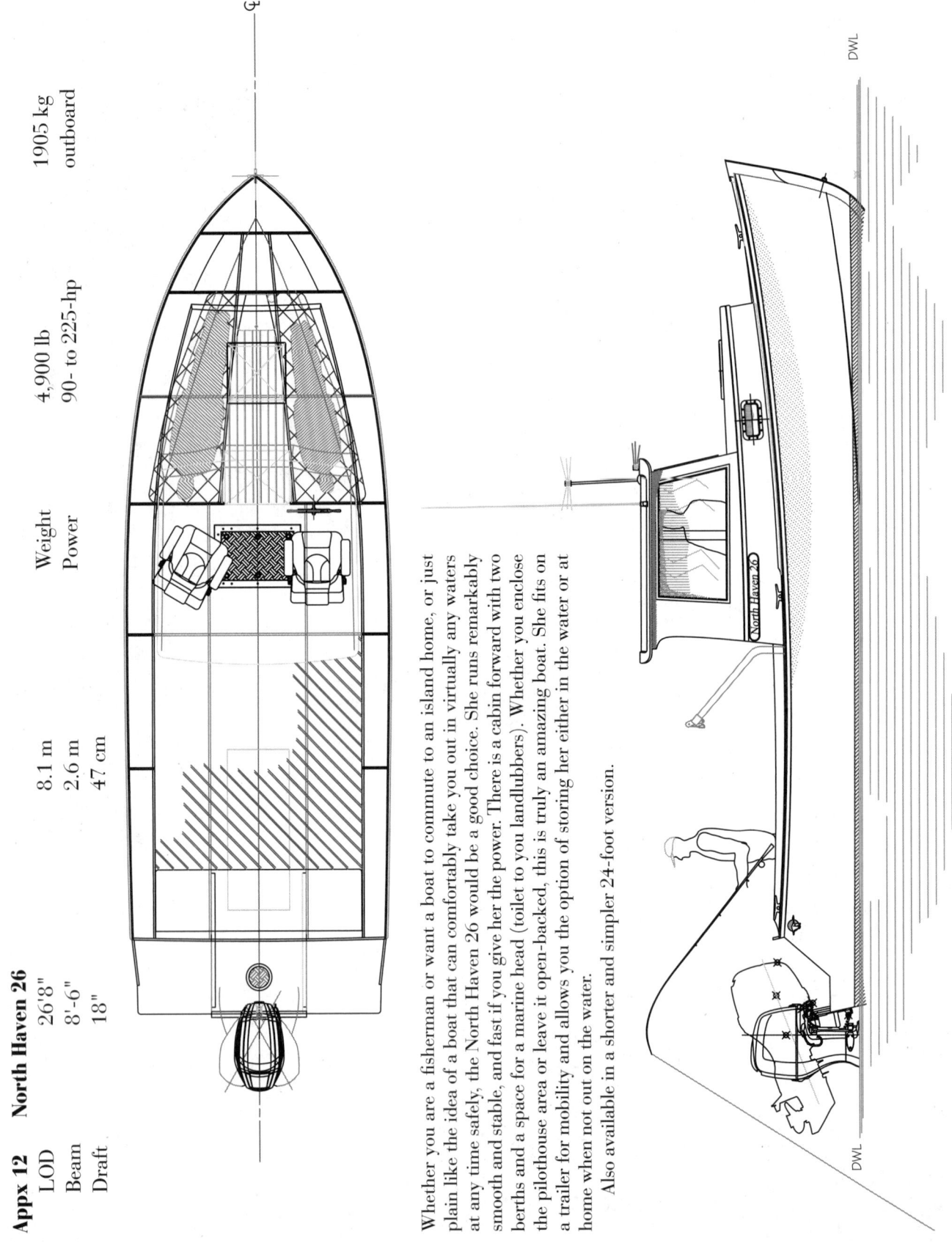

Whether you are a fisherman or want a boat to commute to an island home, or just plain like the idea of a boat that can comfortably take you out in virtually any waters at any time safely, the North Haven 26 would be a good choice. She runs remarkably smooth and stable, and fast if you give her the power. There is a cabin forward with two berths and a space for a marine head (toilet to you landlubbers). Whether you enclose the pilothouse area or leave it open-backed, this is truly an amazing boat. She fits on a trailer for mobility and allows you the option of storing her either in the water or at home when not out on the water.

Also available in a shorter and simpler 24-foot version.

Appx 13 Black Crown 30

LOD	30'6"	9.3 m	Draft	2'2"	66 cm
Beam	10'5"	3.18 m	Weight	10,500 lb	4,762 kg

Our shop has built Black Crowns for more than two decades in many flavors and in sizes from 25 to 30 feet, and it has proven to be a comfortable, powerful, and capable cruising boat. Now we have plans available for home—or barn, more likely—construction of a 30-foot version complete with diesel sterndrive, enclosed head and separate shower, and complete galley. These boats have done cruises as ambitious as Puget Sound to Alaska and back and one even crossed the Gulf of Alaska and makes its home in Prince William Sound. I think these are wonderful boats, stable on the water, economical to run and use, and very seaworthy. She has had sisterships built as far away as Russia, Brazil, Thailand, Korea, and other countries around the world.

339

Appx 14   Red Salmon 33

LOD	33'-5"	Draft	20"
Beam	10'-3.5"	Weight	10,800 lb
	10.19 m		51 cm
	3.14 m		4,89 kg

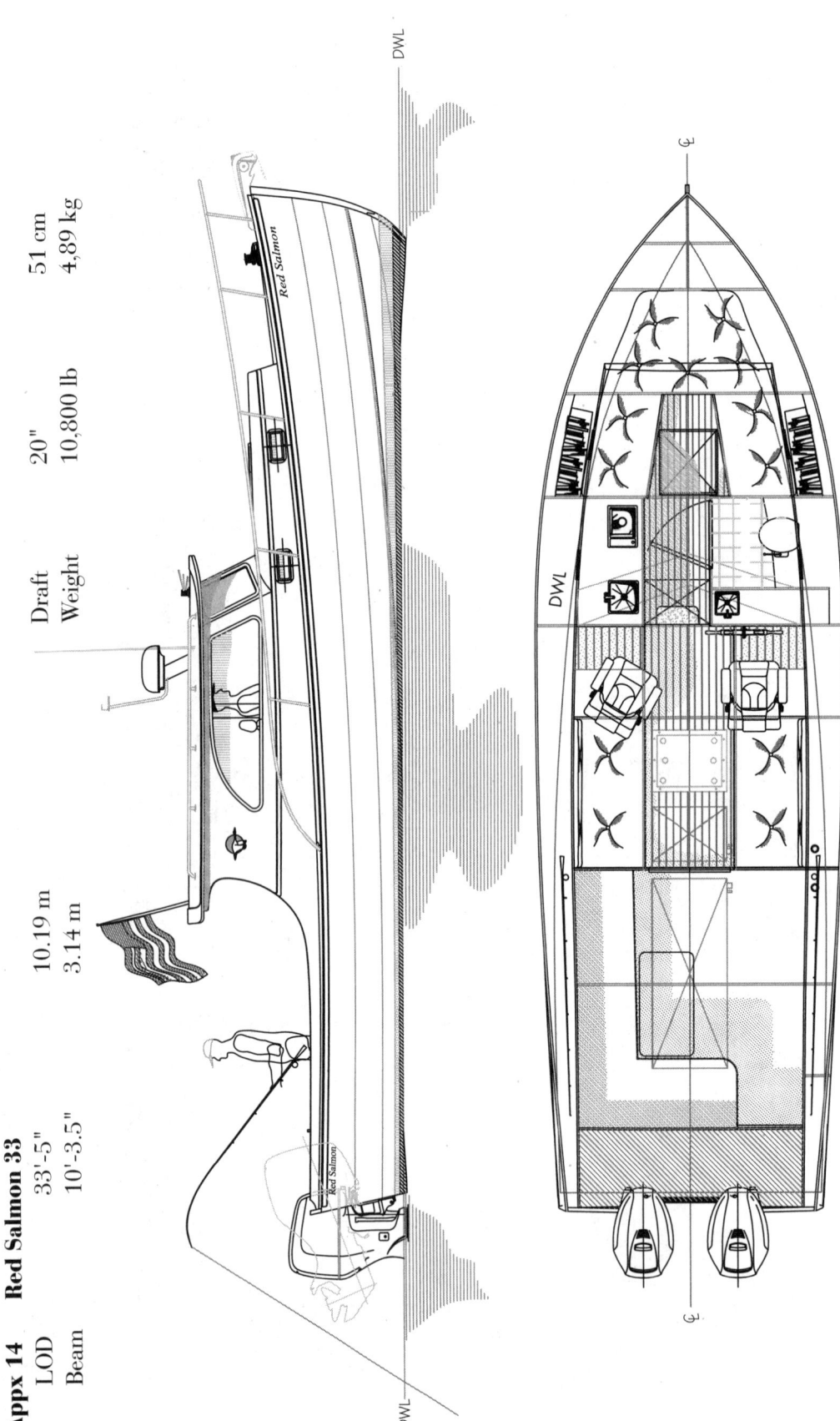

We built this design for a customer who loves to salmon fish on Puget Sound in the Northwest. He wanted lots of speed to allow him to run out to the fishing grounds without wasting time, and with her twin 300-hp outboards we recorded something around 56 mph in early sea trials. Now when I visit him, he admits that while she runs fast very well, he doesn't like the idea of wasting energy and runs her throttled back at a nice 35-mph clip. That allows him to keep a weather eye out for flotsam on the water and stay safe but still get to the fishing grounds in good time. She has a couple of berths and full head below, a full galley and enough seating in the pilothouse for a pack of friends to accompany you for your day's adventures. I also think this design would be very well suited to an island commuter vessel or even wandering far afield for cruising adventures. As for my own dreams, I can see her sitting at the dock on the Washington coast waiting for her crew to arrive, and after a 3-hour run offshore, a days' worth of tuna fishing in the late summer. That way we spend time on the water and reap the rewards of our trip with many fine meals of fresh tuna. That would be way good living!

Available in inboard and outboard versions.

Appx 15   Shanty 36

LOD	36'-0"	10.98 m		
Beam	12'-6"	3.78 m		
Draft	1'-7"	49 cm		
Weight	14,500 lb	6,587 kg		

Who amongst us hasn't read *Tom Sawyer* and mused to ourselves how nice it would be to live on a shanty-type boat on the water. Not tied to shore but anchored up in some secluded cove enjoying the same million-dollar views that our poor-overtaxed neighbors on shore have but without the burden of always paying the big bucks for the view. And if we want to change our view all that is required is to pull up the anchor and move to another location. The Shanty 36 was designed and built for the famed folk singer and artist Gordon Bok, and she serves as his summer home on the water. She is a big boat, most likely too big to safely trailer yourself, but could be docked at a marina that you can rent a slip in or keep her moving throughout the cruising season. She is a true home on the water with amazing mobility with her twin outboard motors in motor wells. Controls are in the forward cockpit and there is a queen-sized berth in the owner's cabin. Make your own Huckleberry Finn stories with a much better base than a raft of logs on the water with a pup-tent pilothouse.

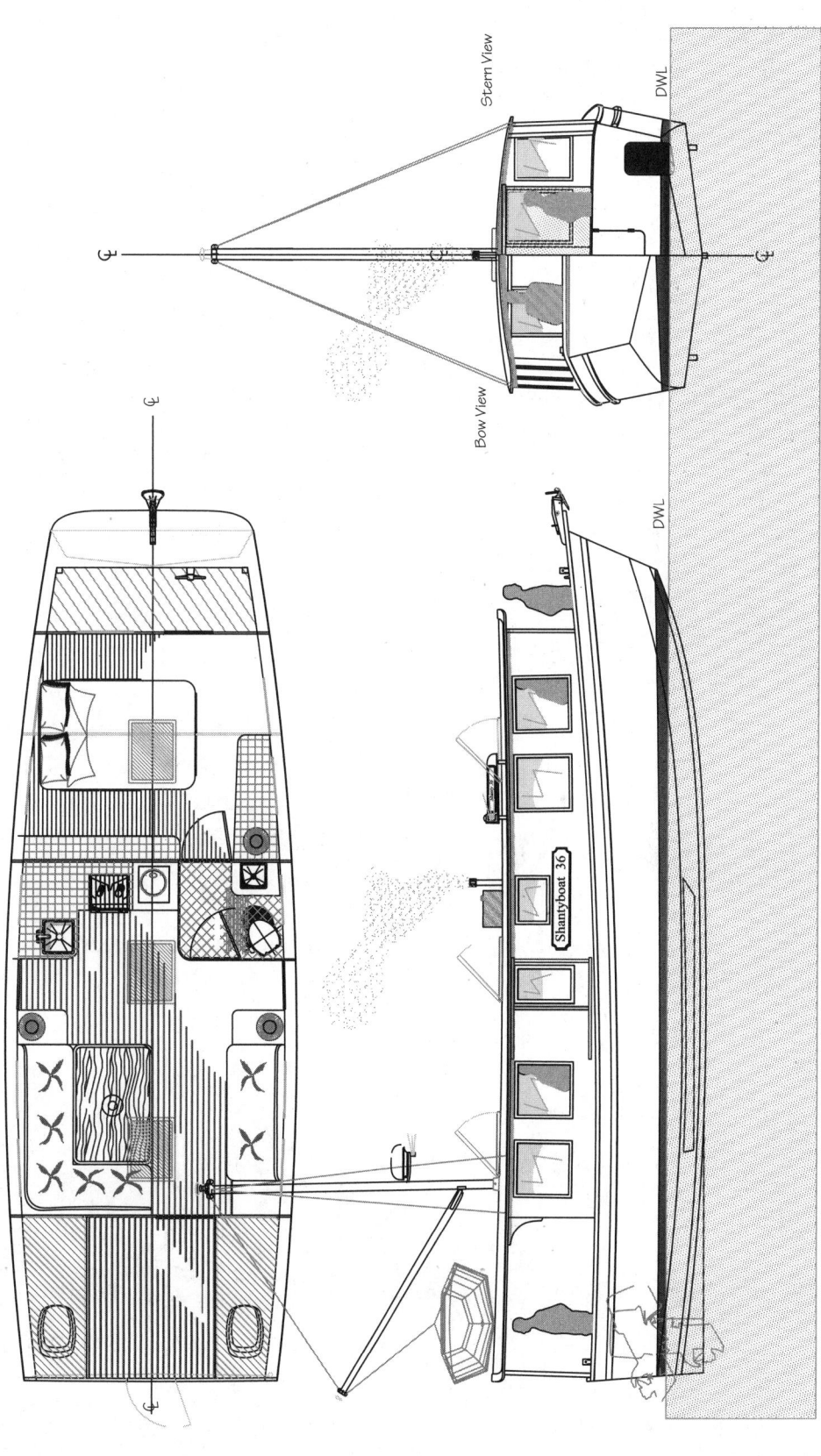

Stern View

Bow View

DWL

DWL

Shantyboat 36

Appx 16   Devlin 37

LOD	37'-1"	11.3 m	Draft	4'-4"	1.3 m
Beam	12'-5"	3.8 m	Weight	26,500 lb	12,020 kg

A design done only for myself and without the input of a customer with their individual wants and needs. I was interested in designing a very capable cruiser of a size that would be very comfortable for extended cruising for a couple hours or for shorter trips with grandkids or friends on board. She is a single-stateroom boat with a huge engine room and plenty of space to not feel too crowded while at sea. A giant covered cockpit protects you from sun and rain and allows a nice crowd to gather for evening visiting and drinks. Ideal for cruising in the Northwest with our potential cruising grounds running almost 1,000 miles all protected and sheltered except for two areas that are only exposed to the open ocean for about 50 miles each. This is a slow but very economical vessel that would cruise burning less than 2 gallons of fuel per hour but moving at almost 8 knots through the water. A summer aboard would be so much more enjoyable than running the lawn mower weekly and sweating through a too-warm summer season. Now I only need to see if my coffers will have enough excess to allow me to build her.

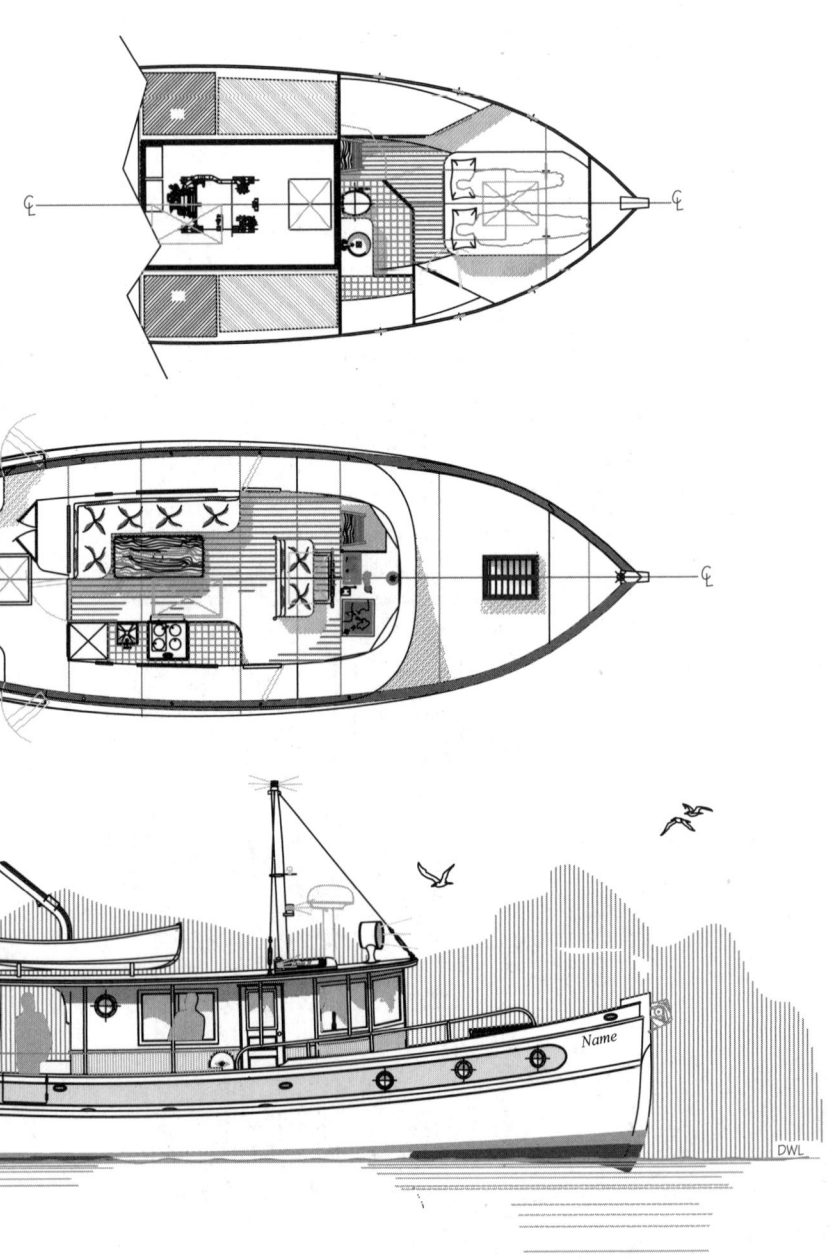

Appx 17   Albacore 40

LOD	39'-11"	12.15 m
Beam	12'-2.5"	3.71 m
Draft	2'-1"	.63 m
Weight	18,900 lb	8,573 kg

A commission from and built in Korea, she won "Boat of the Year" at the Korean International Boat Show in 2019. The Albacore is a lovely and very nimble speedster with twin 350-hp outboards on her stern and an almost 50-knot top speed. Her accommodations are sumptuous for a cruising couple and can easily day handle up to eight people for a wonderful day on the water. She is built strong and light with the Stitch-and-Glue method and showcases how truly versatile this building method can be. Her lines are beautiful, and I am sure she will become a classic that will be admired for decades and decades.

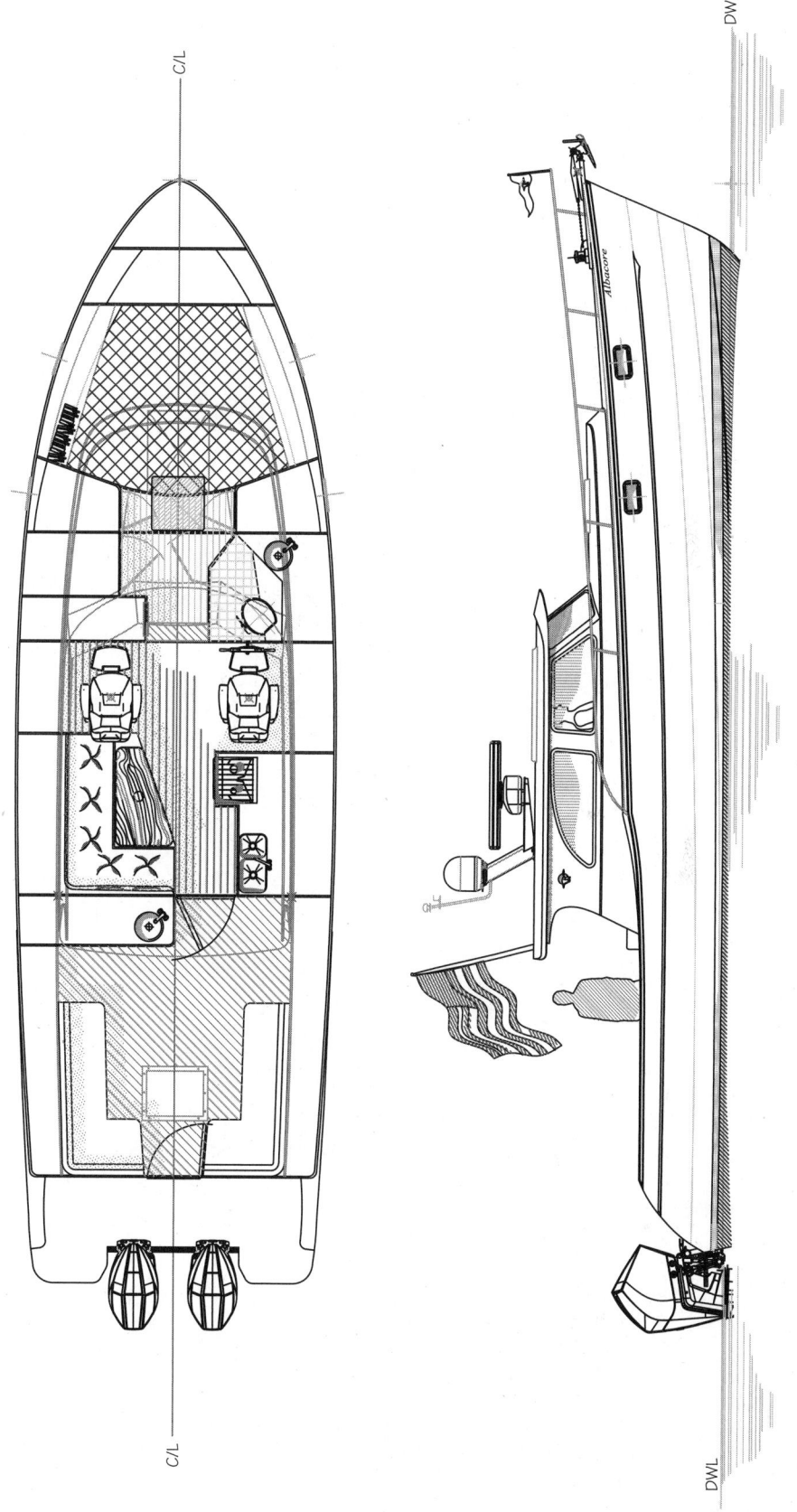

Appx 18   Moon River 48

LOD	47'-9"	Draft	3'-4"	14.55 m
Beam	13'-3"	Weight	32,000 lb	4.03 m
				1.22 m
				14,515 kg

Designed and built for a Northwest doctor and his wife, the Moon River is truly an amazing design and build. She has twin 300-hp diesel inboard engines and tops out at 26 knots. Cruising is done around the 18-knot range with her fuel economy in the 6 gallons per hour range. Moon River is a twin stateroom boat with twin heads to accommodate the crew aboard. An open and very spacious main salon includes large galley, U-shaped dinette table, and true helm and co-helm seats. Her

partially covered cockpit can handle both entertaining under the cover and out of the sun or rain. But if the sun is out and the weather fair, the two stern seats are truly the place to spend the last hours of the day, light breeze keeping things cool, and good friends to visit with and talk about the next upcoming days' adventures.

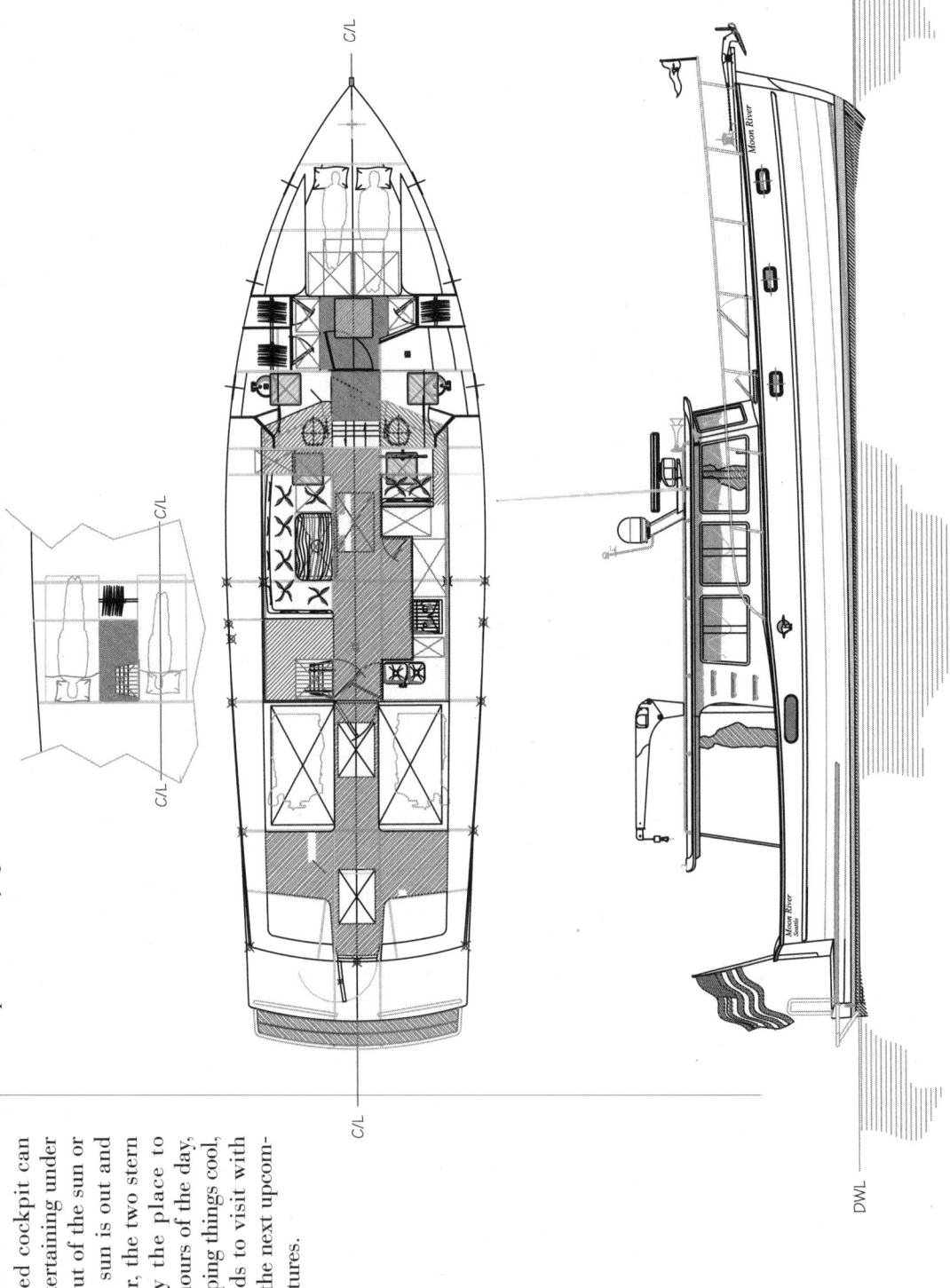

Appx 19 Blue Fin 54

LOA	54'-6"	16.6 m
Beam	12'-7"	3.8 m
Draft	3'-11"	1.2 m
Weight	18,800 lb	8528 kg

Blue Fin is admittedly an unconventional design, as she is an ample 54 feet long but incorporates the layout and accommodations that are more typical of a mid-30s-foot boat. The reason is to provide unbelievably quiet, smooth, and efficient cruising for one couple without the complications and expense of an overmuscled production cruiser with multiple staterooms. This configuration allows us to have an unusually low, sleek, and distinctive profile. A 150- to 225-hp diesel provides an economical 12- to 14-knot cruising speed for the couple who would like to go somewhere with grace instead of hurry. The first Blue Fin 54 was built in Russia by a good friend, Temur Rukhaya, and he has now cruised her more than 16,000 miles. She has been north of the Russian Arctic Circle and has cruised out into the Baltic, visiting Finland, Sweden, and Denmark. A remarkable boat with an amazing number of miles under her keel. She is economical and cruises with a fuel burn of 8.2 miles per gallon, this is a fuel efficiency that is usually not approachable in these days of overpowered boats.

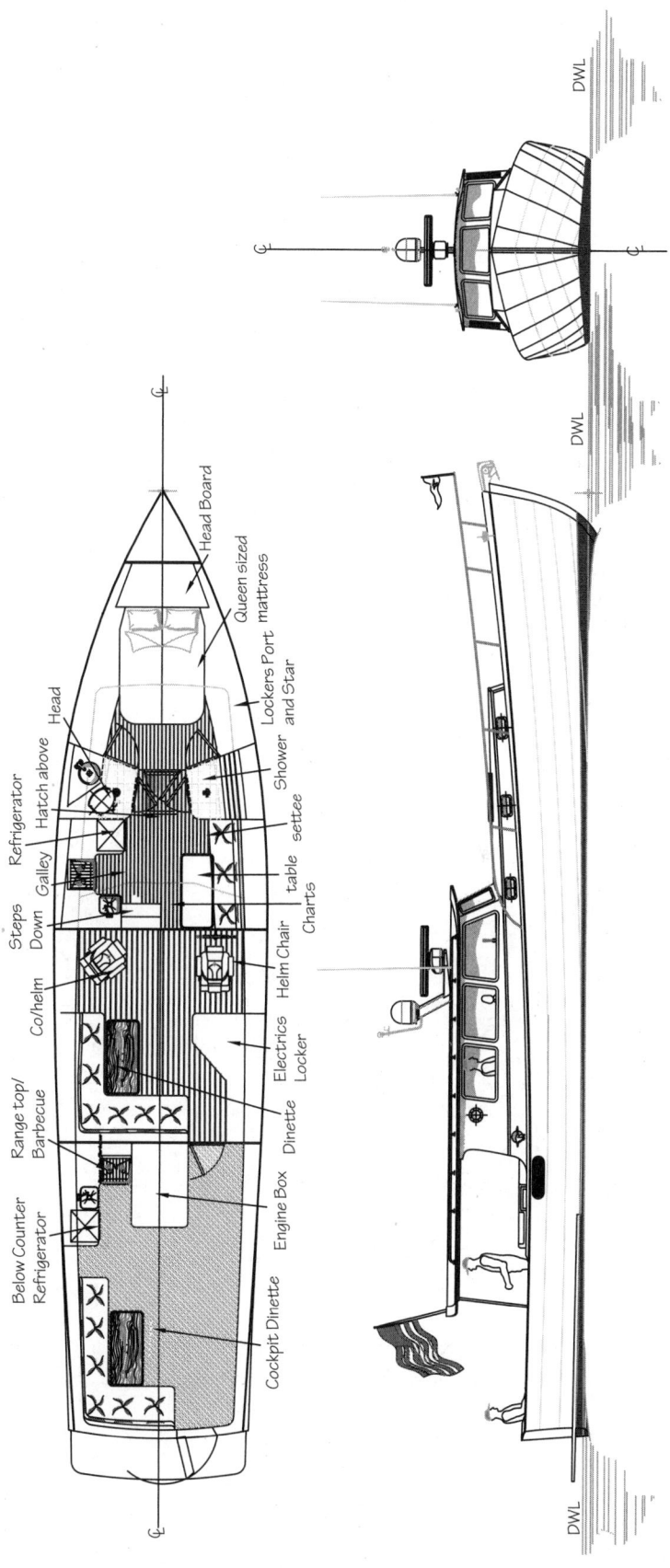

Appx 20   Sockeye 62

LOD	61'-10"	18.66 m
Beam	16'-4"	4.98 m
Draft	6'-5"	1.98 m
Weight	81,900 lb	37,149 kg

The Sockeye has a workboat-type appearance, but it is a displacement-hull cruiser with luxurious accommodations in a 3-stateroom arrangement. A single heavy-duty John Deere 145–hp diesel propels her at hull speed of 9 knots. The fantail stern is distinctive and gives her a sea-kindly motion in a following sea. Unable to leave well enough alone, I have been drawing more workboat-like versions of her and even motorsailer variants with both ketch and schooner rigs. I often tease my friends that if my ship ever comes in, this would be my ultimate dream boat. Hey, even boat designers have a need to dream a little.

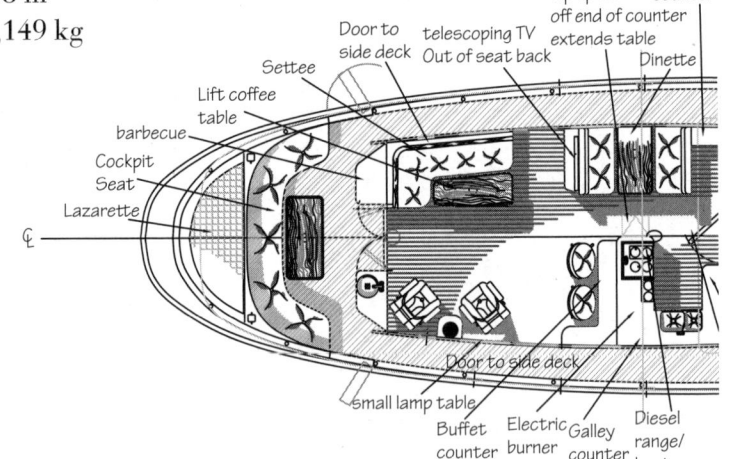

Labels (clockwise): Settee; Lift coffee table; barbecue; Cockpit Seat; Lazarette; Door to side deck; telescoping TV Out of seat back; flip up leaf off end of counter extends table; counter; Dinette; Door to side deck; small lamp table; Buffet counter; Electric burner; Galley counter; Diesel range/heater

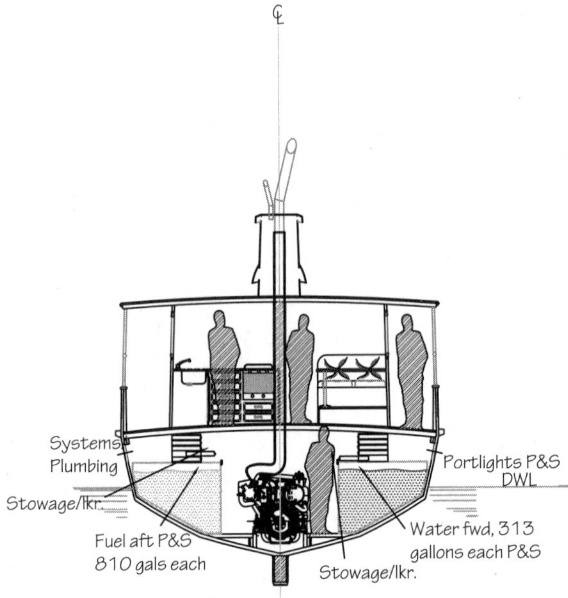

Section Aft, Looking aft

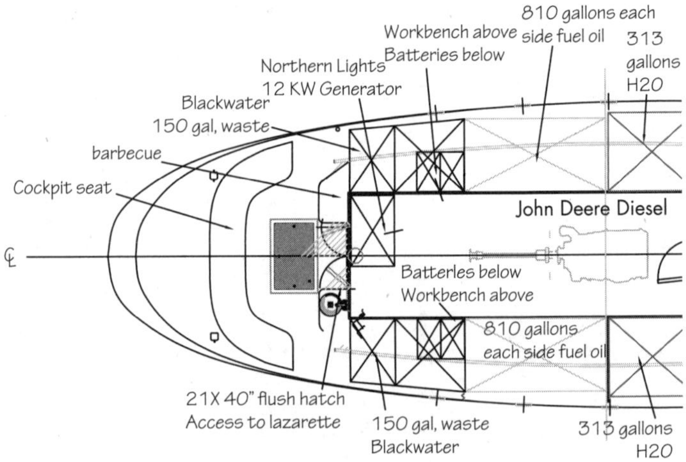

Labels: Systems Plumbing; Stowage/lkr.; Fuel aft P&S 810 gals each; Stowage/lkr.; Portlights P&S DWL; Water fwd, 313 gallons each P&S

Sockeye

Labels: 810 gallons each side fuel oil; 313 gallons H2O; Workbench above; Batteries below; Northern Lights 12 KW Generator; Blackwater 150 gal, waste; barbecue; Cockpit seat; John Deere Diesel; Batteries below Workbench above; 810 gallons each side fuel oil; 21X 40" flush hatch Access to lazarette; 150 gal, waste Blackwater; 313 gallons H2O

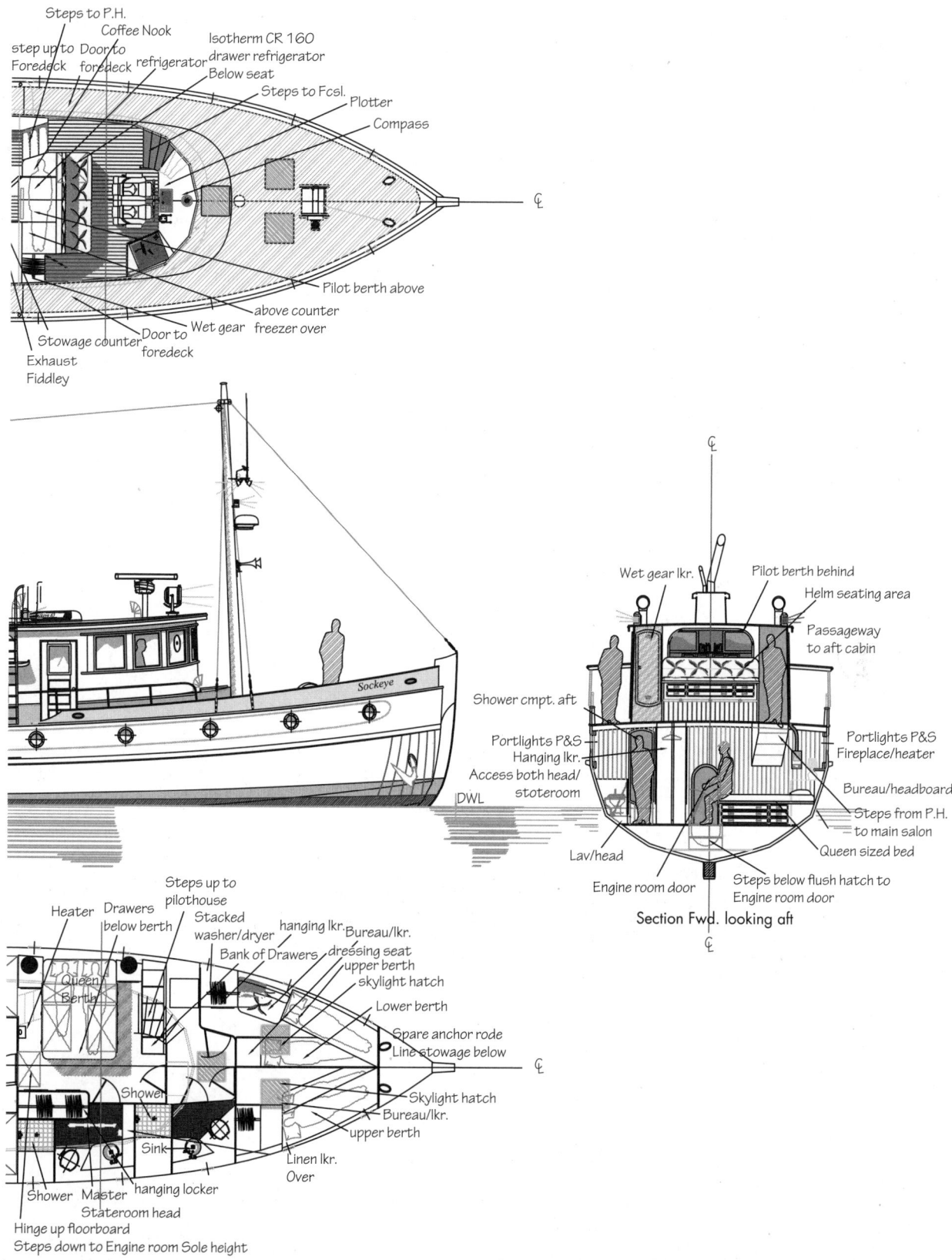

Steps to P.H.

Coffee Nook

step up to Foredeck

Door to foredeck

refrigerator

Isotherm CR 160 drawer refrigerator Below seat

Steps to Fcsl.

Plotter

Compass

Pilot berth above

above counter freezer over

Wet gear

Door to foredeck

Stowage counter

Exhaust Fiddley

Sockeye

DWL

Wet gear lkr.

Pilot berth behind

Helm seating area

Passageway to aft cabin

Shower cmpt. aft

Portlights P&S Hanging lkr. Access both head/ stoteroom

Portlights P&S Fireplace/heater

Bureau/headboard

Steps from P.H. to main salon

Queen sized bed

Lav/head

Engine room door

Steps below flush hatch to Engine room door

Section Fwd. looking aft

Heater

Drawers below berth

Steps up to pilothouse

Stacked washer/dryer

hanging lkr.

Bureau/lkr.

Bank of Drawers

dressing seat

upper berth

skylight hatch

Queen Berth

Lower berth

Spare anchor rode

Line stowage below

Shower

Skylight hatch

Bureau/lkr.

upper berth

Sink

Linen lkr. Over

Shower   Master

hanging locker

Stateroom head

Hinge up floorboard

Steps down to Engine room Sole height

ROWBOATS AND SAILBOATS

Appx 21   Polliwog

LOA	7'6"	2.28 m
Beam	4'1"	1.24 m
Draft	7"	18 cm
Weight	58 lb	26 kg
Useful load	450 lb	205 kg

Polliwog is the simplest Stitch-and-Glue project imaginable, requiring only two 4×8 sheets of ¼ inch plywood, no scarfing necessary! One amateur builder reported completing her in a mere 40 hours. But it's a real boat, useful as a yacht tender, solo fishing platform, or kids' rowboat. We filmed a video decades ago about building her and as an introduction to Stitch-and-Glue boatbuilding. The video has at this writing more than 1.5 million views on YouTube, amazing for such a simple little boat, and at last check, we have had her built as DIY projects in more than 100 countries around the world. Big chops for such a small boat.

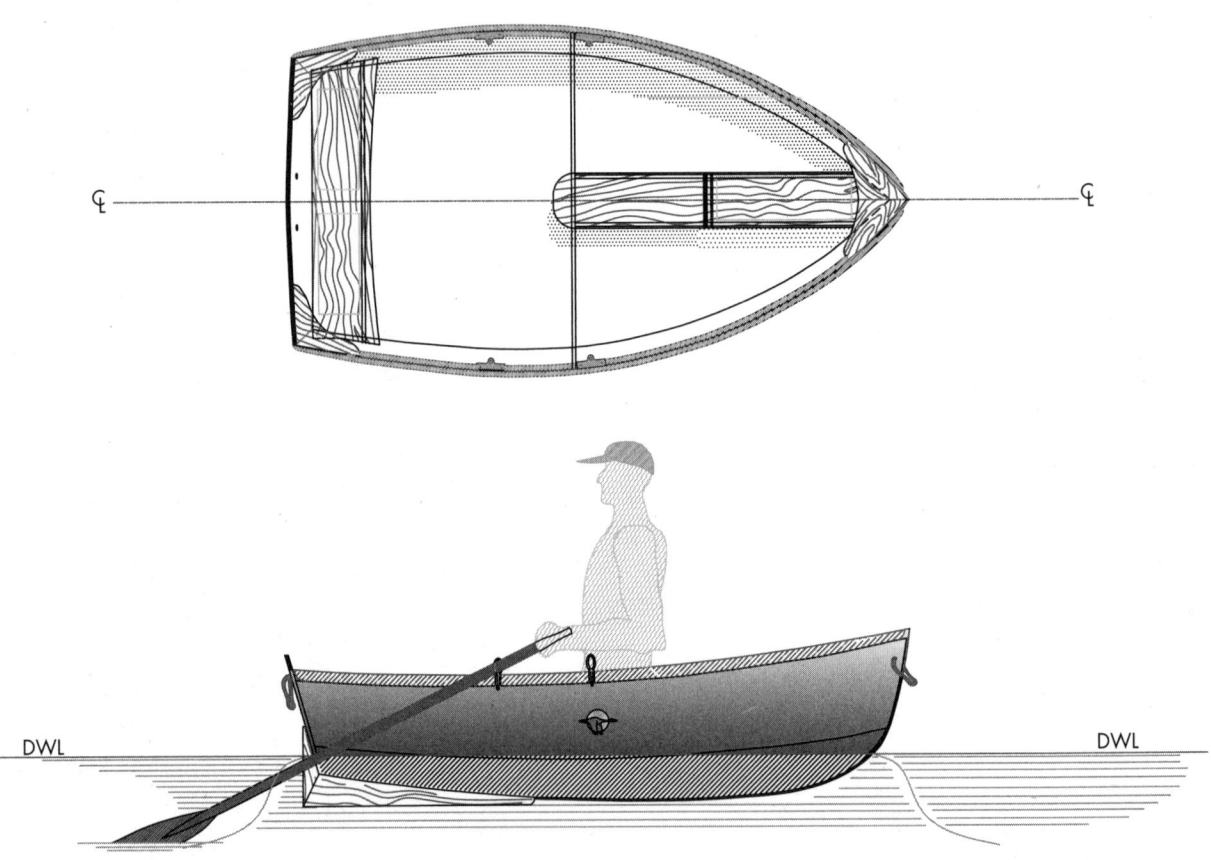

Appx 22   Lit'l Petrel

LOA	8'11"	2.7 m
Beam	4'2"	1.3 m
Draft	5"	13 cm
Weight	68 lb	31 kg
Useful load	340 lb	154 kg

Critics argue, pointedly, that a pram bow makes for a clunky-looking boat. I reply, bluntly, that the great advantage of a pram-bow dinghy is that it affords excellent stability and capacity for the small space it takes up. You can beach this boat and step off the bow, dryly, without tipping her over. You can load her up way beyond design specs: another 110 pounds and she'll sit only an inch deeper in the water. Beauty, I say, is when form perfectly follows function.

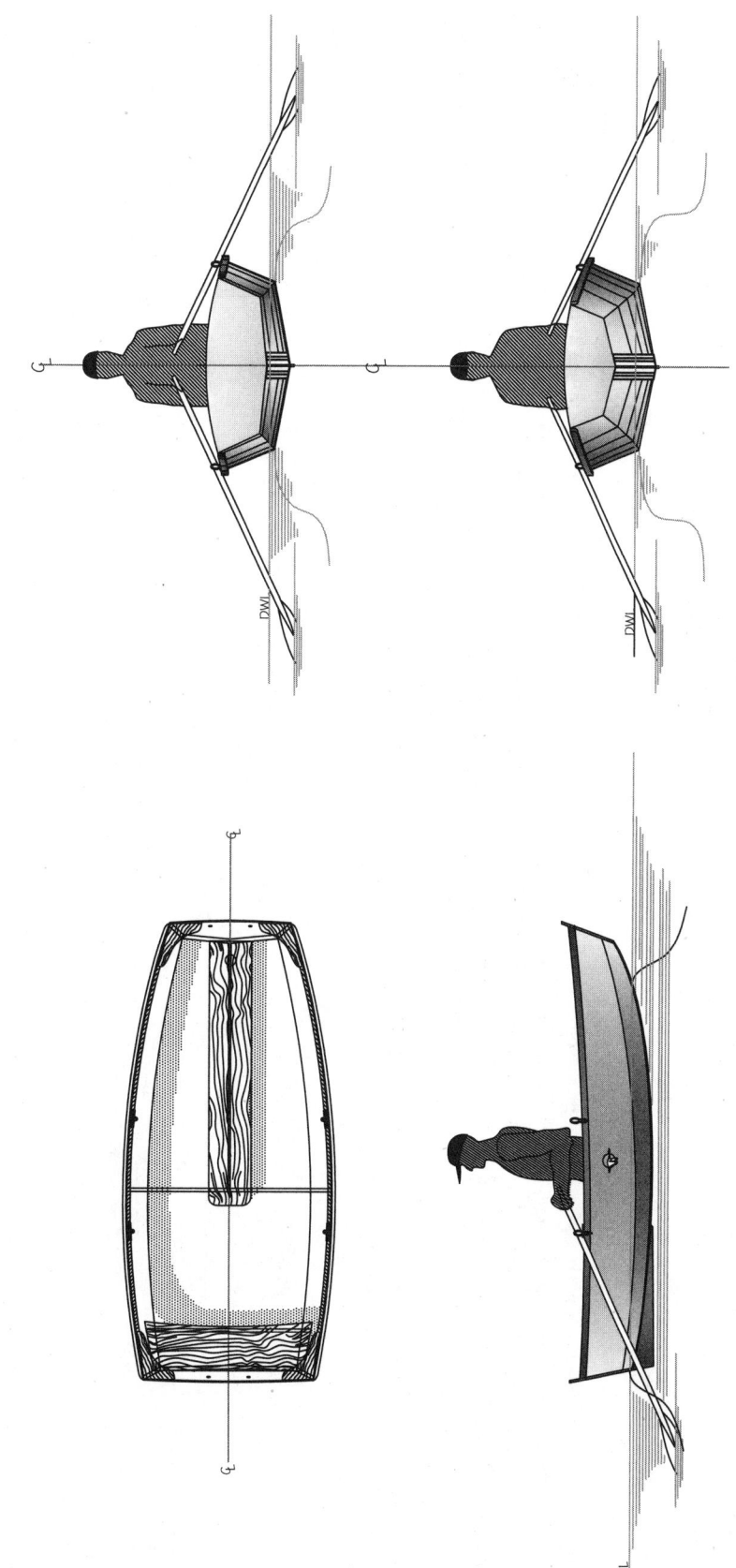

Appx 23 Bella 10 skiff

LOA	9'9"	2.97 m
Beam	3'9"	1.14 m
Draft	6"	15 cm
Weight	52 lb	24 kg

If you called me and asked for an easy-to-build first project small boat to get your feet wet on Stitch-and-Glue construction, I almost always would recommend building a Bella 10 skiff. Formerly named the "5×10 Skiff," this little rowing skiff was originally designed as a teaching tool in my workshops where I teach Stitch-and-Glue construction. She is literally the largest boat you can build out of a single sheet of marine plywood and is all about practical construction: You can build the hull from a single 5 × 10 foot sheet of marine plywood (yes, you can buy 5 × 10 foot marine plywood). The completed boat is also practical because of its very light weight and compact size; she's slender enough to fit in the bed of any pickup. No scarfing and no trailer needed either. But if you can only find 4 × 8 foot marine plywood the building plans cover construction with those smaller sheets also, using scarfing. But whether you tackle her in 5 × 10 foot or 4 × 8 foot, she is always a great little boat and amazing fun to row and use.

The Bella design is available in 10-, 12-, and 16-foot versions.

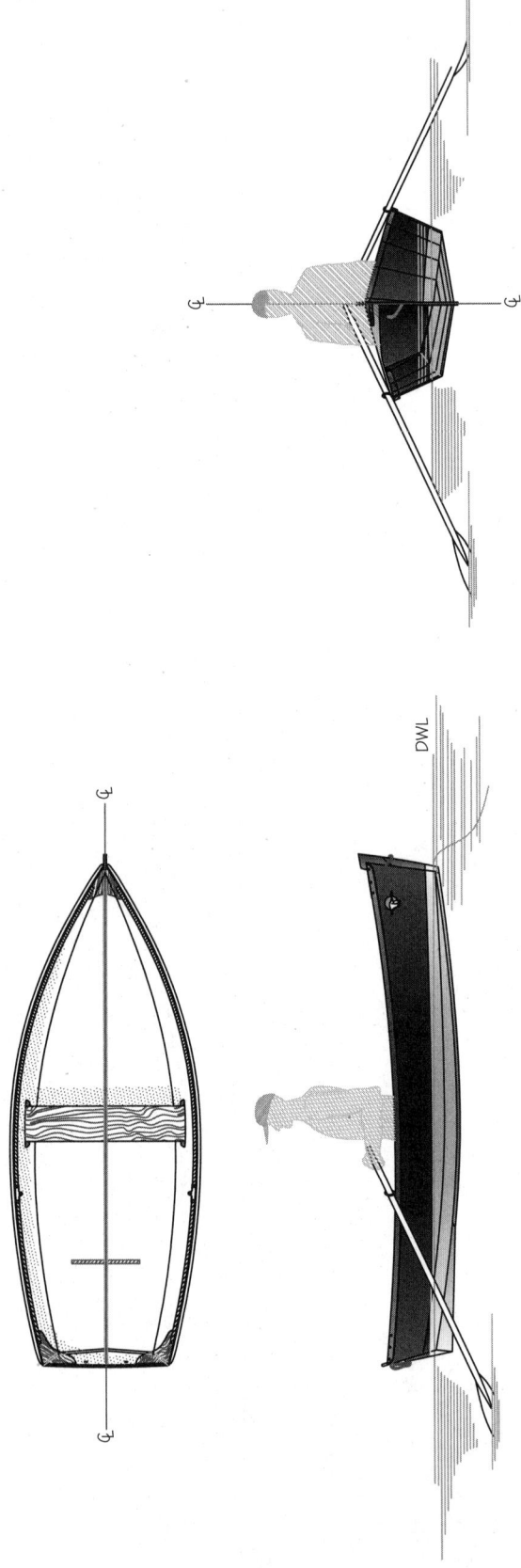

Appx 24   Duckling 14

LOA	14'-4.5"	4.38 m
Beam	3'-5"	1.04 m
Draft	5"	13 cm
Weight	105 lb	47.6 kg

We were building one of our Sockeye 45 cruisers and the owners wanted a rowing skiff that they could launch over the side and go rowing each morning. The owner had rowed in college at Penn, and he had kept the habit of daily rowing in his exercise routine. But we needed the new rowing boat to fit on top of the rear house of his new cruiser and that restricted my length to just about 14 feet long. So, I designed a boat that would fit the cabintop and accommodate Henry and Holly to go for their morning rows. This is the first boat that I stapled instead of conventional stitching because the large mothership was getting close to launching, and we had to have the rowing skiff finished in just a couple of weeks. The result was the Duckling 14, a very nice rowing skiff and we have an optional sailing rig that fits to her for a bit of afternoon fun after the hook is down on the mothership. The Duckling also has a larger sister in a 17-foot length that is a wonderful rowing boat, capable of keeping you safe on larger water for recreational rowing with fixed or sliding seat rigs.

Available in 14- and 17-foot versions.

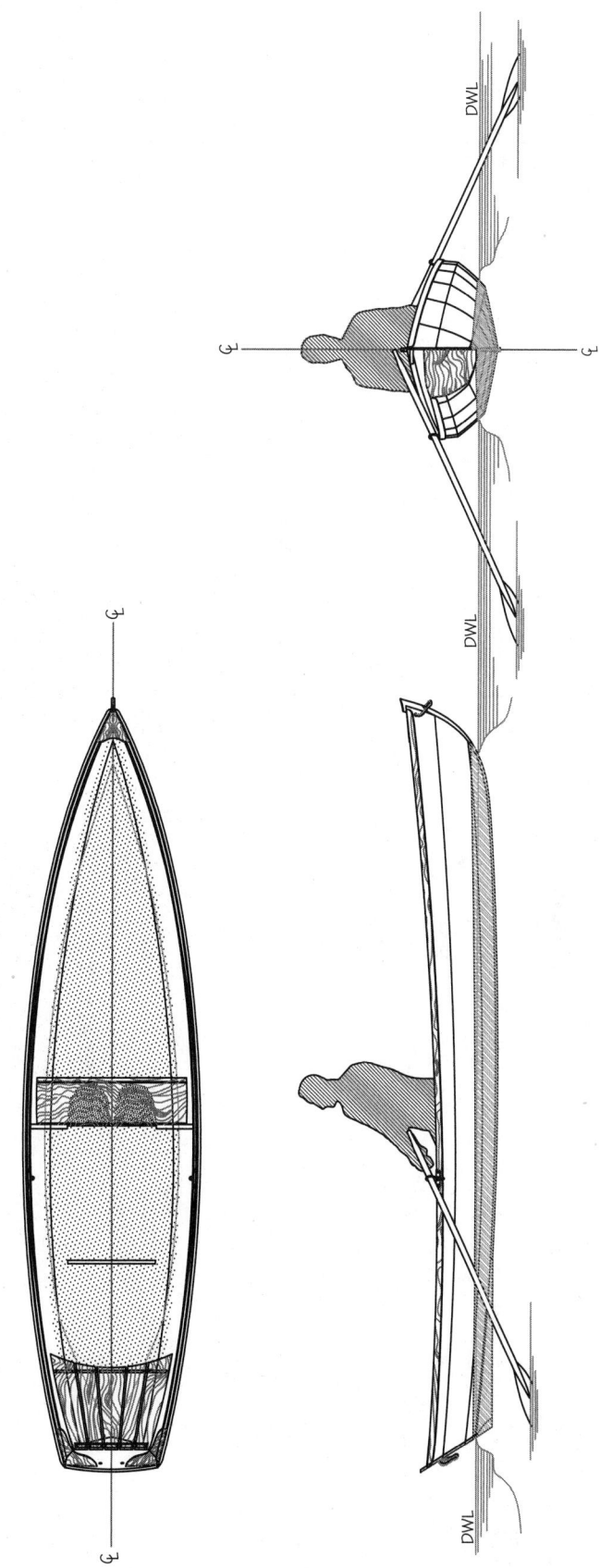

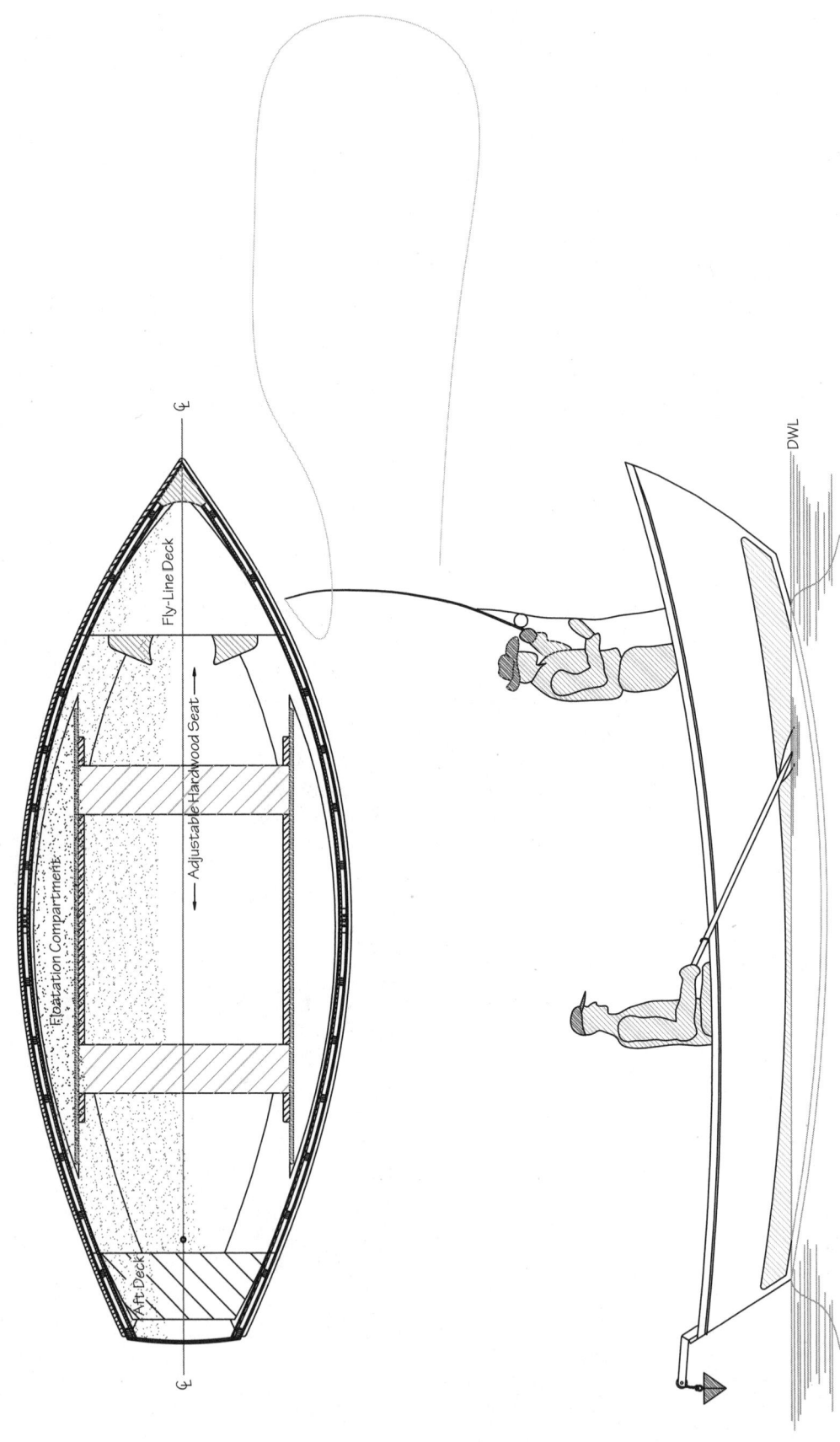

Appx 25 Drifter 15

		Draft	7.5"	19 cm
LOA	15'-1"	4.6 m		
Beam	5'-11.5"	1.8 m		
		Weight	385 lb	174.6 kg

Having grown up in the Willamette Valley of Oregon, I was very familiar with Mackenzie River-type drift boats and even had my hand at building quite a few of them in the early days of my boatbuilding career. I finally designed a Stitch-and-Glue version at the request of my customers, and the result was the Drifter 15. She is a great example of the drift boat genre and does her job very well. The stiffness of the monocoque Stitch-and-Glue hull gives a lot of confidence while negotiating very messy white water, and fishing those calm pools at the base of the rapids gives some memorable hours of recreation on our Northwest rivers.

Appx 26   Fairhaven Flyer

LOA	20'-4"	6.2 m
Beam	4'-2"	1.28 m
Draft	7"	19 cm
Weight	210 lb	95.25 kg

We have for many years had a rowing dory design called the Oarling, and the owners have had nothing but good words about her performance and seaworthiness. Along came one of those customers, a woman named Dale McKinnon; she had built her Oarling from plans and rowed it in several long-distance races on Puget Sound. Dale and I had become friends, and one day she came to me with a dream trip she was considering and needed a proper boat design that would enable her to fulfill the dream. She wanted to be the first woman to row solo from Southeast Alaska to Puget Sound. The Oarling, while a very capable boat, just didn't have the capability of carrying the weight of gear and supplies that would be needed. So, I designed a longer version and named the new design the Fairhaven Flyer. Dale built the first one and shipped the boat up to Alaska and started her trip. It's an ambitious person that tackles an almost 1,000-mile rowing trip solo, and I have done that journey many times in my own powerboats, so I know what an accomplishment it was for her to successfully complete that journey. The Fairhaven Flyer kept her safe and would be a good choice for long-distance rowing or tandem rowing with or without sliding seats.

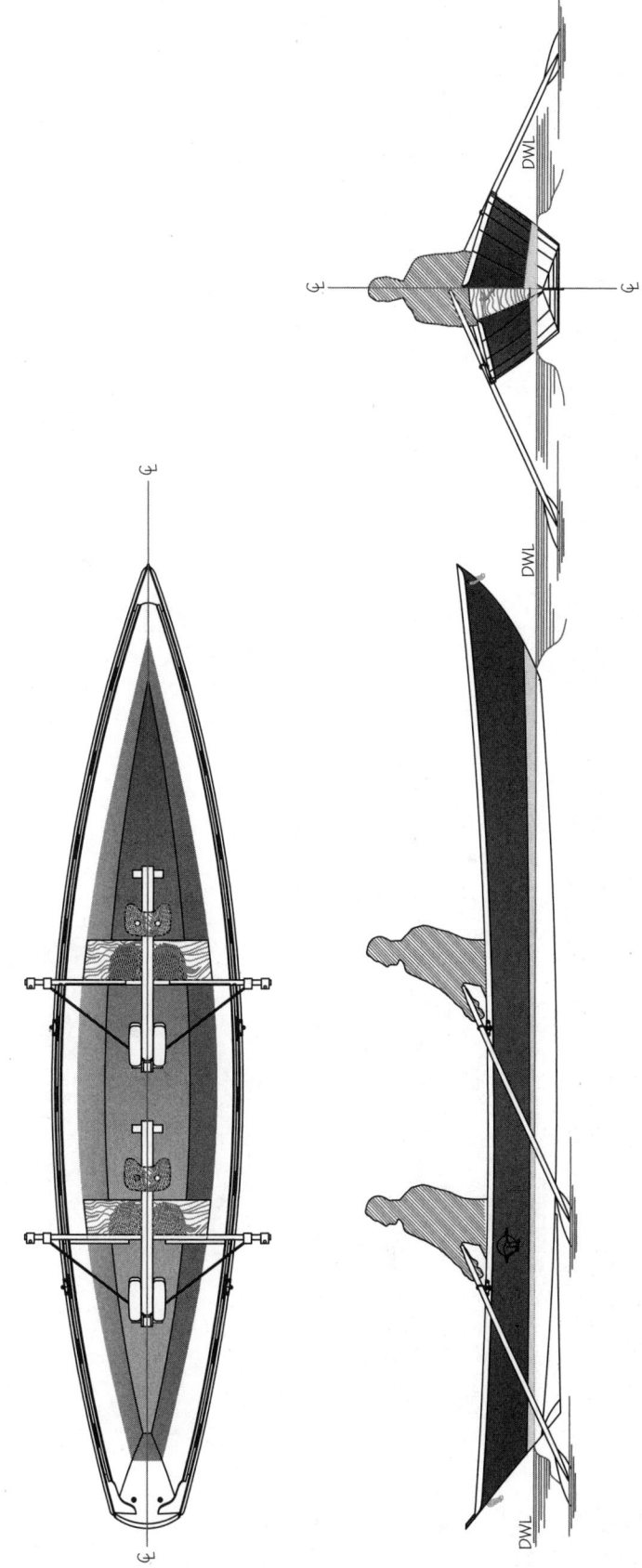

353

Appx 27   Egret

LOA	15'-3"	4.64 m
Beam	4'-10.75"	1.50 m
Draft CB up	10"	24 cm
Draft CB down	2'-8"	81 cm
Weight	350 lb	159 kg
Sail area	72 ft²	6.68 m²

My first design ever done for Stitch-and-Glue construction way back in 1977, and I am very pleased to write that we still sell plans and occasionally build one in our own shop. The Egret really is a very capable small sailboat that also rows very well. She is seaworthy and will handle several sailors aboard without issue. I am slightly amused that I have over the years tried to redesign the Egret but always keep coming back to the original design. It just works so well that there truly seems to not be a true reason to discard her and start all over. If you are looking for a decent sailing boat, and one that you can row also, look at the Egret. I think you might agree she is an amazing boat.

At the time of this writing, I am just finishing up DIY plans for a 20-foot version of this design called the Egret 20.

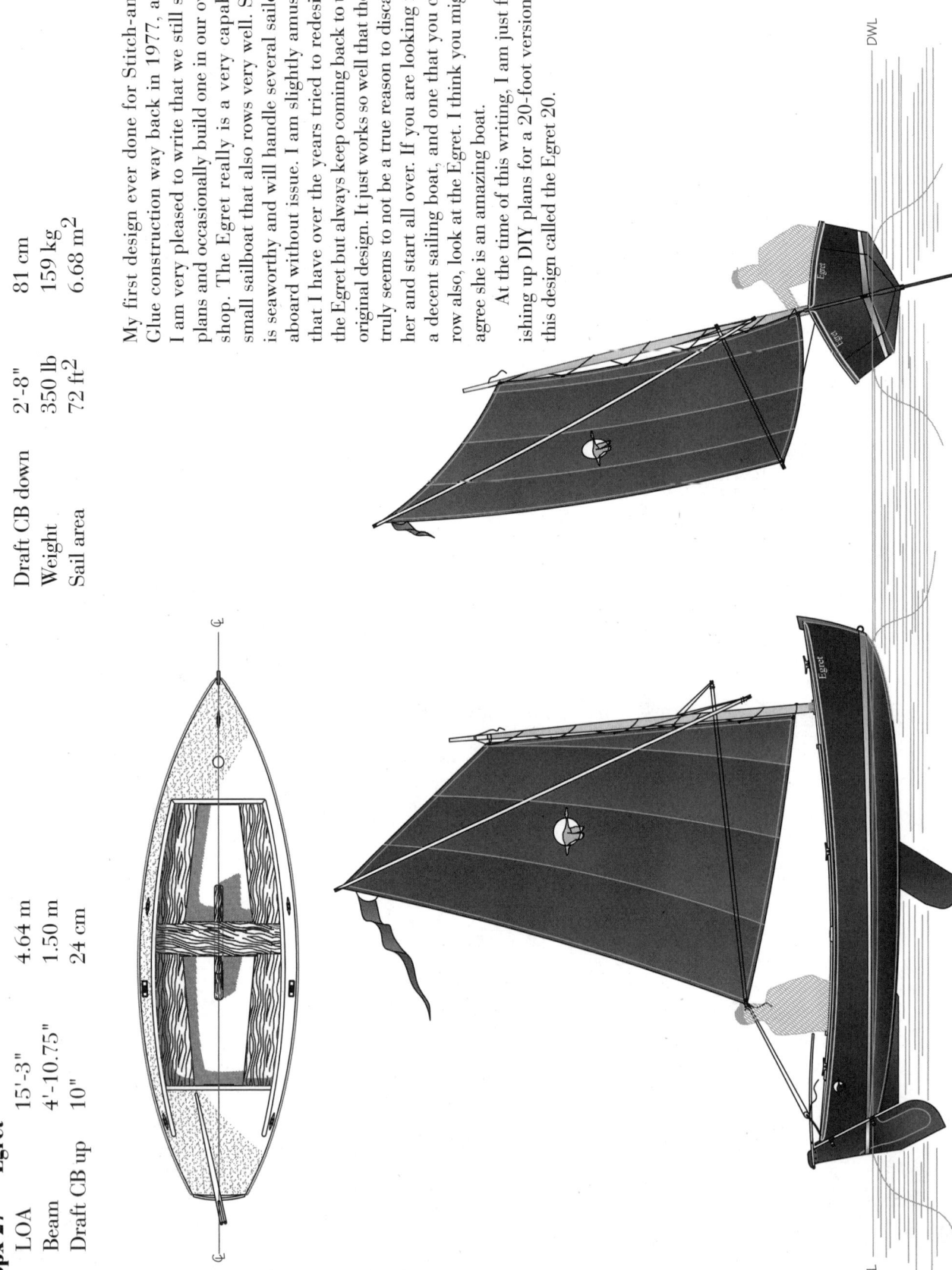

Appx 28 Nancy's China

LOA	15'3"	4.64 m	
Beam	6'2"	1.88 m	
Draft	1'4"–2'11"	43–89 cm	

Weight	850 lb	386 kg	
Sail area	112 ft²	10.4 m²	

I designed this boat in 1980 as an antidote to the high cost and ever-growing complexity of sailing. Her name was inspired by a news report that year about the cost of Nancy Reagan's new china servings that she had bought for the White House; I calculated that an amateur builder could knock one of these boats out for not much more than one of Nancy's $950-apiece place settings. The cost of plywood and epoxy has increased greatly, of course, but I have updated the design and the boat remains very popular. She's practical, pretty, fun, stable, and sports-car responsive to a finger's touch on the tiller. We currently offer three different sailing rigs for this design: a very simple spritsail rig, a Bermuda rig, and a gaff rig.

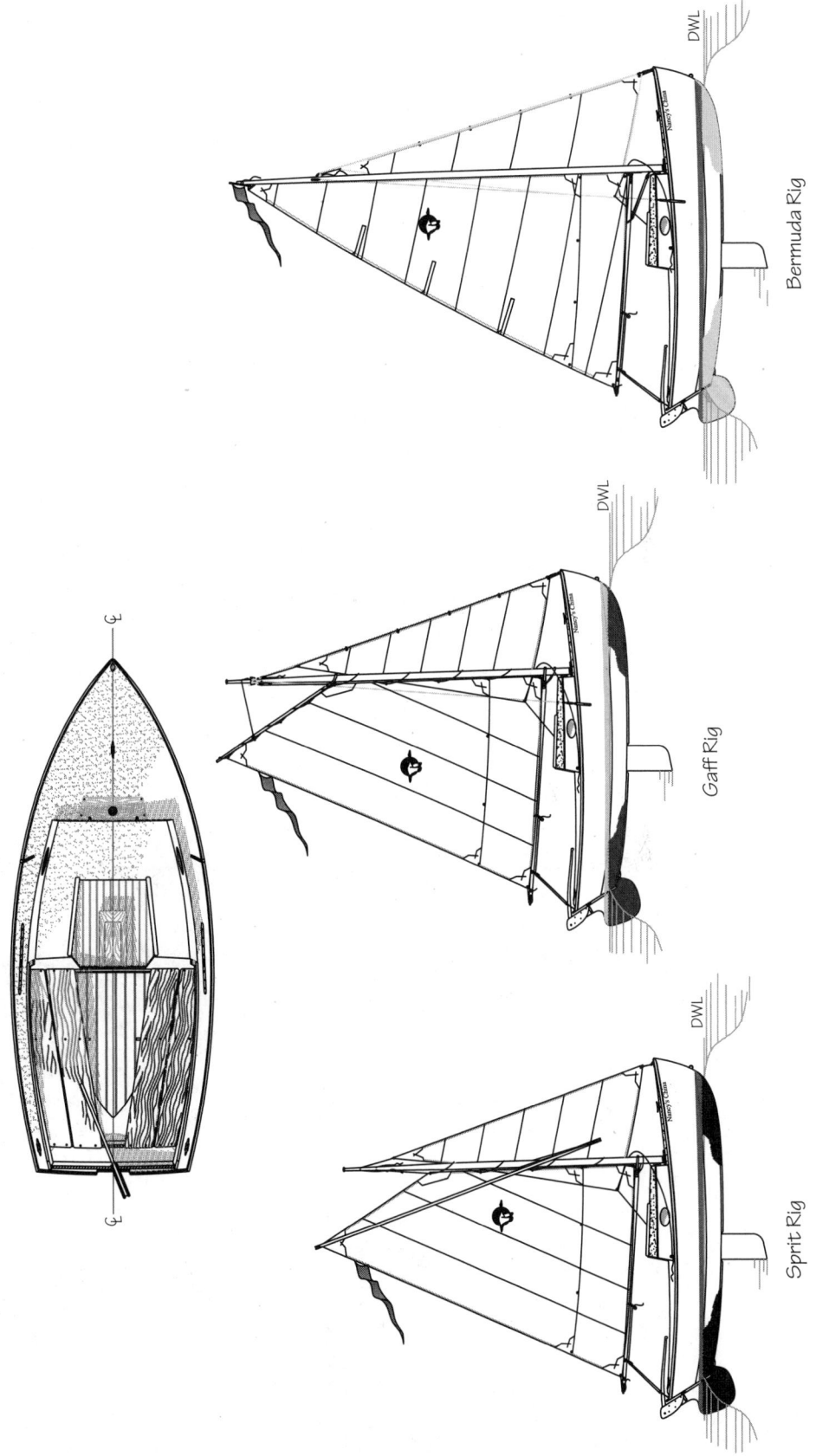

Bermuda Rig

Gaff Rig

Sprit Rig

Appx 29   Eider

LOD	17'6"	5.33 m	Weight	1636 lb	744 kg
Beam	7'0"	2.12 m	Sail area	169 ft^2	15.7 m^2
Draft	1'8"—3'1"	51—94 cm			

A "just-right" size for a pocket cruiser, Eider offers a cabin large enough to sleep two and stow a bit of gear; with her perky bowsprit and gaff rig she sports a traditional shippy look and feel. She's also a thoroughly practical boat, light enough to tow behind a compact SUV and blessed with a self-bailing cockpit—unusual in a sailboat this small.

LINES

Appx 30 **Song Wren**
LOD   21'3"   6.5 m
Beam 7'5"   2.26 m
Draft 2'0"—3'11"   61—120 cm

Weight   2,800 lb   1,270 kg
Sail area 302 ft.²   28 m²

Song Wren, a recent design, plugged a gap in my cruising sailboat lineup between the 18 foot 8 inch Winter Wren and 22 foot 8 inch Arctic Tern. Her gaff cutter rig provides the sailor with plenty of entertaining strings to pull for tweaking performance. She's big enough for a week's cruise for a couple, yet small enough that maintenance won't be overwhelming. The generous sail area is well suited to regions like ours with light summer air; Song Wren provides a good ride in just 5 knots of wind. There is also a fixed-keel version with a 3-foot draft.

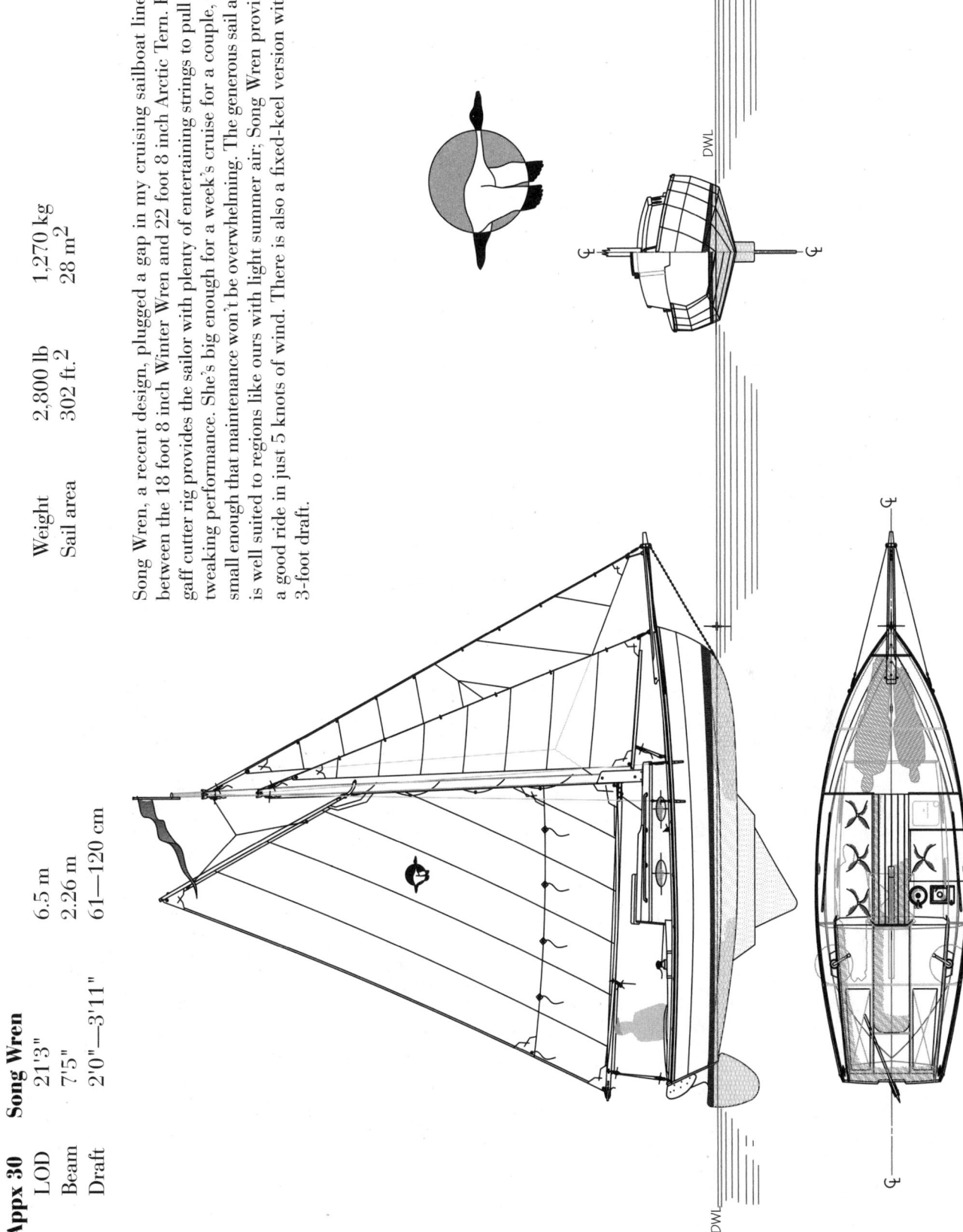

Appx 31   Lichen 26

LOD	26'-4"	8.03 m
LOA	34'-8"	10.58
Beam	8'6"	2.62 m
Draft-up-down	25"-39"	62-98 cm
Weight	4900 lb	2,222 kg
Sail area	401 ft²	37.25 m²

A sailing Garvey (butt-headed) boat boasts a lot of interior room and remarkably good sailing performance. This is a larger version of our smaller Lichen 20 design and has an enclosed head and berths for three in her spacious cabin. The cockpit is large, but the major feature of this boat is the capability to be trailer-sailed. She has a tabernacle mast, and the rig is easy and quick to set up from a trailer. Once you're in the water, her performance is surprisingly quick, and with her twin daggerboards you can sail her into shallower water than most boats of her size. I think this would be a wonderful boat to trailer down to the Sea of Cortez and spend a month or two sailing in warm Mexican waters.

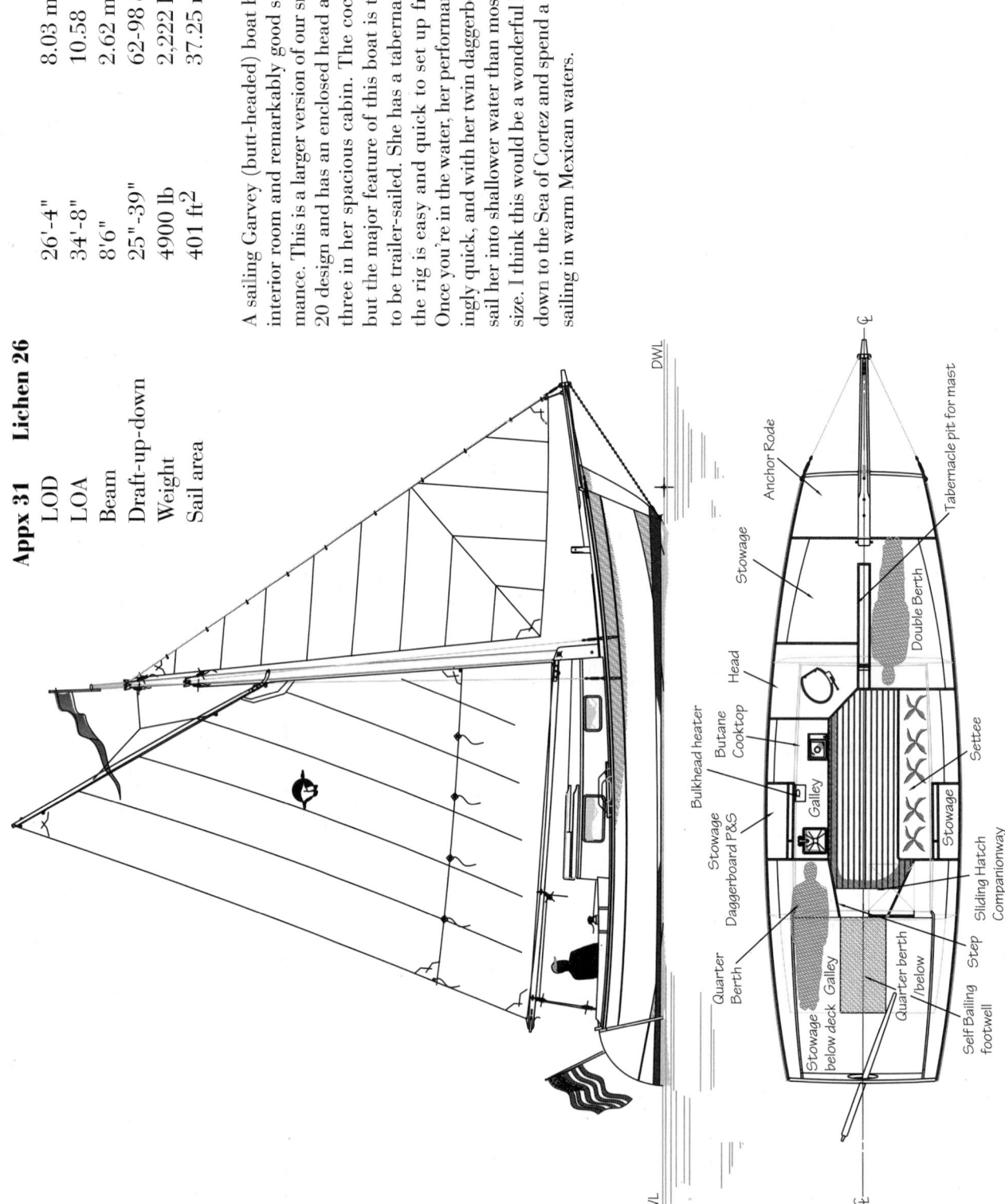

Appx 32 Onyx

LOD	27'7"	8.41 m	Weight	7.350 lb	3,341 kg
Beam	8'6"	2.59 m	Sail area	401 ft^2	37.25 m^2
Draft	4'6"	1.37 m			

Onyx is a multi-chined, full-keel design with the robust construction and rig to make her an offshore-capable sailboat. The cabin is equipped with a full galley and head, and provides a generous 6 foot 3 inch (191 cm) headroom. While this is most definitely not a first boat for anyone to build, plans are available, and for the determined amateur an Onyx would be the accomplishment of a lifetime.

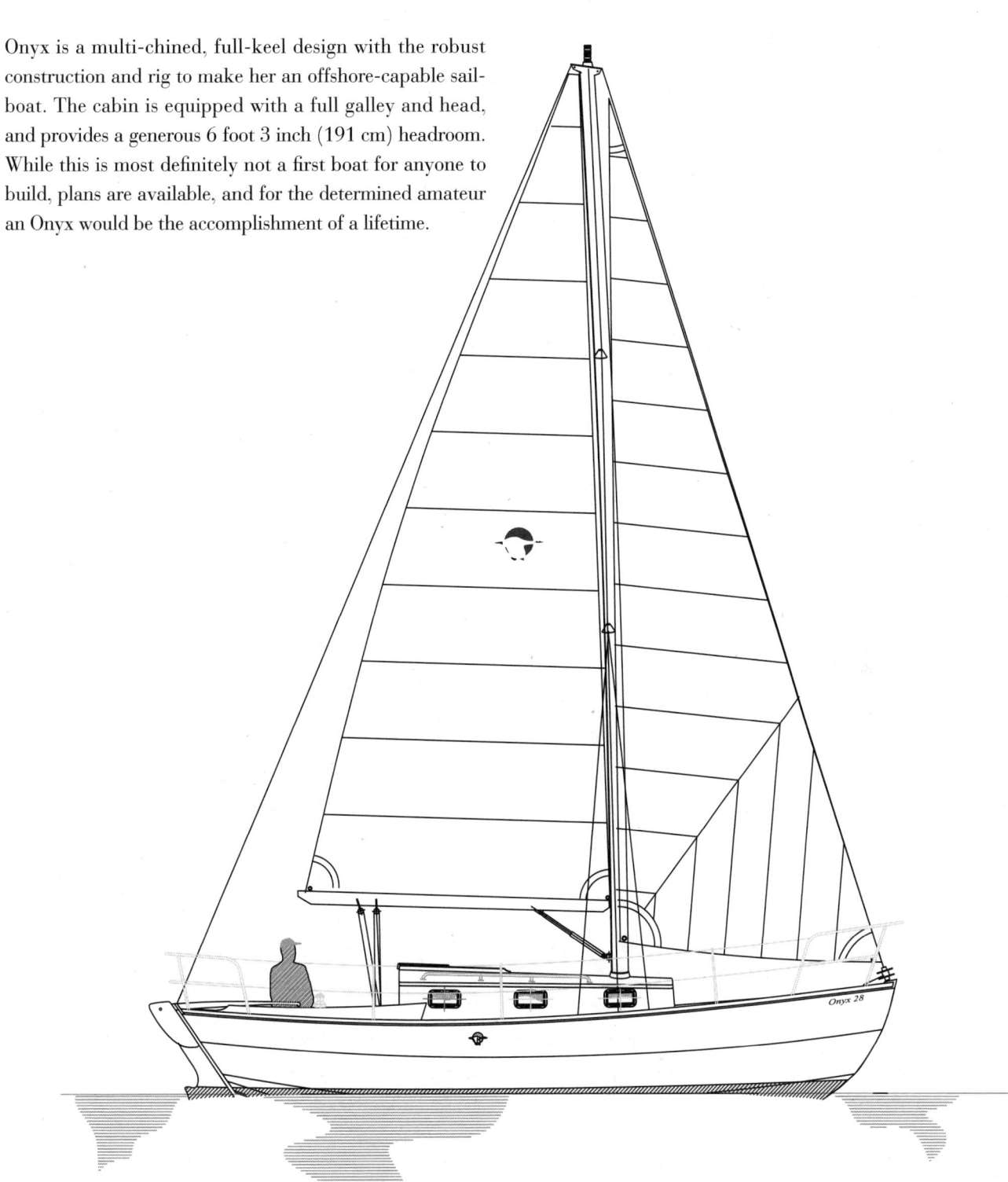

SOLAR–ELECTRIC BOATS

Appx 36   **Solar Sal 27**

LOA	26'-9"	8.15 m
Beam	7'-1"	2.17 m
Draft	25"	64 cm
Weight	3200 lb	1451 kg

The Solar Sal 27 was my introduction to the world of solar–electric boats, and I had no idea it would have such an impact on my design and building life. We built this first boat as a dayboat for some of the large local lakes and sheltered waters of the Puget Sound area of Washington State. But just a couple of years into her life the owners decided in the middle of the Covid 19 closedowns to take her on a very ambitious trip to Southeast Alaska. They spent altogether around 74 days on board and only traveled using the sun as their power source, no additional shore power or generator power, just the solar panels. This was the first trip of record to do the Inside Passage trip under total solar–electric and was an amazing accomplishment. It proves without a doubt that solar is here to stay as a powering option for vessels of all sizes and types.

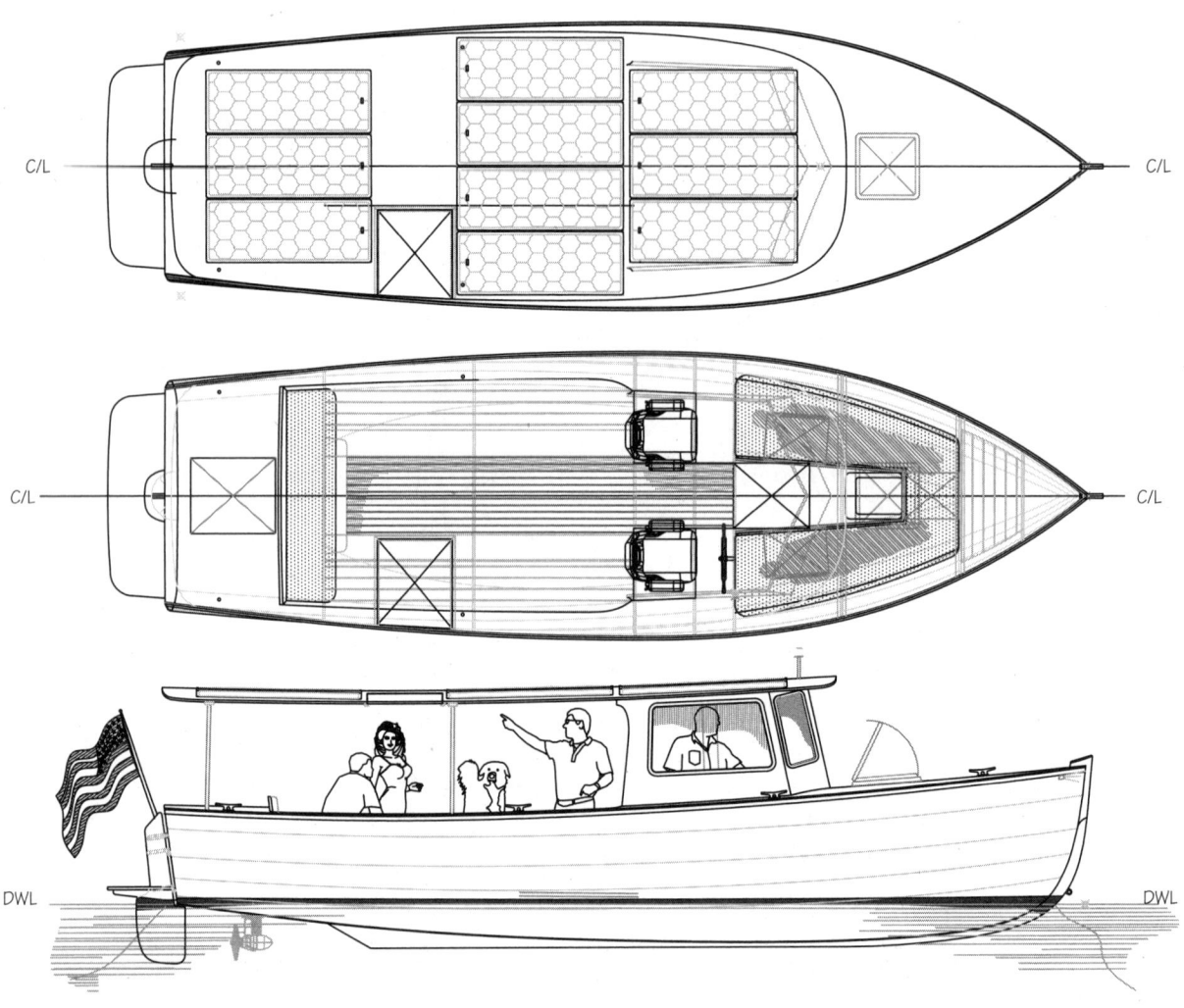

Appx 37   **Electric Philosophy**

LOA	41'-8.5"	12.7 m
LOD	40'-1.75"	12.2 m
Beam	15'-3"	4.65 m
Draft	2'-7"	79 cm
Disp.	24,400 lb	11,111 kg

Ed and Eileen Pauley came to my office one day with the idea of designing and building a liveaboard solar–electric boat. We talked about options for types of hulls that would be a good platform for this project, both a mono-hull design and a catamaran design. I drew two preliminary drawings, and they decided on the cat hull as the best choice. The Electric Philosophy was built in the following year and launched late summer 2021. She now has many thousands of miles under her keel and works exactly as we had designed and wished. To date she has not yet plugged into conventional power for any reason, and though a lightweight generator is on board as a redundant power back up, it has not yet been started. The concept of powering about our northwest waters with only using solar–electric power again is proven with this design. She has almost 10,000 watts of solar panels on her roof and has lithium batteries in her bilge with enough power to run her for 41 hours running time at 6 knots, should the sun not cooperate with their cruising schedule. Her cabin is warm and cozy, and there are walkways on either side of the house, no steps up or down to move about from one end of the boat to the other. A truly fine example of the Stitch-and-Glue construction and modern wood/epoxy boat design.

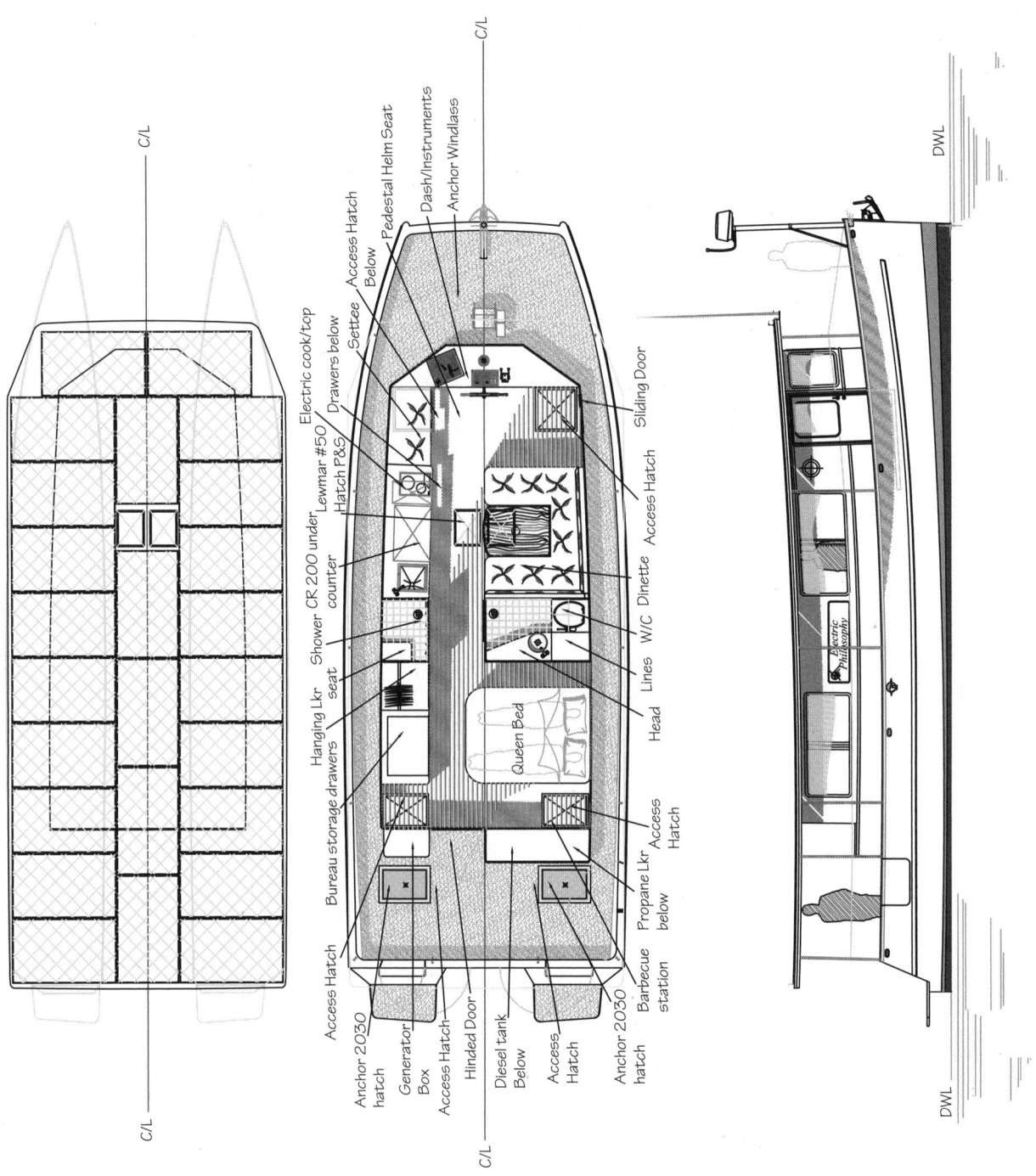

C/L

C/L

Access Hatch Below
Settee
Drawers below
Electric cook/top
Lewmar #50
Hatch P&S
Shower CR 200 under counter
Hanging Lkr seat
Bureau storage drawers
Access Hatch
Anchor 2030 hatch
Generator Box
Access Hatch
Hinged Door
Diesel tank Below
Access Hatch
Anchor 2030 hatch
Barbecue station
Propane Lkr below
Access Hatch
Queen Bed
Head
Lines
W/C
Dinette
Access Hatch
Sliding Door
Pedestal Helm Seat
Dash/Instruments
Anchor Windlass

C/L

C/L

DWL

Electric Philosophy

DWL

MOTORSAILERS

Appx 33   Lit'l Coot

LOA	17'11"	5.45 m
Beam	6' 11"	2.11 m
Draft	2'6"	76 cm
Weight	2,300 lb	1,044 kg
Sail area	160 ft^2	14.8 m^2

Properly designed, a motorsailer offers the best of both worlds, power, and sail, and adapts handily to whatever conditions nature happens to throw at her. This is the aim of my Lit'l Coot, a pocket-sized, gaff-rigged motorsailer. She has enough room inside to sleep one or two plus enough storage for the well-organized cruiser. The self-tending jib allows the skipper to stay inside the pilothouse when tacking. The best thing about this little boat is that in a climate like our Pacific Northwest's, it extends the sailing season well into the rainy season. This design is available in a variety of sail rig designs and keel configurations. My favorite is the cat yawl rig shown with tabernacle masts and twin bilge leeboards that extend up alongside the pilothouse sides. Beachable, very trailerable, and a great micro-cruiser.

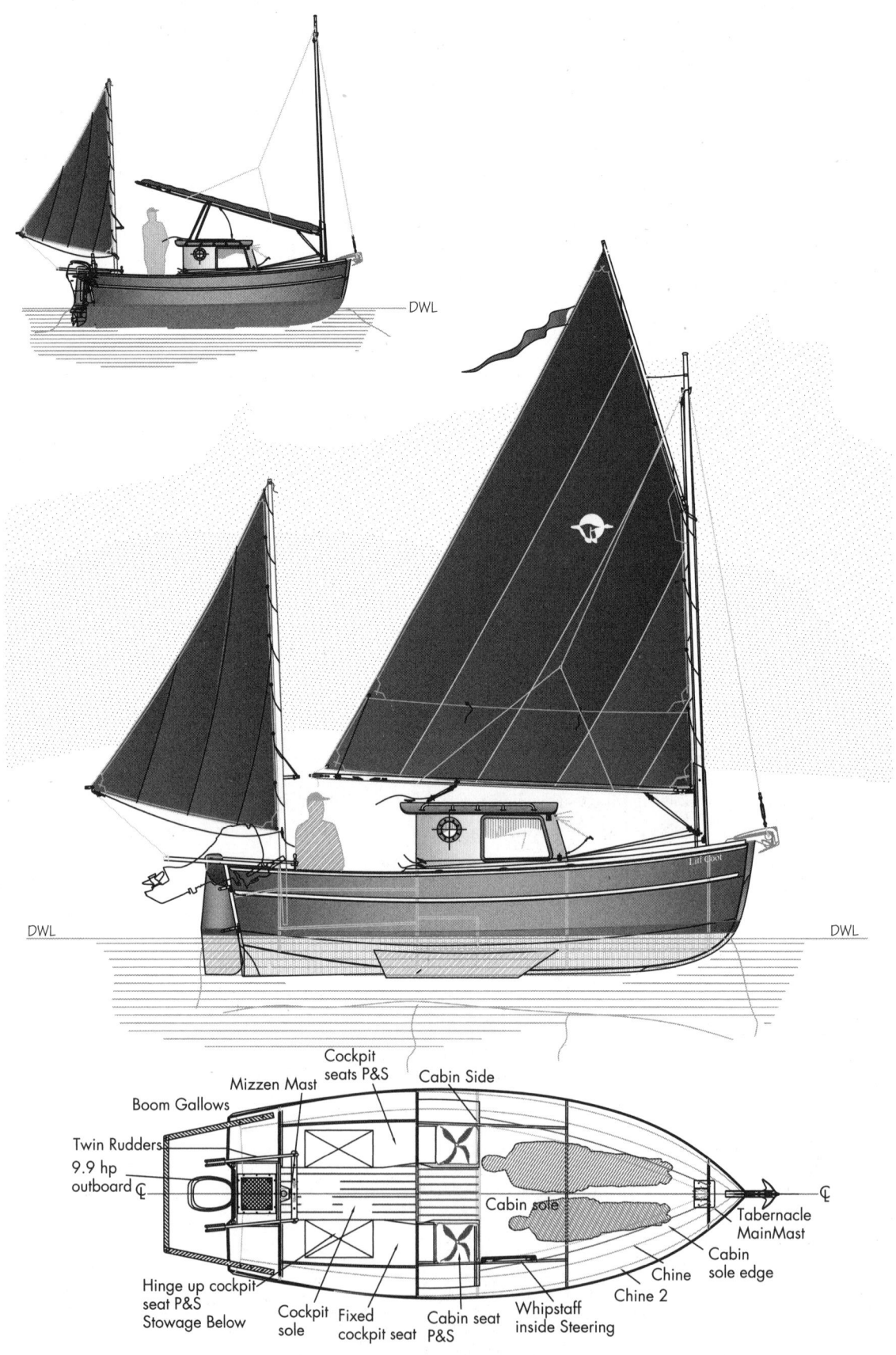

DWL

DWL

DWL

Cockpit
seats P&S

Mizzen Mast

Cabin Side

Boom Gallows

Twin Rudders
9.9 hp
outboard

CL

CL

Tabernacle
MainMast

Cabin sole

Cabin
sole edge

Chine

Chine 2

Cabin sole

Whipstaff
inside Steering

Hinge up cockpit
seat P&S
Stowage Below

Cockpit
sole

Fixed
cockpit seat

Cabin seat
P&S

Appx 34   Camarone 40

LOA	45'-0"	13.7 m
LOD	39'-1"	11.9 m
Beam	12'-9"	3.9 m
Draft	5'- 11"	1.8 m
Weight	30,400 lb	13,790 kg
Sail area	854 ft^2	79.33 m^2

The Camarone 40 is a true 50/50 motorsailer capable of motoring through any sea condition. If considered as an assist, the ketch rig is capable of at least doubling the 1,500-mile motoring range to 3,000 miles. She was originally designed for two Ph.D students for a planned seven-year, round-the-world cruise/research project. She is innovative in that both have private sleeping cabins yet a main salon down below that will be a pleasure to spend dark stormy evenings aboard in a secluded anchorage. This interior merits some study as it provides to my mind virtually all the necessities of comfortable life aboard: good visibility in the pilothouse, great galley area with plenty of room for more than one cook, and a dinette table area that can handle a group of friends comfortably. If you have guests stay over, the dinette area converts to a port and starboard single berth, and there is still a V-berth up high above the dinette. She is being built in Prince Rupert, British Columbia, Canada, and we look forward to seeing her on the water.

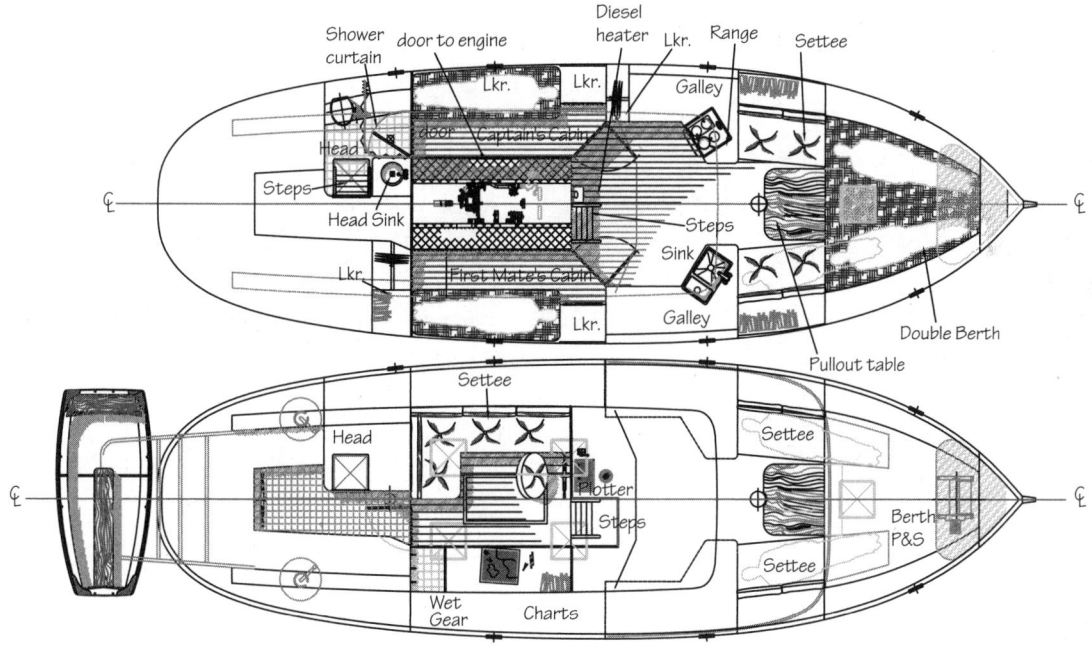

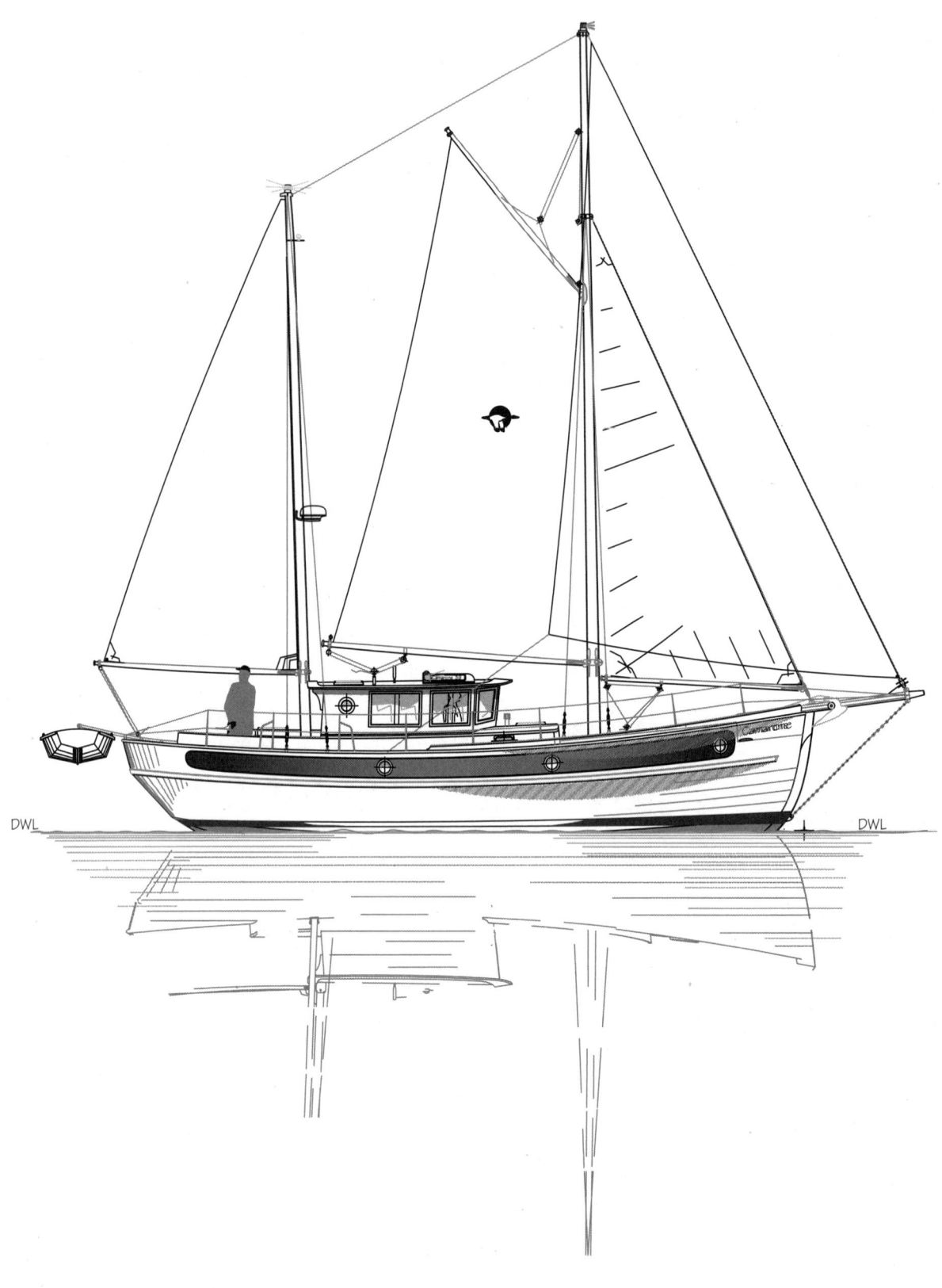

Appx 35   Oso Blanco 54

LOA	57'-2"	17.35 m
Beam	12'-10"	3.90 m
Draft	4'-5"	1.34 m
Weight	43,000 lb	19,504 kg
Sail area	770 ft^2	71.53 m^2

Oso Blanco 54 was designed as an Arctic exploration vessel with a sailing rig to assist with stability and extend her cruising range to anywhere in the world. She has a relatively shallow draft for near-shore exploration, and the box keel will also allow her to stay upright if beached—intentionally or accidentally. There are two staterooms with a separate aft cabin. The pilothouse serves as the main salon, enclosing cooking, dining, and piloting, leaving the private cabins for rest and relaxation. Her split sailing rig provides numerous options for different weather conditions, and there are paravanes rigged on the mizzen mast that assist in providing stability in virtually any sea condition. This is my candidate for best design for around-the-entire-world cruising in comfort and safety, and is a fine ending for our journey into the world of Stitch-and-Glue boatbuilding.

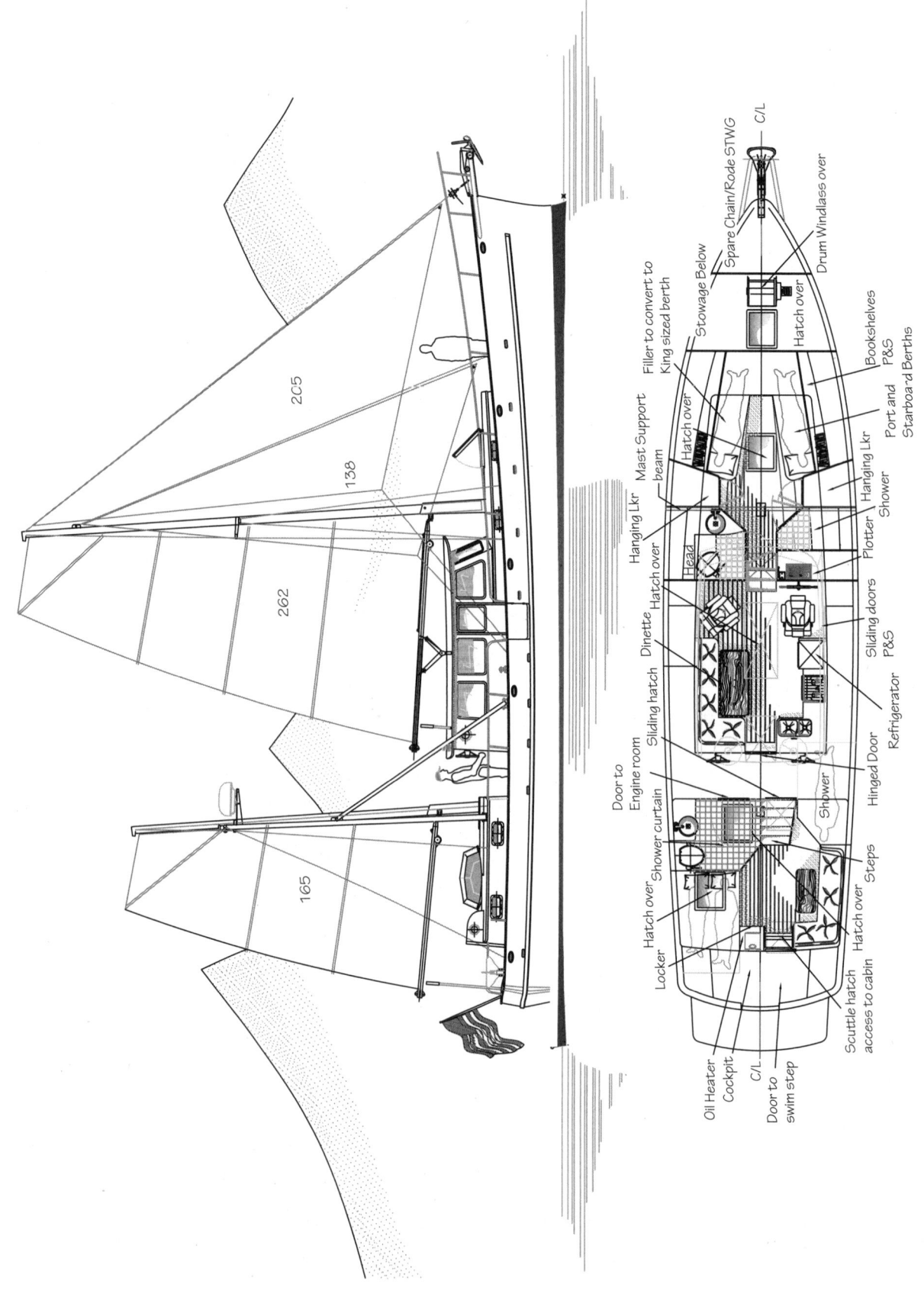

2C5

138

262

165

Filler to convert to
King sized berth

Mast Support
beam

Hanging Lkr

Hatch over

Head

Spare Chain/Rode STWG

Stowage Below

C/L

Hatch over

Drum Windlass over

Bookshelves
P&S

Port and
Starboard Berths

Hanging Lkr

Shower

Plotter

Hatch over

Dinette

Sliding hatch

Door to
Engine room

Sliding doors
P&S

Refrigerator

Hinged Door

Shower

Hatch over

Steps

Locker

Hatch over

Shower curtain

Scuttle hatch
access to cabin

Oil Heater

Cockpit

C/L

Door to
swim step

Index

Note: **Boldface** type indicates a figure.